T0189620

The Antifascist Classroom

The Antifascist Classroom
Denazification in Soviet-occupied Germany, 1945–1949

Benita Blessing

THE ANTIFASCIST CLASSROOM
© Benita Blessing, 2006.
Softcover reprint of the hardcover 1st edition 2006 978-1-4039-7612-3

First published in 2006 by
PALGRAVE MACMILLAN™
175 Fifth Avenue, New York, N.Y. 10010 and
Houndmills, Basingstoke, Hampshire, England RG21 6XS
Companies and representatives throughout the world.

PALGRAVE MACMILLAN is the global academic imprint of the Palgrave Macmillan division of St. Martin's Press, LLC and of Palgrave Macmillan Ltd. Macmillan® is a registered trademark in the United States, United Kingdom and other countries. Palgrave is a registered trademark in the European Union and other countries.

ISBN 978-1-349-53675-7   ISBN 978-0-230-60163-5 (eBook)
DOI 10.1057/9780230601635

Library of Congress Cataloging-in-Publication Data

Blessing, Benita.
    The antifascist classroom : denazification in Soviet-occupied Germany, 1945–1949/ Benita Blessing.
        p. cm.
    Includes bibliographical references and index.

    1. Education—Social aspects—Germany (East)—History. 2. Education and state—Germany (East) I. Title.

LC191.8.G3B58 2006
379.431'0944—dc22   2006045393

A catalogue record for this book is available from the British Library.

Design by Newgen Imaging Systems (P) Ltd., Chennai, India.

First edition: December 2006

10 9 8 7 6 5 4 3 2 1

Transferred to Digital Printing 2009

In loving memory of my father, Don Blessing

Contents

List of Figures   ix

Preface   xi

List of Abbreviations   xv

Introduction: Redemption through Reconstruction   1

1   Antifascism, Unity, and Division   13

2   Setting up the School   37

3   Rebuilding the School   61

4   Rubble Children and the Construction of Gender Roles   91

5   The Antifascist Narrative   121

6   "Vati's home!": From Defeated Nazi to Antifascist Hero   141

7   Reestablishing Traditions   159

Conclusion: Redemption through Reconstruction
and Beyond   187

Notes   203

Bibliography   253

Index   279

List of Figures

1   Excerpt from a Play Written by Pupils at a Girls' School   7

2   "We're building schools. Everybody help out!"   66

3   Pupil's Illustration of Building with Broken Windows
    and Piles of Rubble in 1945 (Top Half); New Panes in
    Window and Cleaned-Up Courtyard a Year Later in 1946   71

4   Christmas, Communism, and Consumerism   97

5   A Pupil's Essay and Illustration of "Rubble Women"   119

6   Pupils' Illustration and Play about the Long Lines at
    the Water Pump after the End of the War   136

7   Pupil's (Probably Boy's) Essay and Illustration of
    Bombing of Neighborhood in 1943. G. Massner,
    January 25, 1946, LAB/STA, 134/13, 180/1,
    no. 524/1 and 524/r   139

8   DEFA Film "Irgendwo in Berlin" Poster   146

9   "What Should I Give My Child?" Poster for an
    Exhibit of Appropriate Antifascist Christmas Gifts   182

10 Christmas 1947   184

Preface

I first visited the East Berlin district of Weißensee in 1985 with a small group of other American high school exchange students and our West German chaperone. We left our larger group hanging around the city center "Alex" to accompany the young woman on her visit to her East German aunt, who worked at a local restaurant. Over a decade later, I stumbled upon the same restaurant while visiting with a friend and colleague who was reading a draft of a chapter I was working on about the Soviet zone's educational system. When I arrived at my friend Henning's apartment and excitedly told him about the coincidence, he asked if that year as an exchange student to West Germany and my two brief visits to the German Democratic Republic (GDR) had influenced my choice of research topic.

Of course it had. I had grown up with the image of the GDR as intransient. Long phone conversations with German and American friends about the meaning of it all in the fall of 1989 and the break-up of the Soviet bloc eventually led me back to memories of that year in Germany and a master's thesis on German educational reforms post–*Wende* (transformation, the period after German unification). An MA in hand two years later, I realized that I still understood little about the forty-five-year history of Soviet zone and GDR schools. I started working backward in time until I landed in 1945, when World War II had ended, and no one could have guessed that Germany would be divided for almost a half century and then, equally surprising to most scholars and politicians, would unify. My eternal gratitude goes to the Congress-Bundestag scholarship program and Youth for Understanding for opening my eyes to another kind of school system in 1984–1985. Negotiating a West German *Gymnasium* as a teenager prepared me well for work as an historian who now negotiates a foreign school system in archives.

My thanks also to the many friends, colleagues, and institutions that supported the research and writing of this book. The *Landesarchiv-Berlin*, the *Bundesarchiv-Lichterfelde*, the *Schulmuseum of the Museum zu Kindheit und Jugend*, the *Deutsches Institut für Pädagogische Forschung*, the *Heimatmuseum-Köpenick*, and the *Archiv der Jugendbewegung-Ludwigsfelder* all offered unfailing support and guidance during the research phase of this project. My dear, departed friend and advisor Sterling Fishman guided me in the right direction while beginning this project, while my coadvisors Rudy Koshar and Bill Reese both kept me firmly grounded in the historiographical and educational scholarship. My colleagues at the Humboldt University and Free University have continued to assist with my questions and requests, even hunting down out-of-print books for me at used-book stores (*Danke*, Henning Schluß). As a postdoctoral student at the Institute for European History in Mainz, Germany, I presented my work to an international audience whose criticisms and questions helped me refine my argument. My year as a postdoctoral fellow with the National Academy of Education/ Spencer Foundation allowed me the valuable time to write and engage with other scholars interested in the topic of education. Maris Vinovskis was an invaluable mentor, while fellows such as Geoffrey Borman, David Gamson, Andy Jewett, Heidi Mesmer, and Lynn Sargeant proved to be crucial to a successful year of thinking and writing. The History of Education Society is perhaps one of the best-kept conference secrets that I herewith encourage other scholars to take advantage of: since my earliest graduate student days, senior and junior scholars there have patiently listened to my conference talks and read versions of my manuscripts and encouraged me unfailingly: Jim Albisetti, Barbara Beatty, Nancy Beadie, Ed Beauchamp, David Gamson, Linda Eisenmann, Ben Justice, Chris Ogren, Craig Pepin, Catherine Plum, Brian Puaca, and the "lunch crowd" from the Madison group have been an inspiration to me over the years. The Midwest German Studies Workshop has also been an important venue for my musings—David Barclay, Erik Jensen, M.J. Maynes, and Glenn Penny have forced me to think in new ways about my manuscript and have always been ready with a helpful citation. Here at Ohio University, I have been fortunate to work with outstanding historians who have always stood ready to listen to my interpretations of German and gender history. The Office of Research and Development generously funded the costs of including photos in the manuscript, which led to a richer final product. My thanks are due to my many students of German and European women's history too, who have read parts of several chapters and asked rational questions that have forced me to rethink many of my premises. My greatest thanks go to my family—my sister, brother-in-law, and mother, for carefully avoiding the topic of when my book would be finished, and to my partner Don, whose interpretations of gender and socialization from a biological behavioral viewpoint have brought about interesting comparisons between lizards and East German boys and girls. As always, all mistakes and misstatements remain my own.

Benita Blessing
Ohio University, Athens, Ohio
March 20, 2006

List of Abbreviations

CDU	*Christliche Demokratische Union*, Christian Democratic Union
DEFA	*Deutsche Film Aktiengesellschaft*, German Film Association
DVV	*Deutsche Verwaltung für Volksbildung*, German Education Administration
FDGB	*Freie Deutsche Gewerkschaftsbund*, Free German Trade Union
FDJ	*Freie Deutsche Jugend*, Free German Youth
FDP	*Frei Demokratische Partei*, Free Democratic Party
FRG	Federal Republic of Germany
GEW	*Gewerkschaft Erzieher und Lehrer*, Educators' and Teachers' Union
GDR	German Democratic Republic
KPD	*Kommunistische Partei Deutschlands*, German Communist Party
LPD	*Liberale Partei Deutschlands*, German Liberal Party
SBZ	*Sowjetische Besatzungszone*, Soviet Occupation Zone
SED	*Sozialistische Einheitspartei Deutschlands*, Socialist Unity Party
SMA[D]	Soviet Military Administration in Germany
SPD	*Sozialdemokratische Partei Deutschlands*, Social Democratic Party of Germany

Archives

AdJB, *Archiv der Jugendbewegung* (Archive of the German Youth Movement), Ludwigsfelder

BArch, *Bundesarchiv* (German Federal Archive), Berlin—Lichterfelde

DIPF/BBF/Archiv, *Deutsches Institut für Pädagogische Forschung, Bibliothek für Bildungsforschung* (German Institute for Pedagogical Research, Library for Educational Research, Archive), Berlin

Heimatmuseum Köpenick (regional museum), Berlin—Köpenick

LAB/STA, *Landesarchiv Berlin/Stadtbibliothek* (Berlin State Archive, Library Branch), Berlin

SM, *Museum zu Kindheit und Jugend-Schulmuseum* (Museum for Childhood and Youth-School Museum

Introduction: Redemption through Reconstruction

No institution stirred more passionate debates in Germany's post–World War II years than did the school system. In the Soviet Occupation Zone in particular—the region that would officially become the German Democratic Republic (GDR) in 1949— German and Soviet decisions about how to educate young people after twelve years of a National Socialist (Nazi) dictatorship became part of a broader social and political discussion about the future of the German nation. At a critical time in the state's development, its educational system both reflected and contributed to a nascent antifascist national consciousness that was uniquely East German.[1] For a half-century, seventeen million residents of the GDR attended a school system that in form and content differed radically from its partner across the border in the Federal Republic (West Germany) and from all previous German educational systems. For the first time, German educators promised all citizens an education that was coeducational, secular, comprehensive, and free. This revolutionary "antifascist-democratic unity school" was neither a product of Moscow nor a tool of hard-line communists in East Berlin. Indeed, German educational reformers in the Soviet zone based their school plans on educational discussions that had taken place during German unification debates in 1848 and 1871 and during the reform pedagogy movement of interwar Germany. The "new school" *(neue Schule)*, as its founders called it, became a distinctly German, socialist-humanist element in the Soviet zone's denazification and reeducation program. The establishment of the GDR in the fall of 1949 signaled the completion of antifascist democratization, or at least the end of German and Soviet administrators' commitment to that program, and the beginning of new debates about the appropriate school for the socialist state.[2] But those first postwar years of educational reforms under progressive

German pedagogues left an important legacy. The "new school" had served its purpose of anchoring the Soviet zone's educational system in German historical legitimacy, even while the official division of Germany ultimately eliminated Moscow's willingness to accept a school system that did not at least reference a Soviet socialist agenda. This book explores the history of this "new school" and its role in communicating antifascism to young people in the Soviet zone during the four and a half years between the end of World War II in May 1945 and the founding of the GDR in October 1949. By introducing this oft overlooked institution of the school in the oft-ignored years before the GDR's official birth, I provide a reperiodization of twentieth-century German history and evidence of surprising East German autonomy in matters of state building and "national," antifascist, consciousness formation.[3]

Antifascist Schools

In October 1945, five months after the end of World War II, the German Communist Party (*Kommunistische Partei Deutschland*, KPD) and the Social Democratic Party (*Sozialistische Partei Deutschland*, SPD) issued a "Joint Appeal for the Democratic School Reform" in the Soviet zone of occupied Germany.[4] Addressing "all parents, teachers, and professors," the two parties summarized the damage inflicted upon German pupils by Hitler's government: "The Nazi regime placed the entire German school system, from nursery school to the university, in the service of fascist party—and racial hatred, of intellectual and physical war preparations, of chauvinistic incitement, and of military drill."[5] The appeal did not propose a return to the "hereditary evil" of the prewar educational system, which had educated obedient "servants of the state" instead of "humans capable of thinking for themselves." Rather, both parties called on "all antifascist and truly democratic forces," including "antifascist parents and teachers, responsible men and women, [and] German youth," to support an educational program of denazification and democratization. The "new school," as educational reformers came to call the new educational system, was to work toward the "democratic unity of the nation" and free "the up and coming generation . . . of Nazistic and militaristic thoughts."[6] This program was the early foundation of antifascist democratic education.

Those postwar educational policy changes were neither completely successful nor entirely a story of disappointment. To be sure, the school reform laws, antifascist teacher training programs, and passionate speeches about antifascist democratic education that comprised the "new school" did not fulfill all the hopes that educational administrators had invested in them. Numerous practical obstacles stood in the way of holding classes in the immediate postwar years, as essays from pupils in the "new school" demonstrated. Problems ranged from reconstructing clean, usable educational spaces to finding didactic material. According to the elementary school pupil Ursula Reimann, for example, who wrote in 1946 about the first days of school after the war, "At first we didn't have any instruction in school, but rather cleaned and straightened up our wing [of the school building]."[7] At fourteen-year-old Eva Schmude's school in the Berlin district of Prenzlauer Berg around the same time period, work units of ex–Nazi Party members were responsible for the physical labor of clearing away the rubble, while pupils and teachers took charge of identifying instructional materials that were not ideologically tainted: "There were big piles of rubble in our schoolyard that Nazis cleaned up. But the sixth and seventh grade classes had to participate in the construction of the school, too. We had on old things and aprons, so that we wouldn't ruin our good things. First we cleaned up the map room . . . We brought all the maps into the hallway and sorted out the Nazi ones from the neutral (*unparteiisch*) ones."[8] From educational administrators' plans for a democratic pedagogy to young people's joy and determination to "learn again after almost three years' interruption in schooling," as one seventh-grade girl exclaimed in 1946, the antifascist democratic school reform program in the Soviet zone demanded the engagement of the entire population, including young people themselves, and it required time.[9] Many proposed radical structural changes in educational policy and practice were not fully implemented; other policies that were implemented often resulted in unexpected consequences. Yet antifascist democratic education, although short-lived in view of the entire half-century of GDR history, was not a failed program of reform. It educated a new generation of antifascist, socialist-humanist Germans. "New school" pupils, teachers, and parents participated in the construction of unique antifascist experiences, consciously building an antifascist society that was distinct from both the Soviet Union and the Western zones.

In order to contextualize this argument, it is helpful to consider recent scholarship that has demonstrated that previous interpretations of the Western zones' educational reforms as little more than a restoration of previous ideological systems are erroneous and lacking in their portrayal of the postwar western German educational systems. Brian Puaca, for instance, has convincingly argued that pupils in the U.S. zone successfully internalized ideas of democracy through such activities as student government organizations, even if the class-based West German system remained largely in place.[10] Wolfgang Mitter's work on postwar reforms in the British and United States zones has echoed this need to look beyond the nearly unchanged structural reforms in those zones' schools and consider instead the initiatives at the student self-government level that contributed to student freedoms to question their society two decades later.[11] In a different vein, Annie Lacroix-Riz has pointed to the surprising role of the strong Catholic presence in the French zone as an impediment to ambitious French programs of educational reform, suggesting further avenues of research for the connection between the Church and the school in that zone.[12] At the same time, Rainer Hudermann has pointed to the historical tensions between the French and the Germans (for example, the post–World War I period) as a key factor in the French military administration's particular attention to cultural policy as a means of "disciplining" Germans toward democracy.[13] As important as these works are to a new historiography on the effects of educational denazification and democratization programs for the Western zones, none of those zones matched the radical structural and ideological reforms undertaken by their eastern neighbors. The Soviet zone school differed radically in content and form from its Western zone counterparts, a development that permanently altered the eastern German postwar experience.

This premise that a new educational system contributed to a new kind of German self-understanding in the Soviet zone rests upon a constructive, complex definition of antifascism. Antifascism was not an empty, totalitarian state ideology. Based upon a message of redemption through reconstruction, it became the centerpiece of collective consciousness in the eastern half of Germany by offering Soviet zone residents hope. The call to rebuild every aspect of society, figuratively and literally, gave eastern Germans a concrete goal, whose realization would redeem them collectively from the guilt of the recent Nazi past. The ideology of antifascism constituted more than a means of socialist oppression, even as the state committed unspeakable crimes against its citizens in the name of antifascism. Antifascism's usefulness extended beyond mythical dimensions, although myths of its glorious past became the pillars in its construction. The antifascist heritage, its disciples claimed, originated in the Enlightenment, long before the birth of fascism. The teaching of antifascism as a popular ideology, however, was the responsibility of the school system, not *philosophes*. Still, antifascist education represented lofty goals: it symbolized a break with the Nazi past and encompassed a broad set of educational reforms that had been under discussion in Germany and other western countries for the previous century and a half. The "new school," its founders believed, would finally bring the French Revolution—and a rejection of German Romanticism—to German soil.[14] The ideological underpinnings of the "new school" were thus not actually new. While Soviet zone educational reformers certainly seized the opportunity after the war to construct their idea of an ideal school system, the suggestion by some historians that the "new school" was a complete break with the past is misleading.[15] "Antifascist democratic education" was a useful term for identifying a number of educational reforms with roots in the French Revolution, the mid-nineteenth century, and the Weimar Republic—reforms that found resonance in the Soviet zone, and almost none in the Western zones. "New school" reformers claimed that previous educational structures had failed to educate the citizenry uniformly. A stratified school system based on class privilege, they insisted, had created a society structurally incapable of resisting National Socialism. Antifascist educational reformers hoped to prevent a return of fascist thought by offering an equal education to all citizens. The resulting educational experience was secular, coeducational, socialist-humanist, national (centralized), comprehensive and unified (nontracked), and German. It created an antifascist consciousness that was indigenous to the Soviet zone.

The "new school" became the physical and ideological space in which official memories and traditions met with group and individual experiences, creating an antifascist consciousness. By analyzing the Soviet zone school from the perspective of nation- and collective consciousness-building, this book offers a unique approach to the study of postwar Germany and the relationship between school and state. It demonstrates how the experiences of pupils in the classrooms of the "new school" affected their relationship to their community and to the German past. Educators and parents both formed and were shaped in turn by antifascist lessons. Schools are not perfect instruments of social control, and the "new school" was no exception.[16] Its influence did not extend to indoctrinating citizens, but antifascist education did affect how individuals and groups saw themselves and the new Germany.

Listening for Pupils' Voices

Two interdependent actors shape the classroom atmosphere: the teacher and the pupil.[17] In this book, I address pupils' voices in addition to those of policymakers, teachers, and parents in order to understand which parts of educators' programs succeeded or failed, employing both traditional and previously unexplored documents. School laws and minutes of school board meetings have long been the backbone of histories of educational institutions. These sources, together with newspaper articles, political posters, and educators' speeches, illuminate one aspect of my investigation: what is it that adults hoped the school would accomplish? Textbooks, curricula, and professional journals identify official, national objectives in the shaping of young people as emerging citizens.[18] Finally, the school day itself comes to life through the use of autobiographies, memoirs, and portraits of schools in films and novels.[19]

Pupils' essays in the period between 1945 and 1949 allow me to address young people's experiences directly. Whether written at home or in class, whether daily assignments or special projects, school essays belong to a gray category of public and private life that clarify the intersection of individual and collective memories and experiences. Essays are the private thoughts written by an individual for a specific, limited readership—primarily the teacher, but to some extent parents as well.[20] Although we must read these essays with an awareness of their context, they provide a unique view into the attitudes and ideas of children. Educators in the Soviet zone did not see them as a simple reflection of the lessons and ideas taught in the classroom but rather as indicators of what their pupils were thinking. One new teacher in 1947 poignantly described her first experience reading pupils' essays: "These children's essays meant more to me than a field for my red pen that looks for mistakes and writes a number for the grade at the bottom. They shook me awake and fundamentally changed my attitude to the children in my class."[21] The effect of pupils' writings on historical inquiry is no less profound.

One essential set of the pupil essays—totaling approximately 1,350— is from the Berlin district of Prenzlauer Berg in 1946. They were later deposited in the State Archive of Berlin (*Landesarchiv*) after originally belonging to the local Prenzlauer Berg archive (see figure 1, below).[22] The School Museum division of the Museum for Childhood and Youth in Berlin houses many more such writings. A dozen essays are part of full assignment books from the same pupil over a period of months or even years, thus permitting a comparison of pupils' writing styles and attitudes over time. Teachers commented on and graded many of these essays, and these dialogues between teachers and pupils added depth and context to the assignments.[23] Occasionally, teachers only marked grammatical or stylistic mistakes. More often, they responded to the pupils' ideas. Pupils were not, however, the only ones who wrote essays in the Soviet zone. The archive at the German Institute for Pedagogical Research (Berlin) includes essay material from new teacher (*Neulehrer*) candidates as they prepared for their teaching careers. The School

Gas, Wasser und Elektrizität im Mai 1946

O r t :   Eine kleine,aber sauber gehaltene Wohnung in der
           Kuglerstr.

Z e i t : Abends 8 Uhr.

Personen: Der Vater, der am Tage vorher heimgekehrt ist,
           die Mutter, der 16jährige Sohn H o r s t und die
           15jährige Tochter I n g e *Inneslel*

Vater: Wie glücklich bin ich, wieder bei Euch sein zu können
           und Euch gesund angetroffen zu haben. Gestern abend,bei
           der Wiedersehensfreude habe ich mir unser Heim gar nicht
           so genau angesehen. Ich war schon froh, dass es überhaupt
           noch da war!
           (Die Mutter schaltet die Tischlampe ein, aus der Ecke
           ertönt leise Radiomusik).

Vater: Es ist mir noch wie ein Traum, das Lampenlicht und die
           Musik. Seit wann habt Ihr denn elektrischen Strom? Nach
           der Zerstörung Berlin's hat es sicher sehr lange gedauert,
           ehe wieder die Stadt mit Strom versorgt werden konnte.

Horst: Gleich nach der Waffenstreckung war es eine der dring-
           lichsten Aufgaben der Besatzungsmächte, die Stromversorgung

                                                            - 2 -

Figure 1   Excerpt from a Play Written by Pupils at a Girls' School.

Source: Christel Novak et al., "Wiederaufbau Prenzlauer Berg," Girls' Middle School, [1946], LAB/STA 134/13, 182/1, no. 105. p. 8 (b/wphoto)

Note: Note the information the young authors have provided: By 1946, there is electricity, running water, fashionable clothes, and homework to be done.

Museum section of the Museum for Childhood and Youth also maintains a collection of teachers' lesson plans and related documents, as well as notes and individual school minutes, providing rich documentation for the classroom atmosphere.

The essays analyzed for this book were written primarily by ten- to sixteen-year-olds. Essays from pupils younger or older than this group did not usually focus on issues of life in the Soviet zone, and were therefore of less interest in answering the questions of this study about the development of antifascism democratic education. All assignments came from German, history, or social studies classes. Four thematic areas dominated pupils' descriptions of themselves and their surroundings: their gender experiences; memory/remembering; reflections on Germany, the German nation and the Soviet zone; and opinions about German unification, including discussions of the Western zones. Not every essay spoke about these themes, but most of them alluded to at least one of these subjects. The range of topics that pupils discussed in essays with similar thematic assignments shows that young people felt free to write about almost all aspects of their everyday lives. Together, the texts narrate a rich story about pupils' experiences in the "new school" and the new Germany.

Einheit versus Spaltung and the Antifascist Nation

Antifascism, the key element in the "new school's" program, was the proclaimed touchstone for all educational reform. *Chapter 1* explores how early postwar ideas of antifascism as a political-cultural approach and dreams of democratic national unification enabled the construction of a loose series of guidelines for everyday life.[24] As Alf Lüdtke has argued, the repetitions in the everyday routines of individuals and groups provide them with a means of appropriating and changing their world.[25] Soviet zone residents developed a collection of ideas and practices that they identified as antifascism and then implemented as the central strategy to construct a new nation through daily acts and thoughts. Soviet zone schools, as institutions of an antifascist perspective, played a curious double function in the struggle to renew German society. The new school system was modeled on the nineteenth-century *Einheitsschule* model, a comprehensive "unity school" whose aim was to provide all citizens with the same education and thereby unify the nation culturally. In spite of the *Einheitsschule*'s emphasis of unity, its unique structure and content offered a discrete set of experiences for Soviet zone inhabitants, encouraging them to see themselves as distinct from the

Western zones. The effect of this tendency was increasing division—
Spaltung—between the zones. The unity school pulled the Soviet zone
farther and farther away from the educational experiences of Germans in
the Western zones.[26]

Chapter 2 examines the limits and possibilities of antifascist
democratic education in the Soviet zone. The most obvious factor in this
respect was the structure of the educational administration itself, which
included a complex relationship between Germans and Soviets. The
German education administration was able to enact its programs with
considerable freedom. Nevertheless, the possibility of the Soviet Military
Administration's (SMAD) irregular but severe reprimands often influenced the administration's decisions about how to present reform proposals. In other instances, potential or actual disapproval from the
SMAD curtailed administrators' plans. Other, basic factors such as cold
and hunger also hindered the school's antifascist democratic lessons.
Lack of heating fuel and poorly insulated classrooms kept pupils from
being able to concentrate and even resulted in school cancellations.
Starving children had a difficult time keeping up with their lessons, if
they were in school at all: many pupils regularly skipped classes in order
to travel to the countryside and gather food (known as "hamstering").
Where feasible, educators and educational administrators enlisted the
help of the SMAD and the Allied Command to find solutions to problems like the lack of windowpanes. Similarly, international relief organizations, the occupation governments, and local residents arranged for
school breakfasts or lunches, one of the aspects of postwar education
most often cited by pupils in their descriptions of their school routines.

Chapter 3 considers the school building as a fundamental symbolic
and physical element in the antifascist reeducation program. As evidenced by teachers' frustrated accounts of temporary classroom solutions, there would be no new learning without a real school building.
This edifice was not just a site in which memories were taught and
learned, however. As one of the largest and most stable physical structures in any community, the school housed many new inhabitants during and immediately after the war. In different cities it served as a bomb
shelter, infirmary, meeting place for separated families, temporary hous-
ing, or administrative space for the Soviet army or German authorities.
A community's school was its local representative of the nation, making
the reconstruction of the school building an important task for the education of young people as antifascist Germans. Educational administrators were not alone in their desire to reclaim the school building from
other uses. Pupils who wrote about their school buildings expressed

relief when they could once again use them solely for educational purposes. This sentiment reflected young people's desire for their lives to return to normal, a wish supported by the comforting presence of the school building. Much of antifascist learning occurred within familiar surroundings, thereby providing the physical foundation for the intersection of continuity and change in Soviet zone education.

Chapter 4 considers the experiences of children in the postwar years and their relationship to the nation. Although other scholars have demonstrated that men and women have different, gendered national experiences, I argue for a deeper investigation that also addresses generational divergences.[27] The process by which boys and girls learn to embrace their nation takes place primarily in the school, where young people learn explicitly to become male or female citizens. In postwar Germany, adults' ambivalent feelings about identifying youth as the future of the new nation provided a further source of social conflict. Although young people were often portrayed as innocent, they were also sentient and sexual beings, whose vulnerability and even complicity made adults uncomfortable. The "new school" offered adults hope for a new Germany, but it was also a painful reminder of the obstacles to social renewal. Considerations of gender and age further contributed to one of the most radical changes in the Soviet zone Einheitsschule and society. Coeducation, the schooling of girls and boys in the same classroom, became a central concern of educators and the public alike. More than a statement about equal educational or career opportunities, coeducation aimed to offer girls and boys the same educational material and experience so that they might be equal partners in the nation. Yet antifascism held different lessons for girls than for boys.

Chapter 5 evaluates how pupils in the "new school" learned to use a specific antifascist narrative in German and history assignments to make sense of their national and individual pasts. This narrative identified antifascists as victims of the Nazi regime and the Soviet zone as an antifascist state. Pupils thus learned to see the Soviet zone, and not the Western zones, as the worthier moral and political successor to the German nation. In order to come to terms with their and their families' recent wartime and postwar experiences, pupils adapted this antifascist narrative in their essays to distance themselves and their families from culpability for the National Socialist regime. In this manner, pupils' own agency in defining antifascism in the "new school" becomes clear.

Chapter 6 looks beyond the classroom to a key partner in the education of children, the family. By identifying fathers as central symbolic figures in the reconstruction of the Soviet zone, I suggest how popular culture attempted to maintain a traditional nuclear family as the ideal family type. Society viewed children in the postwar period as in need of paternal discipline. Thus fathers who returned home as ex-Nazis could be instantly rehabilitated in the public and individual imagination as strong, child-rearing antifascists, the only Germans capable of reconstructing the destroyed nation. In this process of distancing men's wartime activities from their new, postwar roles, antifascism became a primarily male project, relegating women (and girls) in the public mind to supportive roles at home.

But the past could not entirely be laid to rest with political reinterpretations. Ghosts of earlier times were a strong presence in the antifascist classroom itself, as illustrated in *chapter 7*. The tension in Soviet zone educational policies between wanting to fulfill earlier historical plans and creating new instruments of teaching was evident through the end of the GDR's existence. After deciding that only secular material belonged in the new school, Soviet zone educators turned to choosing those historical actors and events that would best relay the desired antifascist lessons to the new society. In this process, educational administrators laid claim to a specific antifascist, German cultural heritage. The commemorations of the anniversary century 1848–1948 and the Goethe Year 1949 exemplify the role that schools played in the Soviet zone's struggle to wrench these icons away from the Western zones. Soviet zone educators repeatedly referred to these "German national events" as part of antifascist culture. The distinction between an antifascist, socialist-humanist Soviet zone culture and a western, capitalist German one made the political division between East and West more distinct. Similarly, the Communist Party's push for traditional Christmas celebrations highlighted the regime's hesitation to abandon earlier German holidays, even those with a primarily Christian connotation. Indeed, by promoting Christmas as a German celebration, the day could be celebrated outside the confines of Christianity.

The *conclusion* considers the meaning of the Soviet zone in the broader context of GDR history. From GDR pupil essays and other writings, it becomes clear that the immediate postwar years had a formative influence on the emerging antifascist nation as a whole. The Soviet zone was more than a dress rehearsal for the GDR, or a time of limbo while the Soviets decided how to proceed with their German territory. Rather, the Soviet zone provided the basis for a distinctly German version of antifascist socialism that would influence how eastern Germans viewed themselves, how they interacted with the Soviet Union, and how they put aside dreams of German unity in order to become a separate German state.

I close this section on a speculative note about writing a history of young people's experiences in the Soviet zone. Reconciling twelve years of Nazi teaching with new school lessons in the Soviet zone posed an enormous challenge for pupils. Children, no less than the adults around them, struggled to reconcile the new enemy status of Nazis—who were also possibly friends, neighbors, and family members—with the new liberator status of Soviets. Like their parents and teachers, young people sifted and weighed their memories against the new narrative structures that they were offered. The traces that young people left behind underline the interactive nature of the school, as the state's representative, and pupils. It is not always clear to what extent these children successfully unlearned Nazi propaganda about the justness of German war victories, or learned to be grateful to their enemies-turned-occupiers. The opinions young people expressed in their writings reflected their often painful attempts to make sense of their past and present, in order to confront and even master their future. As historians, we must address young people's sources with the same critical skepticism and respect that we accord other documents, without treating children's voices dismissively. Only in this manner can we begin to understand how the generation that grew up with the German Democratic Republic made and participated in their transition from Nazism to antifascist-democratic socialism.

CHAPTER 1

Antifascism, Unity, and Division

In October 1945, the Saxon school administrator Herr Viehweg declared in his report on a regional meeting with school supervisors, "Our school should be democratic antifascist. We know what that means. But how do we accomplish it?"[1] Viehweg went on to discuss practical considerations for setting up the "new school," such as the need for new teachers, but, like his colleagues throughout the Soviet zone educational system, he never specified the exact ideology behind antifascist democratic education or even defined its philosophy. Practically, much of what became "antifascist democratic education" referenced earlier school reforms from both the previous century and the Weimar era, but no single educational project was inherently antifascist. There was no one definition of antifascist democratic education in the Soviet zone, just as there was never an explicit definition of antifascism.

Antifascism in the Soviet zone no longer referred to the multitude of political ideas that had accompanied the ideology's development before and during World War II. Soviet zone administrators identified its sole explicit postwar goal of preventing a resurgence of fascism—a concept once clearly identified by communists as the most extreme form of a bourgeois dictatorship, but now a multipartisan reference to xenophobic, militaristic, nationalist, and racist practices. Political groups throughout eastern and western Germany identified any program as antifascist that included rebuilding society, from dissolving the National Socialist party to such practical accomplishments as restoring water and gas services damaged by the war.[2] Antifascism was above all a project of reconstruction, to be led primarily by men returned home from the war. But antifascism in the Soviet zone did not stop at fulfilling basic needs. It meant achieving democracy, international understanding, a healthy pride in the nation, and peace, and political parties agreed early on not to argue much about the specific definitions of these new ideologies. The most important step in reconstructing the nation, Soviet and German political leaders in the Soviet zone believed, was to teach Germans to see and identify themselves as antifascists, as actively working together against the legacy of fascism. Schools carried the responsibility for the majority of formal antifascist democratic education, and educational reformers thus played a significant role in developing a concrete program for antifascism. Antifascist educators in the Soviet zone believed first and foremost that the "new school" should right all of Germany's past wrongs by creating a unified and democratic population that could not be divided by fascist politics again. This chapter explores antifascism as the key foundational concept of the Soviet zone. I first show how educators across the political spectrum used the "antifascist education" program to help define the Soviet zone's "ideological borders," both internally and externally. This "national" system of antifascist education ultimately unified only one half of Germany, pulling the Soviet zone away from the Western zones. I close with a collective biography of the educational administrators involved with the creation of the "unity school" (*Einheitsschule*). The grand national unity project envisioned by Soviet zone educators ultimately fell victim to broader political and everyday developments, both at the national and international level. These would-be unifiers were thus left with the smaller task of the unification of their zone.

The Ideological Map of Antifascism

Defining borders—temporally, historically, and spatially—was fundamental to the evolution of the Soviet zone. Marxist-Leninist historians in the GDR presented the first postwar years as a founding period that consciously and intentionally flowed into the GDR, in line with historical materialist dicta. These years thus comprised the "antifascist democratic period of transition [*Umwälzung*]."[3] In this interpretation, the Soviet zone began in May 1945 with a *Stunde Null*, a zero hour, and ended suddenly in October 1949 with the declaration of the German Democratic Republic and the advent of a socialist state. This description implied discrete spatial and temporal borders for a logical development of antifascism, but the process was more complex.

In striving for an image of societal rebirth, *Stunde Null* proponents rejected blame for World War II and the Holocaust, and ignored continuities between Nazi and postwar programs and policies.[4] Some contemporaries shied away from *Stunde Null* rhetoric in discussions of Germany's future, emphasizing instead the need to be vigilant against the still-present specter of fascism. As Otto Winzer, the city councilor responsible for education in Berlin, warned his colleagues in his report on the state of Berlin's cultural policy in 1946, "the people didn't just waltz away unscathed from the national catastrophe of 1945."[5] The concept of a perceived *Stunde Eins*, a first hour, was more appealing to many antifascist reformers for its sense of promise and hope that many political and social reformers in the postwar period actively felt in both Germanys.[6] Such imagery of a positive first step appeared throughout speeches and essays by political and cultural reformers as well as in the language of pupils in the Soviet zone and GDR.[7] Antifascism gave an identifiable temporal form to the Soviet zone era and region.

The scholar Victor Klemperer, best known for his insightful diaries of his wartime experiences, also chronicled the fascinating postwar years in the Soviet zone. In an early entry, he captured vividly the heightened sense of time and its symbolic meaning for the Soviet zone in his first postwar diary entry: "June 17 [1945], Sunday, Dölzschen [near Dresden]. Chapter TIME. In every broadcast, dozens of times every day, Radio Berlin announces the time, which is a blessing. But when Berlin says it's 8 p.m., it's 7 p.m. here, and 9 p.m. in Bremen: The Russians have Moscow time in Berlin, Dresden has summer-savings time, and the English have central European time in their region. Whenever I am out I ask again and again, what time is it? The standard response: I don't have a watch anymore either."[8] Time had not stopped, and it had not consequently begun anew. It seemed instead to have run amok. Klemperer later complained that he showed up an hour early for the theater because the trains had switched to the new Moscow time change but the opera hall remained on the old time. This event described both a metaphorical and concrete manifestation of conflicting visions of the Soviet zone's actual location in the historical stream—visions not only of officials, but also of train conductors, train passengers, opera directors, and opera-goers.[9] Pupils learning how to tell time likely faced new challenges in this atmosphere of competing clocks. Clearly, placing the Soviet zone on Moscow time, a practice already underway during the war, was more than an administrative convenience for the Soviet Union. It represented Moscow's attempt to place the Soviet zone firmly under Moscow's control. The confusion generated by this redrawing of temporal, and thus physical, boundaries, resulted in the Soviet zone's ultimate return to central European—German—time, and its reaffirmation as a geographic entity that fell outside the realm of Soviet temporal and physical space. Indeed, East German clocks now pulled that region back toward western Europe—adding to the tug of war over the Soviet zone's place in the world. Antifascist time ultimately helped create the concept of an antifascist space, which in turn permitted German agency in those institutions involved in the construction of an antifascist future—like the school.

Pinpointing the Soviet zone's geographical location became even more complicated once some areas changed occupation hands in July 1945. Borders fluctuated as the western Allied troops withdrew from Saxony, Thuringia, and Mecklenburg and Soviet troops replaced them, while U.S. and British troops moved into West Berlin, joined the following month by France. Some districts in Berlin—still treated as a unified city for many administrative issues, including educational policy—were located differently in discussions of eastern and western Berlin according to their economic or political situation. Thus educational administrators often included poorer working-class districts such as Wedding, which became part of West Berlin, in educational statistical analyses of what was geographically eastern Berlin.[10] School officials, assigned the task of helping pupils redefine the German nation, could relate all too well to Goethe and Schiller's classic eighteenth-century *Xenien* satirical question, *"Deutschland? Aber wo liegt es? Ich weiß das Land nicht zu finden,"* (Germany, but where is it? I don't know how to find that country).[11] A speaker at a 1947 Berlin *Hauptschulrat* (central school administrators) meeting observed dryly that all available maps for school use showed the borders of the 1918/19 Versailles Treaty—which identified as German territory most of present-day northern Poland and the westernmost part of Russia. If these maps could not be used, he noted, then there were no maps for the schools.[12] Maps constitute an important means for a people to define and identify itself, making such a problem more than an instructional issue for the Soviet zone.[13] Members of the school community, from school board directors to teachers to pupils, had precious few physical materials at hand to show them the contours of their nation. Without the ideology of antifascism, the Soviet zone's form would have been so vague as to preclude discussing it as anything other than a military-occupied region.

Antifascism was a badge of dignity for Soviet zone educators. Its assistance in mapping out their emerging nation ideologically, even in the face of unclear physical and temporal borders, proved Germany's right to membership in the community of nations. At a time when international public opinion on the German national character was of vital importance to the country's future, antifascism became an important bargaining chip. Educational reformers defined their task of rebuilding Germany consistently in terms of international recognition: they needed to find a course that, as quickly and painlessly as possible, would enable Germans to regain a normal place among nations.[14] As early as 1944, the educational reformer Hans Siebert (KPD/SED), exiled during the war in London, claimed that only a rebirth of German culture would allow the Germans to see the citizens of France, Belgium, the Netherlands, Russia, or England as equals.[15] A country of antifascists, hoped educational reformers, would regain the world's respect for the German nation.

Many individuals and even countries looked skeptically upon Germans' desire to establish normalcy at home and in international relations. One French professor, invited in 1945 to speak for the Soviet zone *Hauptschulamt* (central school administration) on the French educational system, took the occasion to share a few "embarrassing truths" about Germany with his audience. A "Professor Dr. Bousquet," from the University of Algier, stated that he did not believe that Germans would ever again be able to live up to the honorable name that had once been associated with Germany. His judgment was severe: "It is quite clear for Europe and for the rest of the world and for me as well that the name of Germany today is equated with an unspeakable barbarism and immeasurable atrocity which is despised by the entire world. And Germany deserves this contempt."[16] Bousquet then explained that "the number of victims" in Nazi Germany did not change the fact that the majority of Germans had been determined to follow Hitler. "We, the victors, have understood this quite well and we will not forget it."[17] Creating his own antifascist narrative with this statement, he unconsciously imitated the Soviet zone formula of replacing the Nazi past (in this case, the collaborationist government of Vichy) with a heroic, postwar image (here, France as victor). His audience members kindly did not question his reinterpretation of France's role.

After the presentation, the director of Berlin's central school administration, Ernst Wildangel, responded to Bousquet's critiques by pointing to his own and his colleagues' unsuccessful attempts during the war to recruit support abroad against Nazi Germany, including in France. Wildangel closed by stating his optimism that Germany would once again be worthy of its "ancestors such as Bach and Beethoven," continuing in the antifascist canon of connecting antifascist Germany to its purportedly unscathed classical past.[18] Bousquet interrupted him— "impatiently," as the minutes noted—with protests that Germans were politically inept: "You have never done anything like we did, like chopping off the head of a state leader. You should try it. It offers a real feeling of security, and the clouds part in the heavens to let the sun shine through again. This is why I see no possibility of understanding between us (*Verständigung*) . . . I could just rip Germany's throat out."[19] Here he unwittingly alluded to a theme that antifascist reformers had already embraced: the need to bring the lessons of the French Revolution to Germany—even if no one advocated the guillotine per se. Bousquet's closing violent fantasy of literally disgorging Germany could not have been a source of optimism for Soviet zone educational administrators hoping to find like-minded thinkers among their western colleagues.

Bousquet's visit indicated the direction in which Soviet zone reformers hoped to guide the formation of a new German national consciousness as well as how other countries would perceive this political and ideological course. The genuine antifascist wartime experiences of education reformers such as Wildangel allowed these social reformers the relatively privileged position of credibly rejecting responsibility for Hitler and the Holocaust and embodying the new Germany. In Wildangel's response to Bousquet, he did not downplay Nazi atrocities or German crimes. He instead alluded to his own recent, unsullied history, which permitted him to impose a shared international responsibility for the paltry German resistance onto France and the world.

Wildangel provided an ideal profile of a collective, antifascist German resistance fighter during the war. In 1933 he had escaped to France, where he directed a school for German Jewish emigrants until he was interned in southwestern France. After escaping that camp, he lived on a farm under the pseudonym Pierre Delorme as an informant for the French Maquis. He was eventually captured by the SS and shuttled throughout various concentration camps until he succeeded in going underground with Berlin communists. After the war, he became one of the first "activists" to participate in the development of the new school system.[20] With this pedigree, he could claim to have been the exact opposite of the picture that Bousquet drew of Germans. Wildangel grew up as a devout Catholic and worked with Jewish children during the war in France, so he could neither be associated with anti-Semitism nor with Victor Klemperer's worry of appearing to act as a "*jüdischer Rachegeist*" (Jewish angel of vengeance).[21] Nor could he be dismissed as a Moscow party hack. Furthermore, he sought help from abroad during and after the war and had worked with the French resistance—implicating those in France who had collaborated with the Nazis and strengthening his own position as the representative of antifascist action.[22]

Nonetheless, Bousquet had expressed unequivocal disdain for Germany's past—including damning the country for its lack of a democratic revolution, devalued the myth of German antifascist activities during the war, and deflected culpability away from France. Calling attention to his status as a Frenchman, and thus in his mind as a true heir to the French Revolution, he flashed his democratic credentials and alluded to Germany's failure to mature historically. Wildangel, in response, reiterated his and his colleagues' innocence. He claimed to represent a different Germany, one that retained the right to someday inherit Germany's past cultural greatness. Finally, he warned Bousquet away from protesting his and his country's innocence too loudly. The exchange ended in a stalemate. Bousquet promised to arrange for future speakers. Wildangel and his colleagues left the meeting with a clear reminder of the obstacles ahead of them in rebuilding the German nation, and of the international community's hesitance to accept a Germany that claimed to be antifascist and democratic.

Such incidents illustrated the residual hostility between Germany and its neighbors and attested to unresolved national alliances. As mentioned earlier, in the fall of 1945, one central school administration was still responsible for all four zones of Greater Berlin. With the brewing political divisions in Berlin between the Soviet Union and the western allies, the central school administration (located in the Soviet half of Berlin) found itself slowly excluded from decision making powers for western Berlin. For the moment, however, because Greater Berlin was jointly controlled by all four occupying powers, the central school administration felt the need to draw on advice from all its occupiers' countries and from around the world. The growing division of East and West had not yet advanced to the stage of acknowledging only socialist countries' educational systems as acceptable role models in the Soviet zone. Similarly, the Soviet zone professional teachers' journal *die neue schule* regularly featured articles on schools in many different countries in the first postwar years, including positive evaluations of schools in western countries.[23] Inviting speakers such as Bousquet signaled a desire on the part of education reformers to anchor the Soviet zone in the international community. It also offered a means of retaining the ties established by German exiles during the war to western countries. Wildangel, after all, was not the only educational reformer to have spent time during the war in western countries. Hans Siebert and the prewar youth movement leader Eberhard "Tusk" Koebbel, for example, were both active in anti-Nazi activities in London.[24] But such inclusiveness did not last in antifascist discussions. Approving articles about western countries eventually disappeared from Soviet zone journals, as did mentions of exiles' work in England or France.

Defining Antifascism

Literature on antifascism presents a clear partitioning of forces. On the one hand, there is a "good" antifascism, a popular alliance of various groupings on the left for the tactical defeat of reactionary forces, as associated with the Spanish civil war. On the other hand is a "bad" antifascism, a state-sponsored ideological mask that covers a totalitarian system. This insidious antifascism is most strongly identified with the postwar Soviet Union's sphere of influence, particularly in the German Democratic Republic (GDR). The first version has few critics; it is accepted, if not always admired, for its constructive possibilities of popular democratic political action and its grassroots base.[25] The second version, a "web of lies" whose economic interpretation of fascism "ultimately stripped Auschwitz of its core," has become the target of historians' wrath.[26] The historiographical "antifascist legacy" has retained the polarized character of antifascism's origins, a desperate fight that Anson Rabinbach described in 1997: "In the global life and death struggle against fascism there could be no middle ground, no neutral space, and no non-combatants."[27] With the struggle against fascism over, historians have turned their attention—albeit less frantically—to denouncing the most totalitarian forms of antifascism, which they link directly to Soviet-dominated communism. But there is room for a third category of antifascism, one that cannot easily be described as a democratic "good" or socialist "bad" ideology. German antifascism in the Soviet zone, as a political, ideological, and educational program, was an ongoing process of rebuilding the German nation and national consciousness around a new political and cultural ideal. It must be understood within its many functions, which included rhetorical, symbolic, doctrinal, ideological, idealistic, and cultural uses.

Before the eastern German population could be united behind this project, the "antifascist democratic bloc" of parties in the Soviet zone had to find a means of resolving their ideological differences. This bloc comprised four parties: the two labor parties, the Communist Party (KPD) and the Social Democratic Party (SPD), which joined forces in 1946 to become the Socialist Unity Party (SED); and the two smaller bourgeois parties, the Christian Democratic Union (CDU) and the Liberal Democratic Party (LDP).[28] Under pressure from the SMAD and the Communist Party, but also out of a genuine commitment to working together for Germany's reconstruction, the SPD, CDU, and LDP accepted the Communist Party's claim to leadership within the bloc.[29] Although the initial impetus came from the KPD and its protector, the

SMAD, the other parties quickly rationalized the bloc system of institutionalized cooperation because they saw it to be the only viable means of political action available to them. Their agreement to work together also resulted from a desire to avoid divisive ideological disagreements. Political leaders from all parties, especially the workers' parties, deplored their failure to have created a united front against Hitler's National Socialists. Their vow not to repeat interparty fighting, however, entailed a decision not to examine too closely their sometimes conflicting definitions of antifascism or democracy. All parties nonetheless successfully maintained their individual political lines in a number of areas; even if the CDU and LDP were not able to significantly alter SED policies, they did offer strong resistance to the centralizing tendencies of the SED.[30]

The Communist and Social Democratic Party leaderships' postwar decision to put aside decades-old animosities and join forces was a significant new policy direction for both parties. In 1924, the Comintern (the Communist International, the central international communist decision making organ) had declared that social democrats were obstructing the true interests of the working classes.[31] Repeating a sentiment formulated earlier that year by the Soviet Bolshevist Grigori Sinowjew, Stalin had angrily claimed that social democracy and fascism were "not antipodes but twins."[32] By the end of the Weimar Republic, the KPD leadership had eschewed all cooperation with the SPD, using the accusation of "social fascism" to implicate social democracy as a midwife to fascism.[33] True democracy, claimed the KPD leadership, could only be found after "the revolution" under a socialist system.[34] The SPD, in turn, upon comparing the Soviet government with the totalitarian structures in the increasingly powerful National Socialist party, suspected communism and fascism of being "twin brothers" and coined the equally dismissive term "communazis."[35] The social democrats maintained a parliamentary vision of democracy, a system they planned to reinstate after the war. They believed that the numerous internal factions among the workers' parties had weakened parliamentary democracy in the Weimar Republic, and thus called for a unification of the splinter groups.[36] The social democrats' wartime plans in this regard did not include negotiations with the KPD.[37]

Given this historical background of hostility, the gap between social democrats and communists might have been unbridgeable after the war. But major areas of concord among party leaders and members, such as in school reform, facilitated general agreement on larger programs of reconstructing the German nation as a united front. This cooperation emerged as the educators within these parties, who had long shared similar educational ideas, came together again after the war to set up schools. KPD and SPD educators agreed on a program for restructuring society because they agreed on the best way to teach antifascism as a unifying political ideology: by creating a unified school system, the *Einheitsschule* on social democratic terms. This membership level of SPD–KPD cooperation is not surprising. The KPD was never a monolithic, uniform organization. Even before the war, membership often had not seen eye to eye with the party elite's hard-line interpretation of policies or events.[38] And after the war, in order to maintain its image as a leading party of compromise, the KPD leadership omitted almost all references in its postwar programs to socialism as a political goal, and to Marxist-Leninism as its fundamental ideology—a policy that remained generally in place until the founding of the GDR.[39]

In the first years after the war, therefore, antifascism connoted a framework of parties and individuals working together for a new Germany, an ambivalent alliance of convenience and conviction. Educational reformers in the Soviet zone in particular generally avoided specific theoretical and philosophical discussions about the future of Germany, instead demonstrating broad ideological consensus and addressing concrete educational tasks at hand. A party that did not work in this spirit made itself vulnerable to public criticism by the other parties. For example, when Anton Ackermann (KPD/SED), cofounder of the cultural association *Kulturbund* (Cultural Union for the Democratic Renewal of Germany, established July 1945), presented the joint KPD–SPD program for "Democratic School Reform" in November 1945, he began with the announcement that the bourgeois LDP party also supported the proposal.[40] The obvious absence of a reference to the CDU here allowed Becher to imply that the Christian Democrats refused to work with other parties for the good of Germany. Later in the speech, Becher called for a secularization of the school curriculum, an issue the CDU vehemently disagreed with. He then emphasized his party's support of the right of all individuals to follow any religious confession or philosophical ideology, thereby portraying the CDU as a dogmatic party that was intolerant of other belief systems. To emphasize this point, he further explained that even a "Catholic worker, farmer or teacher" who accepted the communist program could be a member of the KPD.[41] Noting pointedly that "I believe that we are in total agreement on this point as well with our comrades of the Social Democratic Party,"[42] Ackermann's implied criticism of the Christian Democrats was damning: they were allowing their religious beliefs to become a divisive political issue, and were therefore not working toward the reconstruction of the German nation, and were even prejudiced against other ways of thinking. He reassured his audience that true Marxists and Leninists would not lower themselves to belittling other beliefs, and noted that his colleagues and he expected the same respect from others. This official line insisted on cooperation among individuals and groups from different ideological traditions to work together, and marginalized those critics of specific political differences by repeating the need to focus on the larger picture of moving ahead.

Yet antifascist democracy was not only a strategic ideological battle cry for parties to work together. Antifascism was at the core of emerging new concepts in the Soviet zone about the German nation and German culture. Antifascists in the Soviet zone advocated a political-cultural definition of the new nation, in which culture would be the basis of a political consciousness that could be *acquired*; and not an ethnic-cultural one, in which culture constituted an essential, congenital aspect of ethnicity.[43] Antifascism, in theory and practice, was not an ethnic program that focused on a biological definition of the German nation.[44] In direct contrast to National Socialist racial policies, as well as earlier conceptions of Germany as a natural, biological, cultural community, antifascism connoted voluntary, achievable membership in the nation.

The state's responsibility herein was to provide all citizens with equal opportunities for active participation in the nation, an undertaking in which the school played a key role. Because the "new school" would teach all pupils the same knowledge and cultural tenets, classrooms—theoretically—equipped young people with identical tools for political agency. By smashing the "educational monopoly" of the upper classes, antifascist educators dreamed of a nation comprising groups that had learned to be equal, using their common cultural knowledge to work as a unified collective. The architects of the new German nation defined antifascism as a political and cultural program that could be taught and thus learned, thereby implying that the Western zones were free to participate in this version of the German nation if they so desired. A large number of Soviet zone educational reformers in fact hoped to convince their colleagues in the West to cooperate at least in cultural realms, especially in restructuring the school system throughout Germany.

Antifascists used their cultural-political definition of the nation to emphasize the German character of their ideas. Although inspired by the movement's heritage as a protest movement in Italy and a unifying ideology for troops in the Spanish civil war, antifascism in the Soviet zone had neither internationalist nor Soviet pretensions.[45] Assumptions of "Sovietization," that Moscow and its German agents quickly stamped out German political culture, deny that Germans in the Soviet zone had agency in the governance of their zone.[46] The power relationships between the occupier and occupied did not occur unidirectionally.[47] Moscow feared strong German and Allied resistance against a Soviet-style political system for the Soviet zone, and it also supported the KPD leadership's belief that a left-wing coalition and parliamentary democracy were logical, temporary steps to achieving traditional communist programs at a later date.[48] "We are of the opinion," stated a June 1945 declaration of the KPD, "that it would be wrong to impose the Soviet system on Germany, because this way does not correspond to the present conditions of development in Germany. We are rather of the opinion that the decisive interests of the German people in the present state call for another way for Germany, the way of the construction of an antifascist, democratic government, a parliamentary-democratic republic with all democratic rights and freedoms for the people."[49] The SMAD oversaw all of the German Education Administration's (*Deutsche Verwaltung für Volksbildung*, DVV) proposals, but German educators in the Soviet zone reacted with hostility to suggestions that their ideas actually emanated from the SMAD.[50] The "new school" did not merely appease Moscow, even if the Soviet zone took care to maintain positive relations with its occupier. As a constellation of perceived democratic and socialist-humanist practices, based upon international cooperation but national consciousness, antifascism's references were in Germany and not in the Soviet Union. The school gave Germans in the Soviet zone an opportunity to define their nation for themselves—in explicitly antifascist, German, political terms.

Just as the Soviet zone was not completely dominated by unilateral directives from Moscow, neither was it bound to conform to the Western zones' policies or ideologies. Antifascist educators carried a unique definition of the German nation into Soviet zone classrooms. Again and again, antifascist reformers claimed that the lessons of the French Revolution about civic equality and rational self-government had finally been brought onto German soil—with specifically German, socialist-humanist improvements. These aspects generally focused on Germany's "cultural heritage," which included scholars such as Johann Wolfgang von Goethe and Friedrich Schiller. By laying claim to this cultural heritage, antifascists named the Soviet zone, and not the Western zones, as the true Germany. KPD and SPD antifascists strengthened this position by pointing to their own resistance activities during the war, and to their new antifascist political system, which had ostensibly eliminated fascist ideology. Then they accused the Western zones of still being trapped in Germany's fascist past.

Antifascist claims that the Soviet zone was the only legitimate heir to the German nation contained two mutually exclusive goals. The first was to convince the Western zones of the need to unite all of Germany under the banner of antifascist democracy. The second was to discredit the Western zones if the first goal should fail, and to blame them for German division. In this vein, one of the constantly proclaimed objectives of Soviet zone educational reformers was to work for the unity, *Einheit*, of all of Germany, even after a reconciliation of the two halves seemed impossible. In November 1948, the Soviet zone Saxony's educational minister Helmut Holtzhauer reported on "Today's Political Problems." He lamented the fact that German unification was no longer under discussion in the Western zones.[51] The topic had not by any means disappeared among educators in the Soviet zone, not even after the June 1948 London Conference. Attended by the United States, Great Britain, France, Belgium, the Netherlands, and Luxembourg, the conference recommended the creation of a separate West German state. Shortly thereafter, the four-power joint command of Berlin ended due to unbridgeable ideological differences between eastern and western military administrations, dividing the capital city into East and West (and necessitating the creation of a new West German capital, Bonn). The western powers and the Western zones had chosen at least a temporary division of the nation.

Antifascist educators in the Soviet zone continued to hope, in some cases into the early years of the GDR, that their plans for an educationally and culturally unified Germany might one day erase the political lines of division between East and West.[52] These aspirations hinged upon the Western zones accepting the Soviet zone plans for restructuring social institutions, such as the school, according to an antifascist model. But the use of antifascism as a political orientation and as a set of guidelines for structuring everyday routines and attitudes found almost no support in the West, particularly after the western powers' plans for Germany were announced. This situation was complicated by the increasingly unstable antifascist democratic bloc system of the two labor and two bourgeois parties. The SMAD sought to increase its authority in 1948 by allowing new political parties to join the bloc, weakening the bourgeois parties' voices by reducing their percentages.[53] Thus, while the Western zones gradually pulled westward, the Soviet zone presented itself even more emphatically as a unique, non-western (nonbourgeois, nonimperialist) entity on the German landscape. Antifascism became even more clearly identifiable as a Soviet zone ideology. The Soviet zone's focus on antifascism and antifascist education, as a cultural and political program not accepted by the West, thereby contributed to German division, *Spaltung*, between East and West.

The focus on culture, in terms of language, literature, and philosophy, as a cohesive function alluded to a further, emotional aspect to the debates about Germany's future and the evolving definition of antifascism. With obvious shame, politicians and educators referred repeatedly to the collapse of Germany's previously much-admired educational system. Speakers referred to the National Socialists' abuse of the school system as a "national disgrace," claiming that the "Hitler years" had brought down the educational level to record lows, undoing the educational work of decades.[54] The 1945 joint call of the Communists and Social Democrats for a democratic school reform closed with the plea that the German school should help the Germans regain the respect previously afforded them by "the world."[55] Germany had once enjoyed the status of a nation of well-educated and knowledgeable citizens, a *Bildungsnation*. After twelve years of Nazi rule, it had become a nation whose educational system was no longer in the service of knowledge, and this status embarrassed its reformers.

Most antifascist educational reformers went further in their criticisms and claimed that postwar reforms had to correct the structural inequalities in the German school system that had existed even before the National Socialists' reign, drawing particularly on nineteenth-century educational reforms for their postwar goals. The first sentence of the 1946 Law for the Democratization of the German School expressed respect for achievements of the earlier school system, but blamed its class-oriented character: "The German school was—in spite of its considerable heights before 1933—never a site of real democratic education of youth to become free citizens, aware of their responsibilities and self-confident. It was a school divided by class."[56] The "new school" would return to the tradition of locating Goethe and Schiller at the core of the curriculum. But now these lessons would be made available to everyone in the "unity school"—not only the elite classes—further proof that social democratic ideals continued after the war to dominate pedagogical thought.

In fact, revolutionary rhetoric all but disappeared from the KPD's postwar speeches, replaced by consensus with the SPD on the need for reform within legal structures. Official allusions to fascism's evolution focused primarily on rewriting militaristic and nationalistic school lessons, and much less on the traditional communist rhetoric of capital or class issues. Antifascism offered socialist humanism and democracy, communicated to the population through schools, as political solutions to overcoming fascism. In the first Soviet zone years, "democracy" and "fascism" referred to concepts that the Allied Command and the general population could easily accept and that provided legitimation for the new political structures. Democracy did not mean a Soviet people's democracy but rather a limited parliamentary democracy with elected provincial assemblies (*Landtage*)—although by the spring of 1947, the centralizing maneuvers of the SED and SMAD began to strip the assemblies of their powers.[57] Correspondingly, the traditional communist interpretation of fascism as a consequence of economic inequalities found almost no mention in speeches or policies, especially in cultural-political discussions. Instead, KPD and SPD leadership most often referred to fascism in terms of its totalitarian racist and militaristic politics. Typical of this strategy was Ackermann's observation that "Hitler trickled [the idea] into the heads of the German youth that the supposed higher Germanic race was called upon to exterminate the so-called inferior races."[58] Similarly, when Paul Wandel (KPD/SED), director of the DVV, spoke at the First Pedagogical Congress in August 1946, he demanded that educators eliminate all the elements from the German school that had prevented the nation from becoming a "land of peace, progress, and freedom."[59] Such discussions about antifascist democratization focused on a program easily accepted by all parties: create the conditions for Germany's peaceful existence domestically and with other nations. This discursive strategy of focusing on objectives like peace and normalcy allowed for a broad multipartisan political consensus on very general goals that could be specified later as needed in specific political and institutional areas.

The use of the school system in this sense implied a partial triumph, especially in educational circles, of the social democratic interpretations of fascism, antifascism, democracy, and school reform that had developed during the Weimar Republic and the war years. There had been no need for a communist revolution in the SPD's prewar view, and this belief remained unchanged in the postwar years. It was enough to reorganize and stabilize existing social structures, they insisted, such as the educational system. Schools that institutionalized inequalities and militaristic thought, according to the SPD, had been one of the factors in the rise of fascism. Neither the pedagogy behind antifascist democratic education nor the schools themselves were revolutionary or socialist; they were social democratic. This situation would change in the early 1950s, but for the moment, the existence of historical social solutions such as the *Einheitsschule* determined the political-cultural landscape of the Soviet zone.

Educators in the Soviet zone proposed to use "antifascist democratic education" to teach young people and the larger school community to lay claim to their German past and to promote a Germany that was more unified socially and culturally than it had ever been. From the structure of the unity school to the foundational myths for the nation it communicated, antifascist democratic education was a self-consciously German creation. Like all institutions in this period, the school was subject to a complex set of relationships between the Soviet Union and the Soviet zone, and it did not have full autonomy.[60] But when German educators outlined the role of the school in constructing Germany's future, they wanted to create a "national consciousness" (*Nationalbewußtsein*), insisting that a "German democratic school is part of our unified democratic Fatherland."[61] Relations with the SMAD were a constant source of worry to German administrators in the Soviet zone, but their vision of the "new school" and the new, antifascist society remained a German one.

Unity and Division

With the establishment of two separate states looming on the horizon, antifascist educators had presented a common school experience for all of Germany's young people as an antidote. By offering all Germans the same cultural knowledge through a comprehensive school, educators in the Soviet zone hoped that the school would create a single, antifascist, democratic society. But antifascist educators failed to realize one of their key political objectives for the "new school": instead of helping unify the German nation, the development of the "new school" contributed to an educational and cultural division of Germany. Antifascist educators hoped that a collective cultural consciousness, transmitted through a unified school system, would pave the way for a common political consciousness that had not yet existed in German history. "We thus call out to all Germans," wrote one official emphatically, "Fight with us for a school from which young Germans will go forth who will transport the *idea of unity* into the political, cultural, and economic life of Germany and thereby serve the *firm establishment of German unity*."[62] Young people's ensuing sense of a common national culture would facilitate the creation of a future, undivided Germany. Of course, educators' calls for German unity assumed a specific kind—one based on socialist humanism. This development, Soviet zone social democrats and communists believed, had been the true will of the German people since at least 1848. The next revolution, they claimed, would succeed because it would be purged of deceptive bourgeois elements.

The appeal to work toward a single German nation belied a reality understood all too well by many educators in the Soviet zone: German unity had disappeared with the end of the war. Educators in the Soviet zone had already seen a number of events that informed them of newly drawn lines in the sand. From the January 1947 official bizone conglomeration of the British and American zones into one economic zone, to the February/March 1948 London Six-Power Conference mentioned earlier (United States, the Great Britain, France, and the Benelux states) and its announcement of the massive infusions of money into western states with the Marshall Plan, many Germans in the East and the West saw division as a certainty. If some Germans in East and West viewed the creation of two German states as a temporary measure that would later enable unity,[63] others did not so easily accept this explanation. Thus a March 1948 editorial for the American zone *Frankfurter Hefte* envisioned the possibility of the "two halves" growing so apart politically and ideologically that after ten years they would be permanently separated.[64]

Educational reformers in the Soviet zone were still willing to continue the fight, even as they often acknowledged the near hopelessness of the cause. By the fall of 1947, educational administrators had already begun to discuss the "battle for Germany's unity" (*Kampf um die Einheit Deutschlands*). Educators hoped to achieve this objective through the "creation of a truly democratic national consciousness, and with this the ideological support for the struggle for a unified, undivided German republic."[65] Some educators believed that they could still influence the future of Germany's internal borders. As Erwin Marquardt (SPD/SED), vice president of the DVV, announced at an educational conference in December 1947, "We are faced with the question of how the unity of Germany will come about, which is also a decisive moment for the construction of culture."[66] National destiny and the reconstruction of a national culture were interdependent in Soviet zone reformers' views. Other educators reacted to the developing political situation with more alarm. A January 1948 educational presentation, for instance, discussed the threat of the nation's division and the ensuing damage to German culture and education, concluding that the crippling effects of division would be permanent.[67] At the end of the year, the tone remained the same. Attendees of a December 1948 school administrators' conference listened to a sobering speech about how Germans in the Western zones had stopped discussing the possibility of a single Germany. The speaker added that Germans in the new western capital of Bonn were planning to negotiate with the western Allies for a West German state, one based on a divided Germany and that would exist "for many years."[68]

Educators throughout the Soviet zone despaired. Even those Soviet zone Germans who foresaw the split as temporary doubted that their zone would benefit from this situation.

Antifascist educators were not ready to accept just any unity, though, nor unity at any price. Although they feared that the Western Allies and western Germans had already laid in an irreversible course away from the Soviet zone, they believed that the time for compromises had long been over. The campaign for German unity was thus launched unequivocally as a German socialist-humanist project, whose goal was fundamental social upheaval. The failure of Soviet zone educators to convince western Germans to participate in the revolution did not force educational administrators in the Soviet zone to doubt the validity of their cause; they saw instead proof that the imperialist, reactionary forces in the Western zones remained quite strong. As the former Weimar educational activist Paul Oestereich insisted, "As long as we are obliged to live in a zone, we at least want to be the more progressive part and change the world as much as we can."[69] If the Western zones were lost to the cause of a unified socialist German nation, then educators in the Soviet zone planned at least to continue with their plans as best they could. The very idea of unity, the *Einheit* so ardently hoped for by antifascist educators, in this manner became one of the main factors of division, *Spaltung*.

The conflicting goals of unity and division had not only ideological dimensions but also practical ones. Obstacles to constructing one Germany through the educational system plagued educational administrators from the first days of school. Some of these problems were purely bureaucratic, but they were more often at least partly the result of ideological differences. For instance, educational journals licensed only in one zone separated teachers' abilities to exchange pedagogical ideas between the West and the East. Furthermore, the Western zones' individual *Land* school administrations did not automatically accept the school-leaving certificates in the Soviet zone as valid in the West, thus eliminating incentives to keep the course offerings or requirements of the school systems similar. Moreover, the 1948 currency reform slowly eliminated the previously unifying effect of teachers living in one zone but working in another—especially in divided Berlin.[70] Whereas those teachers traveling between zones had served a tangible function of permeating a supposedly rigid border, they were now paid in a currency invalid for buying groceries or paying rent in their own neighborhoods. The economic and political division thus became too difficult to ignore. Antifascist educators believed that a unified classroom experience would lead the nation toward unity, but the unified educational system was still subject to the movements of actors in the Western zones, from military administrations to educational reformers. Ultimately, the "new school" contributed to the construction of a different kind of zone from the Western zones. The people and their daily educational practices—attending or graduating from a school with unique, identifiable Soviet zone characteristics—drew the line of division between East and West more distinctly. Antifascist democratic education, transmitted through the *Einheitsschule*, remained a product of the Soviet zone.

The Bearers of Antifascism

In 1947, the educational reformer Johannes "Hans" Siebert wrote in his report on the previous year of the culture and education division of the SED that it was not enough to eliminate Nazi teachings from the "new school's" curriculum. When antifascists spoke of the "catastrophe of Nazism," he insisted, they were not merely speaking of the twelve years of the Hitler dictatorship. Rather, they also meant the damages done by the "centuries-old Prussian-German authoritarian state."[71] He called for nothing less than radical change: "the foundation for our entirely intellectual (*geistiges*) life must be fundamentally altered and entirely rebuilt anew." These stirring words were not empty phrases. Siebert had immigrated to England in 1936, where he spent eleven years convincing the German exile community and sympathetic local audiences that only by reeducating the young in antifascist democratic thought would postwar Germany recover from National Socialist ideology.[72] Returning to Germany in 1947 after helping other German exiled communists emigrate home, Siebert joined a large group of antifascist educational reformers who tirelessly worked for the construction of the "new school" and a new Germany.

These key individuals responsible for defining antifascism in the schools exercised political and pedagogical influence through the creation of curricula and school texts, teacher training, and the construction of educational policies from the zonal down to the regional level. They did not, however, conform to the typical profile of most party leaders in the SED inner circle, who had often been in Soviet exile and returned as part of an elite party cadre.[73] In this sense, the historiographical assumption that the real political leaders in the Soviet zone were exclusively Moscow puppets is inaccurate and misleading. Persistent claims that Stalin entrusted only a small group of German communists to fly in from Moscow and "[take] over the reins of power on the ground" ignore the reality that "power on the ground" extended far beyond Moscow's or even East Berlin's reach, and that many social reformers throughout the Soviet zone were not necessarily "like-minded comrades" in the narrow sense that such a phrase implies.[74] Most antifascist educational reformers, from teacher instructors to pedagogical theorists to educational policymakers, had originally been trained and worked as teachers before the war. They were first and foremost educators who saw their primary responsibility in teaching young people, and who genuinely had long believed in the potential of schools to construct a better society. These individuals had seen a link between education, society, and politics when they were teachers in progressive Weimar classrooms, and this experience forever marked their later discussions about the future of the German nation and the "new school." Without recognizing these reformers' preexisting attitudes toward educational and national reform, the historical scholarship uncritically repeats the GDR's own forty-year evolving representation of its early days. Antifascism was not the easily coopted, meaningless founding myth of a few party schemers, as too many historians have insisted, even if historiography written in the GDR itself was constantly rewriting the origins of antifascism.[75] It is a gross misrepresentation of those postwar years to conclude that "most East German communists" had nothing more from antifascism than a mere superficial knowledge of empty symbols.[76] A more critical approach in particular to these early years of Soviet zone antifascism forces a renewed, flexible, and even positive definition of antifascism and its proponents.

Later biographies written in the GDR about the "activists of the first hour"—those personalities who played an important role in bringing about societal reforms in the period of "antifascist democratic transition"—insisted that these educational reformers came from the same working-class background that then dictated a later single vision of mission and antifascist purpose. Yet, a close examination of the rhetoric reveals a picture of antifascists whose experiences and worldviews did not always comport with the official version of events.[77] The 1989 GDR work *Wegbereiter der neuen Schule* (those who paved the way for the new school), for instance, profiled thirty-five education reformers active in the immediate postwar years. Even at that late date in the GDR, the editors had difficulty introducing the individual life stories as a convincing harmonious collective: "The life sketches portrayed here make clear the breadth of the alliance of those involved in the struggle for the new creation of the educational system under the direction of the Socialist Unity Party of Germany. As different as the backgrounds, educational experiences, and path toward socialism were of those persons presented here, they were all rooted in the antifascist democratic movement."[78] The admission that not all activists shared the same working-class credentials and KPD membership was tempered by claiming that these individuals all eventually overcame their misguided ideas, as in the case of one of the first recipients of the "Honored Teacher of the People" award, Käte Agerth: she had worked as a Red Cross nurse in World War I, "initially influenced by a false sense of patriotism." In another instance, the educational researcher and teacher trainer Wilhelm Heise had passed his *Abitur*, the final examination of the elite upper secondary school, the *Gymnasium*, and had followed a traditional academic career. This background precluded his credentials as a member of the working class. The authors therefore could not follow the standard pattern of making introductory remarks about the formative nature of his family background, and merely skipped over that part of his troubling bourgeois life. They opted instead to emphasize his postwar communist activities.[79] Here, as in other examples, even the determined Marxist-Leninist historians of the GDR could not successfully force the divergent pasts of postwar antifascists onto a narrow path of working-class family life and unfailing communist loyalty. More important for these Soviet zone educational reformers was the continuity in educational and ideological orientations that they had shared before the war.

A belief in education's ability to reform society along social democratic and antifascist lines united the individuals responsible for the codification and dissemination of antifascist education. Since their past experiences and backgrounds differed, points of intersection in their thinking and experiences are thus striking. For the most part, education reformers in the Soviet zone had worked in schools before World War II, frequently as teachers. They were often social democrats, sometimes communists, and usually male if they held an administrative position. Many of them had frequently switched back and forth between the two parties before and after the war, occasionally more than once, underlining their ambivalent stance toward the KPD's goals of socialism and people's democracy. They had often taught during the Weimar period at so-called experimental schools (*Versuchsschulen* or *weltliche Schulen*), which generally practiced some form of reform pedagogy. In addition to being secular, these schools educated boys and girls together (the practice of coeducation), and often had a commitment to educating workers' children.[80] The revolutionary and international nature of such reform programs should not be underestimated. If it is tempting now to read the "new school" as a socialist or communist precursor to later GDR educational reforms, it is imperative to remember the roots of such educational directions in a worldwide discussion of education and society—one need think only of the early twentieth-century work of pedagogues John Dewey in the United States, the Swede Ellen Key in Germany, Maria Montesorri in Italy, or the anthroposophist Rudolf Steiner in Austria.[81]

As former teachers in the Weimar period, antifascist Soviet zone educators had not begun their careers with the intention of someday playing important roles in larger societal reforms.[82] Typical of this profile were individuals such as Robert Alt, who taught at Fritz Karsen's Karl-Marx-Schule in Berlin-Neukölln until 1932 or 1933, after which he worked as a teacher at the Jewish schools in Berlin until his internment at Auschwitz in 1941. Alt had joined the SPD in 1924, switching to the KPD in 1933. In 1945 he rejoined the SPD, and became an SED member in 1946. Active in teacher training and the creation of instructional materials after the war, Alt was one of the key educational theorists in the Soviet zone and GDR.[83] Marie Torhorst, one of the few women active in postwar educational politics, represented another variation on this biographical theme. She also had worked at the Karl-Marx-Schule in the same period as Alt, even cofounding a local branch of the teachers' union with him in 1929.[84] Torhorst was best known in the GDR for heading the education ministry in Thuringia, the first female political minister in the Soviet zone.[85] Like Alt, she had first been an SPD member, in 1928; she joined the KPD in 1945 and the SED in 1946.

The differences in these reformers' backgrounds underline the wide variety of experiences they brought with them into the "new school." Alt was an assimilated Jew who was interned by the Nazis in Auschwitz and later survived the bombing of the concentration camp prisoner ship *Cap Arcona*.[86] He maintained lifelong international contacts to other educators, including serving as president for the "Society for Cultural Ties with Other Countries." Significantly, though, he expressed almost no interest in the Soviet Union.[87] In contrast, Torhorst was a pastor's daughter with a strong affinity for the Soviet Union. She left for her first of many trips to that country in 1932, remaining there for six months.[88] She stayed in Germany during the war, holding odd jobs in kitchens and businesses and working in the communist underground in Berlin. In 1943, she spent two months in a work camp after hiding a Jewish communist in her home.[89]

The heroic leitmotiv of antifascist educators from diverse backgrounds, working together to avoid the mistakes of the past, surfaced repeatedly in subsequent historical accounts concerning the rise of the GDR school. Yet these heterogeneous backgrounds demonstrate the need to rethink the uses of antifascism in postwar education in the Soviet zone. As educational administrators and theorists, educators such as Alt and Torhorst were not part of the SED elite inner circles such as the Moscow-trained "Ulbricht Group" around the apparatchik Walter Ulbricht. Nor did they fit neatly into a category of working-class communists. They were committed educators who based their definitions of antifascism upon consensus and compromise. Political ideology played an integral part in their educational philosophies, instead of a leading role. In this way, educators both helped develop, or "elaborate," the applications for antifascism and transmitted these ideas in the classroom.[90] The inclusion of their many viewpoints accounts for antifascism's successful development in Soviet zone schools as a narrative framework to explain the redefinition of the German nation.

* * *

Antifascism was at the heart of the Soviet zone's program for reconstructing the German nation; the "unity school" was of central importance to this project. As the core of a new political-cultural national consciousness that was German and social democratic, antifascism provided a meaningful narrative of self-expression and self-definition to residents of the Soviet zone.[91] It became a meaningful founding myth of the new nation that created a sense of common purpose among parties and individuals who otherwise might have begun to focus on their differences, while providing a unified sense of destiny and direction to eastern Germans.[92] Antifascism, particularly as it was articulated by educational reform programs, represented the sort of "creative political action"[93] necessary for the metamorphosis from multiple fragmented experiences to a sense of collective national consciousness. Antifascist education became an accepted strategy for identifying the ideological borders of the nation, even though at the outset it was not clear where the internal geographical boundaries of the new nation would lie. By contributing to the perception of a unified experience for that part of Germany in areas such as education, it pulled the Soviet zone mentally and experientially away from the Western zones. Antifascism, defined by its architects as a German movement, permitted reformers in the Soviet zone to create a safe sphere for the development of a German consciousness that was not dictated by Moscow. Furthermore, because it based itself upon a German cultural heritage, antifascism allowed the Soviet zone to proceed in a policy of claiming its status as the worthier inheritor of the German nation and the "better Germany."[94] Finally, antifascism provided Germans in the Soviet zone with an argument for full membership of their half of Germany in the community of nations by creating an educational system that would eliminate fascist ideology from the minds of young Germans.

CHAPTER 2

Setting up the School

Constructing the best educational system for the implementation of antifascist democratic ideas posed a significant challenge to educational reformers. At this intersection of theory and practice, educational administrators discovered the structural limitations on educational reforms. The school's purpose, organization, and administrative infrastructure helped determine how antifascist democratic education was transmitted in the classroom. The ensuing educational system, the "new school," gave practical dimensions to the theoretical concept of antifascist, democratic education. The core of the resulting educational system was a comprehensive, nontracked "unity school" (*Einheitsschule*) of eight years, which left the optional, elite four-year upper secondary school intact. This system, which had its roots in nineteenth-century egalitarian ideas about expanding citizens' access to education and thereby their participation in the nation, was at the base of antifascist educators' understanding that the school was a key instrument in constructing German national consciousness. Similarly, new faculty reflected the primacy of democratizing the educational system. Regional school boards fired or did not rehire the majority of teachers with National Socialist affiliation who had taught during the war, necessitating the immediate training of thousands of so-called new teachers (*Neulehrer*). Also, the German educational administration oversaw the creation of curricula, textbooks, and teacher training, although the Soviet Military Administration's (SMAD) department of education had the final authority on all matters. After all these details had been realized, educators then still had to ensure that they could attract the Soviet zone's two and a half million children to class and keep them there.[1]

These practical considerations defined the possibilities and limitations of antifascist education. Many of the daily aspects of antifascist education resulted from issues universal to any school system; others were specific to the peculiarities of the postwar years. A number of factors affected antifascist educators' visions of what the "new school" could accomplish, from German mistrust of Soviet-style pedagogy to unheated classrooms, often altering administrators' plans. The national system of antifascist education ultimately unified only one half of Germany, pulling the Soviet zone away from the Western zones.

Historical and Cultural Legitimacy of the *Einheitsschule*

The idealistic visions of German unity rested upon the central structural element of the antifascist educational program: the *Einheitsschule*. Like so much of antifascist education, the school's creation and legitimacy drew upon previous German educational and social reform movements. The first major German debates connecting egalitarian education to national political agendas had surfaced in the early nineteenth century, when Wilhelm von Humboldt unsuccessfully proposed a unified horizontally tracked school system in 1810.[2] German progressive educational reformers began referring to the concept of a horizontally tracked system as an *Einheitsschule* in the mid-nineteenth century. They linked the educational structure to political objectives, claiming that a citizenry with the same "unified" educational basis was a prerequisite to a unified German nation. The Weimar period saw partial implementation of this system in some reform pedagogy schools, although those versions of the *Einheitsschulen* retained significant internal tracking structures. This clear German pedigree helped ensure the *Einheitsschule*'s construction after 1945.

The antifascist democratic "new school" evolved in stages during the Soviet occupation, a product of interactions between Germans and their occupiers, between different political traditions, and between adults and children. Those who pushed to reopen the schools had varying motives. Many teachers and parents, for example, wanted the school to begin its work of denazifying pupils right away in order to structure young people's turbulent lives and keep them out of trouble. Other educators realized that being among the first group of social reformers offered potential job security. During the war, many Germans in exile or in concentration camps had developed plans for the postwar school system. After the war, competing interest groups again clashed over educational aims and ideologies. Some of these tensions reflected prewar discussions; others resulted from where the various planning groups had spent the war—in concentration camps, "internal" exile in Germany, abroad, or underground.[3] Returning German exiles ("reemigrants") from the Soviet Union made up the largest single group of future party cadres, but Moscow was not the only wartime factory for blueprints of a better postwar Germany.[4] The antifascist committee in the concentration camp Buchenwald included an educational faction, and had proposed a detailed multipartisan program for postwar Germany as well. Many of these members, such as the KPD activist Walter Wolf, went on to become central figures in educational policymaking for the Soviet zone.[5] Communist groups exiled in London also worked throughout the war on plans for future social programs, often enlisting the support of other countries' citizens and politicians to help realize these postwar German plans.[6] Yet at war's end, no official blueprint for antifascist democratic education existed.

Still, neither schools, teachers, nor pupils waited patiently for permission to begin instruction. By the time the SMAD issued a decree on September 13, 1945, for school to begin throughout the Soviet zone on October 1, many groups and individuals had already been organizing classes.[7] Popular opinion strongly favored opening schools as soon as possible after the war, as in the case of a Berlin-Prenzlauer Berg school in which parents put up posters to ask educators to begin classes again.[8] Contradictory and spontaneous directives came from local SMAD and German educational authorities as well as from individual teachers and school directors. In some cases these mandated schools to open their doors immediately after the war, only to close them by another decree. Particularly after such unorganized beginnings, public acceptance was a precondition for a successful implementation of educational reforms. All the occupying powers had agreed to involve Germans in the reconstruction of their nation, so that none of the military administrations had enough structural power to introduce educational changes without the support of the Germans in their zones. Germans in the Western zones in fact maintained the tracked structure with some minor changes, in spite of American protests. SPD and KPD educators enjoyed broader, cautious support in the Soviet zone for more sweeping changes, but educators throughout eastern and western Germany viewed recent Soviet culture, including pedagogical developments, with skepticism. Aside from organizational considerations, the SMAD also did not have the resources to force families to send their children to school. Antifascist education thus offered an apparent rupture with the Nazi past without demanding that Soviet zone residents accept another country's educational system. SPD and KPD educational policymakers in the Soviet zone generally agreed with Soviet (and American) postwar recommendations to eliminate the vertically tracked system, but neither Western nor Soviet zone Germans were ready to accept an entirely new, foreign model of schooling. Antifascist educators won support for the *Einheitsschule* because they could point to its German origins. The Soviet zone public accepted the school reform because they did not see it as the product of Moscow.

Germans have historically identified strongly with their schools, and have demonstrated an interest in foreign educational systems primarily in order to affirm German traditions.[9] In the Soviet zone, this ambivalence toward other educational systems was particularly high because of anti-Soviet attitudes. Long before the war's end, Germans began to express anxiety about Soviet treatment of a defeated Germany. Antifascist educators in exile addressed these fears, resulting for instance in a 1942 London counterpropaganda broadcast that reassured "German youth and its educators" that the Allied powers were in "absolute agreement" on all points, such as how to punish war criminals and annexation issues, "and that no different treatment can be expected from the West as compared with the East."[10] Nazi propaganda had contributed in large part to the images of the Soviets as monsters, and the KPD exiles in London worried that antiwar propaganda would be ineffective if Germans were terrified by a potential Soviet victory. Germans also suspected that the Soviets would be more interested in revenge than U.S. or British troops. When it became clear that Germany would lose the war, this fear turned into panic and resulted in a mass exodus to the Western zones.

More specifically, although German educators had once admired Soviet pedagogy, this was no longer the case after the war. From 1917 to 1930, the educational theorists Anatolij Vassilevič Lunačarskij and Nadešda Konstantinovna Kruspkaja (Lenin's wife) had further developed the "work school" ideas of the German reform pedagogue Georg Kerschensteiner. But Stalin's ascent to power brought about a change in the pedagogical winds, ringing in the era of "pedagogy without children," so called for its rejection of reform pedagogy's central focus. From 1931 until the death of Stalin in 1953, the work of Anton Semëovi Makarenko dominated Soviet pedaoical practice. Makarenko, most famous for his development of work colonies that "resocialized" alienated, orphaned boys, elaborated a theory of the "social organic" nature of the collective. According to these ideas, "productive work" within the collective would create "new men" who would then bring about the socialist revolution.[11] Germans in East and West balked at the rigid nature of these developments in Marxist-Leninist pedagogical theory.

Reform pedagogues in the Weimar period had made significant strides toward a child-centered, flexible approach to teaching. Makarenko's opposition to spontaneity and his emphasis on discipline did not find a receptive audience in a nation with fresh memories of authoritarian Nazi pedagogy. The postwar atmosphere thus acknowledged the faults of the National Socialist school system while desiring to rediscover untainted German traditions. During a speech at the first Soviet zone Pedagogical Congress in August 1946, the director of the DVV, Paul Wandel, assured his listeners of the German nature of school reforms: "The basic ideas of the new school law and the reconstruction of the German school are not discoveries from our days and not plans thought out by us . . . In its basic ideas, our new school is founded on very old ideas."[12] With the 1946 school law, therefore, antifascist educators in the Soviet zone agreed upon their version of the clearly German eight-year, nontracked *Einheitsschule*, followed by an optional, more traditional, academic secondary school, divided into three "branches."[13] The school reform thereby strengthened a sense of Soviet zone unity, because it responded to German fears of Soviet control of cultural and political realms.[14]

Although the upper secondary school fell under the responsibility of the DVV, the school's function as an elite academic institution kept it out of the central discourse surrounding schools in the Soviet zone. Because only a small percentage of the school population continued to the upper secondary school, educational reformers devoted more of their time to the *Einheitsschule*. When administrators mentioned the secondary schools, they most often criticized their conservative and reactionary faculty. A 1947 Thuringian report, for instance, noted that 183 teachers belonged to the CDU and 57 to the LDP at the beginning of 1947, compared to 96 members of the SED; SED membership outweighed the combined CDU and LDP numbers in the lower grades. The report concluded that "only a barely perceptible democratic wind is blowing in the upper secondary schools."[15] Antifascist educators' disdain for the upper secondary school also stemmed from other philosophical convictions. Because these schools were more academic, the upper secondary school needed teachers with specialized knowledge. In a time of teacher shortages, this meant that its ranks were overwhelmingly filled with teachers who had previous teaching experience and training. This image did not fit in with what educational administrators wanted to present as the new educational system, which was supposed to be constructed with new teachers. Antifascist educational reformers therefore excluded the upper secondary school from "new school" discourse except as a negative example.

The curriculum for the *Einheitsschule* did not represent a significant break with past academic traditions. Antifascist education in this sense was not revolutionary, in part because public support did not exist for more radical measures. Besides a basic core curriculum, many schools offered a specialized course selection of either classical languages, modern languages, or sciences in order to facilitate pupils' transition to the upper secondary school, which offered those three subjects as "branches" or "tracks." This format appeared to fuse the idea of equal education at the basic course level with an academic curriculum.[16] However, the course selection was not available at all unity schools, and educational administrators discontinued its practice in 1948 after deciding that it did not fulfill the principle of a comprehensive, unified education. Furthermore, the lack of adequately trained instructors or insufficient classroom space meant that many pupils in the Soviet zone did not attend the full thirty-two-hour school week mandated by educational regulations, so that discussions of a uniform education often remained a theoretical ideal. Still, arguments surrounding the new system were strident despite the continuation of many curricular traditions. Three main issues became the most tangible points of contention in both public and educational circles: the numbers of years that the *Einheitsschule* took away from the upper secondary school, the implementation of coeducation, and secularization of instruction. Additionally, educators and the broader public alike accused the *Einheitsschule* of offering substandard academic education.[17] This skepticism accompanied the school throughout its existence until 1959, although there was never a significant movement to revert to the previous tracked system.[18]

Structurally, therefore, antifascist education referenced German educational history. At least as important as the national unity project behind the *Einheitsschule* were the school's identifiable and acceptable German roots. Soviet zone residents accepted a modified version of the school that allowed for a classical academic education in the upper secondary school, but did not support further reforms. Educators in the SPD and KPD, who continually pointed to the wisdom of 1848's campaign for social unity, believed strongly in the *Einheitsschule* concept of educating all citizens equally and uniformly. The bourgeois parties remained suspicious of the new system, but they focused their attacks on specific elements that could potentially be altered within the *Einheitsschule* structure, such as pushing for religious instruction. The compromise of the *Einheitsschule* with a traditional curriculum was one that everyone could accept, and one that Moscow supported as part of a transitional period of antifascist democratization.

Running the New School

Who was in charge of the lessons that pupils actually learned in the class-room? Certainly, the *Neulehrer* were an enduring symbol of the Soviet zone educational system, but they were only one element in the organization of teaching and learning.[19] An entire hierarchical structure with decision making powers stood behind them, one that helped define the contours of antifascist education. One important aspect of the antifascist educational system was its dual administrative bodies. Soviet and German authorities had their own educational bureaucracies for the schools, whereby the Soviet administration retained final authority in all matters. But official administrative, zonal, and regional lines of authority did not necessarily reflect actual practice. Antifascist education developed through the interactions of these different bodies.

On July 27, 1945, the SMAD decreed the organization of several new central administrations, among them the *Deutsche (Zentral)verwaltung für Volksbildung* (German [Central, eliminated from the title in 1946] Educational Administration, DVV), led by Paul Wandel.[20] The son of a working-class family, Wandel had joined the KPD in 1926. During his emigration to the Soviet Union from 1933 to 1945, he had served as the personal secretary to future SED chair and GDR president Wilhelm Pieck before returning to the Soviet zone in 1945.[21] The DVV's responsibilities extended from schools and universities to the department of propaganda. It advised and coordinated decisions between the SMAD and the schools, but it did not enjoy official autonomy. As in all the zones, each German-run administrative body had a mirror SMAD "parent" organization to which it ultimately answered.[22] Both the SMAD Educational Department and the Information Administration regularly evaluated the German administration's performance, often reprimanding the DVV for not doing enough to implement antifascist democratic education. A 1946 memo from the SMAD Education Department to the DVV, for example, complained of pictures of Hitler in schools and unacceptably run-down classrooms: "The German departments of education are not doing enough to ensure that schools are being appropriately repaired, which increases the discontent of teachers and pupils. In a number of schools, there is no heating, the windows have no panes, and the rooms are not kept clean."[23] The tone of such complaints always reaffirmed the SMAD's position of final authority. Still, its decisions to intervene directly were random and its overall influence far from omnipotent.

The common historiographical interpretation of a communist-run school system, buttressed by the fact that former KPD members held key administrative positions, portrayed the school system as dominated by a party loyal to Moscow, and that aimed to turn young Germans into Soviet-style communists. This conclusion is misleading. Internal school structures were much more complex and capable of accommodating different German viewpoints. Furthermore, SED school policy suffered from a divided communist membership. Nor did differences of opinion in the SED fall only along communist-social democrat lines. Extensive KPD in-fighting characterized the Soviet zone and GDR school system to a degree unacknowledged by those who have claimed that the party maintained a military-like "commando" hierarchy of rank and file members following the leadership's orders.[24] The KPD was prone to different visions and rivalries that at times hindered the creation and implementation of policies, and of antifascist education.

Aside from in-party bickering and power struggles, the *Land* regional educational ministries made real zone-wide policy implementation even more difficult. Each region had its own ministry of education that functioned autonomously, a return to the pre-National Socialist governance structure. The DVV did not have any central law or decision making mandate from either the SMAD or the regions. It regularly tried to coordinate and control regional decisions with some success, made easier in 1947 by the mutual signing of the educational ministers' agreement to cooperation with the DVV.[25] No formal methods existed to force an educational ministry to comply with decisions, however, and the SMAD had the final word in all matters. Representatives from each regional administration met monthly for zonal meetings, usually in Berlin, and a considerable amount of correspondence went back and forth between Berlin and the regions in the form of statistics and reports.

None of this gave the German Educational Department the illusion that schools in the Soviet zone were actually doing what Berlin had requested. Indeed, many regional administrators consciously rejected control of their implementation of antifascist democratic education. For example, the Saxon education minister Wilhelm Schneller (KPD/SED) strongly defended his autonomy, warning anyone against making major structural changes in the decision making process. At a meeting of the regional school administrators, he insisted that reform attempts needed to be focused on curricular forms, and not on organizational concerns: "There is no methodical school reform, there is no organizational school reform. There is only a democratic school reform that has to give our school system a different face and especially a different content."[26] Individuals who stood between the DVV and pupils had their own ideas about what should be happening in the antifascist classroom, regardless of DVV plans.

Lower administrators and teachers contributed at least as much to the political atmosphere of the antifascist classroom as district supervisors did. In those areas where regional administrators did not have transportation to observe events in different districts, a typical complaint voiced by a 1947 Thuringian report, schools relied even more on directors and teachers for orientation and guidance.[27] Teaching at the same school did not, however, guarantee a cooperative atmosphere. Faculty in the schools fought their own battles, often shaped by ideological issues, almost always based on pedagogical concerns. The "unity school," like the SED "unity party," suffered from fissures.

Two different and overlapping types of SPD members influenced the structure of the "new school." Some refused to join the fusion of the SPD and KPD to the Socialist Unity Party in June 1945; others became SED members but remained suspicious of the KPD. This absence of unqualified trust in the KPD concerned the SMAD as well as communist school administrators, and GDR historiography would later have a difficult time describing this lack of unity. At the time, however, the KPD and the SMAD saw the advantage of gaining the cooperation of the SPD, whose skeptical membership could have prevented the KPD from gaining the support of the Soviet zone population.[28] Schools could also be left to SPD influence, since political education in the extracurricular youth organization, the Free German Youth (*Freie Deutsche Jugend*, FDJ) fell almost entirely under KPD control.[29] Moreover, the majority of those individuals who entered the SED with teaching experience usually had an SPD background dating to the SPD's educational heyday of the Weimar Republic.[30] This SPD pedagogical influence continued throughout the entire GDR period.[31]

Later GDR accounts identified an SPD "right wing" that "undertook to intrigue against a reform of the schools," including demanding key administrative positions.[32] This sort of "right wing SPD" explanation collapsed two different events together. First, the SPD in the Western zones under the leadership of Kurt Schumacher rejected the idea of an SPD-KPD fusion.[33] This decision permanently splintered the SPD between East and West, furthermore weakening the KPD's position in western Germany by depriving it of unquestioned social democratic support. Second, many Soviet zone SPD members in the SED began to voice discontent soon after the fusion because they did not trust the KPD's willingness to treat them and their demands fairly.[34] Postwar communist and social democratic determination to work together, and thus make good the historical failure to oppose National Socialism, quickly fizzled. Instead, a mixture of negotiation, communist strong-arming, and compromise characterized school politics. Thus, a report to the Saxony KPD in September 1945 noted that SPD members already held the majority of regional school administrative positions. The report emphasized that, with some exceptions, the SPD would therefore not need to be considered for further key positions. The author recommended giving both the Christian Democrats (CDU) and the Liberal Democratic Party (LDP) a few positions to appease them.[35]

At the classroom level, KPD reports always complained that SPD teachers wanted only to pick up where 1933 had left off—that is, a return to reform pedagogy, the early twentieth-century progressive educational theory that was losing favor in the Soviet zone because Soviet-oriented educational administrators believed the practice to be too bourgeois.[36] The reform pedagogy movement represented the worst of many evils for the Soviet Union and many members of the KPD. It had its roots in turn-of-the-century Germany and had quickly become part of a larger western international discussion of a more child-centered, less authoritarian pedagogy.[37] Its theorists included the Chicago professor John Dewey, once courted by the Soviets between the wars but now persona non grata there because of his bourgeois-imperialist heritage.[38]

Communists implied that Germany had fallen behind international education developments, and this prevented the Soviet zone from adopting a communist educational program more suitable for its Marxist-Leninist needs. As county education administrator Herr Richter explained at a conference in Dresden-Wachwitz in September 1949, reform pedagogy did not have the interests of working-class pupils at heart. "But this reform pedagogy could never be the expression of the proletarian class or have any effect in that direction, because the ruling institutions of the state, which was governed by a single class, knew to prevent any attempt in that direction. And the overwhelming majority of reform pedagogues never had the intention to fight the ruling state with pedagogical means."[39] The only appropriate pedagogy for the antifascist state was one whose method and content sought to definitively prevent a return to fascism.

Reform pedagogy remained unpopular with the KPD leadership and Moscow for other reasons as well. It had pushed its way into the curricula and ideology of Weimar schools, an era firmly associated with social democratic teachers. Allowing it to dominate pedagogical practices meant once again permitting the SPD to gain ground in the schools. Moreover, reform pedagogy's emphasis on individuality and self-expression worried the KPD and Moscow. It was a pedagogical practice that rejected authoritarian structures, and it could be identified as the intellectual property of social democrats. Since it was also presumably practiced in the United States, the SED leadership found it unacceptable for a new Germany. But the large number of SPD teachers and the survival of reform pedagogy practices in postwar Germany forced the SED and the SMAD to accept a pedagogical compromise or lose social democratic support in educational and political areas. This was a decision that carried consequences for educational practices over the next half-century.[40]

Meager Soviet zone statistics have resulted in incomplete information about political affiliations and loyalties in the antifascist school. However, the KPD's absolute majority did not mean that everyone supported or even wanted to cooperate with its members. KPD/SED members even complained that some educators refused to deal with those who had been in the KPD.[41] Educational documents did not usually note previous affiliation before entry into the SED, concealing political cleavages. In the 1947 totals reported for party affiliation of teachers throughout the Soviet zone, 27,357 teachers belonged to the SED; 7,300 to the LDP; and 6,111 to the CDU. However, 19,374 did not belong to any party at all.[42] Even assuming that the SED teachers firmly believed in either the SPD or KPD programs can lead to false conclusions. For instance, some teachers joined the SED to ensure better job security.[43] Pronouncing that the vast majority of teachers belonged to the SED hardly provided an accurate picture of their political leanings.

Other educational organizations proved in some cases of equal or greater importance to the political landscape of schools, and thus created a more complex administrative and policy situation than one dictated by communists. The teachers' union (*Gewerkschaft Unterricht und Erzieher*) belonged to the umbrella organization of the *Freier Deutscher Gewerkschaftsbund* (FDGB), one of the first mass organizations to be legally permitted in the Soviet zone in the first year after the war. The teachers' union initially enjoyed considerable autonomy through the early years of the GDR. Paralleling general FDGB developments, different political parties within the teachers' union weakened its supposed unified character.[44] Particularly in the first postwar years, the teachers' union took critical stances on a number of educational issues, although the SED slowly achieved greater influence in the union's ranks. Likewise, the parent-teacher organization *Freunde der neuen Schule* (Friends of the New School) provided a forum for individuals of different political leanings to discuss issues about their children's education in a relatively open atmosphere.[45] "Friends" helped schools in other ways as well, from organizing school parties to repairing damaged buildings. A number of institutionalized opportunities thus existed for parents and politicians to influence educational policy in the Soviet zone, none of which ever exactly followed the programs of the SMAD or SED leadership.

Another administrative problem that affected antifascist education involved hiring and retaining enough teachers who had an acceptable ideological stance and adequate knowledge of their subjects. Of the 39,346 teachers at the end of the war, over 20,000 were fired or not rehired because of NSDAP membership, although this number reflected wide variations in regional practices. Regions that had enough new candidates to train in crash courses for new teachers (*Neulehrer*) could afford to dismiss those Nazi Party member teachers who had been employed before 1945 (*Altlehrer*). Thus a school report in 1946 from Mecklenburg-Vorpommern proudly claimed to be the only *Land* school system to have fired all Nazi Party members, even if they had been deemed merely "nominal" party members; that is, those who had joined after 1937 in order to save their jobs or for similar reasons.[46] Later in the report, however, the district noted that it would soon have to rehire nominal party members to fill vacancies in certain subjects.

Saxony also claimed to have fired all of its former Nazi Party members, but the chronic lack of teaching staff and constantly underenrolled teacher training courses suggest that these reports might have been inflated.[47] An administrative report in 1947 noted soberly that Mecklenburg-Vorpommern's and Saxony's strict policy of firing all former Nazi Party teachers meant that approximately 80 percent of teacher positions had to be filled by *Neulehrer* without any kind of training at all. Furthermore, these two districts caused a chain reaction by pushing off their former Nazi Party teachers onto other districts who would hire them because they needed staff, creating a permanent sense of "unrest and uncertainty."[48] A region such as Thuringia could not have even dreamed of dismissing so many of its former Nazi Party teachers. After calculating former Nazi membership, an administrator explained that this practice literally would have meant firing 97.5 percent of the employees, causing the collapse of the school system in that region.[49] To complicate matters even further, the public did not always support policies of firing suspected Nazis, who were often respected educators in the community. In an article in the *Sächsische Volkszeitung*, the Saxon education minister Wilhelm Schneller (KPD/SED) defended the decision to prohibit former Nazi members from teaching in the schools. Schneller noted with dismay that the critics of this policy included non-Nazi members as well.[50] Pupils, too, often disagreed with the mass firings of Nazi teachers. When asked by an educational administrator to identify the problems of the "new school," one eighth-grade boy from the Thuringian town of Altenburg complained that not all teachers should be fired, only the "real Nazis . . . The teachers from back then shouldn't all be lumped together!"[51] Even where schools could afford to dismiss Nazi teachers, this aspect of denazification was one that did not necessarily meet with public approval.

The chaotic personnel situation led to a number of decisions made in autumn 1945 about hiring or firing teachers who were often overturned at a later date, aggravating school directors' attempts to keep teaching positions filled. Such incidents plagued school administrators throughout the Soviet zone era, well into the first years of the GDR. The educator Charlotte Diesel-Behnke, for example, had worked in a school in Berlin-Neukölln before the war and had joined the SPD in 1927. She did not specify if she had her own class or if she worked in an assistant or administrative position. According to her résumé, the Nazis had required her to take early retirement because of this background, but had rehired her in 1940 because of a desperate shortage of teachers. She apparently kept this position after the war.[52] Her materials do not state the circumstances that allowed her to continue teaching, but her SPD background likely gave her a seal of political reliability. Then came a whole new set of career problems for Diesel-Behnke. In the summer of 1949, she and her colleagues had arranged a faculty trip, and she wanted to combine this with a visit to her ailing mother in the Western zone. She had waited weeks to receive an interzone pass. The Soviet administration then rescheduled a teaching exam that she had to take, interfering with her plans. She wrote the Ministry of Education in Weimar, requesting to change the date of the teaching exam. Her next letter in October protested being fired for this attempt to change the exam date. Furthermore, she complained, she had not been allowed to demonstrate that she would have obeyed orders had she been denied permission.[53] Her résumé indicates that this "daily and hourly" public servant of the democratic school, as she described herself, did eventually receive her job back. Other documents show that the incident was not an isolated one and that other victims of the increasingly politicized bureaucracy were less fortunate. Diesel-Behnke experienced first-hand that, as in her own case of being dismissed and then rehired under the Nazi regime, only to experience a similar pattern after the war, educational policies in the Soviet zone oscillated between exaggerated ideological enthusiasm and administrative expediency.

Many educators found such incidents too trying. A 1949 report on the number of teachers who left the profession entirely (known as

Lehrerfluktuation) stated that every region lost approximately ninety–hundred teachers every month.[54] Forty-five–fifty of these teachers left for personal reasons, twenty-five–thirty simply broke their contract without obvious grounds, and the rest departed because of illness, antidemocratic behavior, or general "lack of ability." Those who left for personal reasons listed problems such as not having shoes or shirts to wear, the route to school being too long, tensions between locals and teachers, or a lack of support from local school administrators. A report from an educational administrator in Berlin, responding to the school administration in Mecklenburg, countered that unhappy teachers who had been identified as "reactionary" or caught in a Nazi philosophy were in fact only frustrated and disappointed with the lack of support from educational administrators and their pitiful financial situation.[55] Professional dignity was not always easy to maintain under these circumstances: In addition to standard items of clothing needed by teachers, a 1948 report from Berlin-Köpenick listed a desperate need for underwear.[56] Teacher retention was not unique to the Soviet zone, but the structural problems that teachers had to endure there as well as a shortage of newly trained reserves made continuity in the teaching staff—and thus in classroom lessons—even more difficult.

* * *

The question of who was actually responsible for teaching antifascist democratic lessons to pupils has multiple answers. Administrative lines of authority in the Soviet zone educational system were blurry, and subject to practical obstacles of policy implementation. The organizational structure ultimately answered to Moscow, but a whole battalion of administrators, teachers, and parents criticized SED directives and had enough autonomy to influence educational policy. This arrangement varied depending upon Moscow's attempts to control political and cultural developments more or less closely. The SMAD at times violently demonstrated the limits of its patience, and examples exist of closed Soviet military tribunals meting out work camp and death sentences to pupils and teachers for subversive political behavior.[57] Harsh punishment did not curtail protest actions throughout the zone, though, and pupils and teachers who observed brutal reactions by Soviet authorities continued to discuss their disapproval of policies among themselves.[58] The SMAD could not and did not control all aspects of school communities' actions, even if its shadow was always present. Yet individuals learned to adapt within this system, often formally insisting on their

right to be part of antifascist democratic education. The extent of autonomy that existed and that was used by individuals and organizations
provides further evidence that Moscow lacked a vision for the future of
the Soviet zone or the GDR in the first years after the war, still imagining that a unified, neutral Germany was possible.[59] Antifascist educators
thus had limited but clear ideological space in which to negotiate and
develop antifascist democratic education to correspond to their needs
and desires.

Creating School Routines

Deciding upon the appropriate type of school system was only one step
in the task of defining antifascist democratic education. The next
involved getting pupils to school and then producing instructional
materials and curricula. Not only the quality but also the quantity of
school lessons in the "new school" varied greatly, depending on a host of
factors. Some were specifically linked to the postwar situation in the
Soviet zone, and some were familiar ones to any school system at any
time. For example, regional conditions in the Soviet zone varied; some
areas had more severe damage to school buildings than other areas.
A region like Mecklenburg-Vorpommern, in which school buildings suffered only minor damage, was in a better position to offer regular classes
than a region like Brandenburg, which had seen more serious bombing
to its buildings.[60] A 1948 school report for Brandenburg noted significant progress in the region's ability to offer instruction as compared to
the previous year, but added that the lack of adequate school buildings
and teachers meant that pupils went to school in shifts, so that they
received fewer hours of instruction.[61] When confronted with the actual
number of hours that pupils received instruction in the "new school,"
antifascist educators admitted that their ability to democratize and
denazify pupils was limited.[62]

The subject of reduced school hours or even cancellation of classes
and the effect this had on democratizing pupils surfaced throughout the
Soviet zone in both official reports and pupils' essays.[63] One of the most
common reasons for not holding classes was the weather. The harsh winter of 1945/1946 created life-threatening food and heating shortages for
Germans, and this affected schools. Some schools remained open and
heated, so that one boy, Paul wrote of how lucky he was to be able to
enjoy a few hours of warmth during the school day while his parents
were exposed to the cold all day long.[64] Teachers also asked pupils to
donate some of their household coal to the classroom in an effort to keep

schools heated and open. This situation implied a classroom with at least a source of heat; young Ursula wrote of not having had an oven in the classroom. When the weather turned really cold her class moved over to a different school.[65] Another pupil, Margot, noted that the windowpanes they replaced in her classroom had unfortunately not made it any warmer. She then wrote amusingly that all the pupils' hard thinking would cause their heads to smoke like ovens, and thus warm up the classroom a little.[66]

Where school could be held, the lack of adequate classroom equipment made instruction very difficult. The conditions could not have been worse for pupils and their teachers. The fifth-grader Sonja remembered having to bring something to sit on. Otherwise she had to stand during her lessons—hardly conductive to concentration, especially for starving and weak children.[67] Gerda wrote without irony that having to sit in coats, hats, and gloves made writing awkward.[68] She was lucky to be able to write, since pen and paper were also in scarce supply. Pupils wrote their essays on the backs of sections of Nazi posters, Nazi administrative forms, or continued in the notebooks that they had used only months ago in the Nazi school. One Thuringian new teacher proudly reported her discovery that broken bits of the plaster of Paris molds used by a local porcelain factory could be used as a chalk substitute.[69] Teachers and children, bundled up in whatever rags and odd clothes they could find, writing on scraps of paper with odd bits of pencils or pens, created a curious sight in the "new school." They also demonstrated a determination to find creative solutions for continuing their antifascist democratization.

Arranging for school materials did not guarantee that pupils would be in school to use them. In difficult times, pupils either missed classes, or school was cancelled. Hunger kept both teachers and pupils away from the classroom. The *Hamsterfahrt* (hamster trip), a semilegal foraging trip taken out into the countryside to "collect" vegetables and fallen wood, or *Kartoffelbuddeln* (potato gathering) became the only means of survival for many urban families and a common excuse for pupils' absences. Young people accompanied their parents on over-filled trains, hauling back wheelbarrows full of mostly illicitly acquired food and fuel and, incidentally, material for future school essays. Without a doubt, school and home overlapped in young people's lives, but not always in ways that educational administrators anticipated. The *Hamsterfahrt* often left little room for school, assigning the classroom a minuscule role in pupils' formal instruction. On the other hand, pupils did not hesitate to bring stories of home, including the *Hamsterfahrt* into the classroom. They thereby took an active part in determining the content of their school experience, and used their school essays as a means of making sense of their world. Some even blamed the "hamsterers" for taking food that would otherwise have gone to the larger population.[70] Thus, an unanticipated effect of the shortages of heating fuel and food was the redrawing of boundaries between school and home. This development extended beyond the mere fact of young people spending less time in school and more at home or with families. It also helped determine how involved parents were with their children's homework. As noted in pupils' essays and teachers' comments, in most homes, the primary room in which pupils completed their homework was the kitchen. Children sat at the table reading or writing, mothers stood at the stove cooking, and other family members hovered nearby.[71] The explanation for this was simple, as Paul-Ernst wrote. "I could only complete my schoolwork in the kitchen, and even then only when the one small table there was free. My fingers would have frozen if I had written in the living room."[72] Teachers continually worried about this arrangement, insisting that pupils needed a quiet working space to study at home.[73] Little could be done, though. While mothers prepared food for their families and boarders, the cooking stove gave off heat. Even in the undamaged, larger homes of wealthier families, limited fuel supplies made the kitchen the central and often only room in the house in which family members could sit in relative comfort.

Under these conditions, children's homework could become a family affair, as evidenced by pupils' own accounts of completing their assignments. For example, a middle-school group of girls wrote a play in 1946 to describe the postwar situation. The first scene shows Christel discussing her homework with her family members, who gave suggestions and also reflected upon the meaning of the assignment for their own lives, a typical situation for families with school-aged children.[74] Parents or siblings who consciously or unconsciously participated in working through difficult questions with the pupils about the family's own Nazi past, or who spent time trying to come to terms with the horrors they had experienced together, generally did not dictate how children should view the past. More often, adults and children participated in a mutual endeavor to organize their thoughts about the past, present, and future.[75] "Assisted" by material circumstances such as lack of rooms, Soviet zone school administrators had constructed a school that extended its reach through the classroom into the population at large. This broader audience was not passive, though, and the school did not function as the mouthpiece of the state.

School administrators had a difficult time exercising authority over children when children attended school irregularly, and when the consequences of missing class were minimal. Many areas competed for pupils' time, including the search for food or fuel, or entertaining distractions. As noted at a regional educational conference in October 1945, teachers needed to be concerned with more than the denazification of their pupils: "Because today it is not at all only about saving our youth intellectually. For the coming winter especially it is also about saving them physically. This saving of our children must be the task of the new democratic school community that we will build, which must include youth organizations, teachers, parents, political parties and the administration. Their duty must be to help our youth get through this difficult time with warm classrooms, school lunches, and warm clothes."[76] Disastrous everyday conditions furthermore distracted young people from their studies. As one observer poignantly reported, "The children secretly steal bread from each other or fight over it out of hunger. They are so hungry that during their lessons they often eat their dry piece of bread or cry from hunger. In spite of teachers' warnings, the children often eat their breakfasts right away and then starve during the rest of the lesson."[77] The effects of extreme hunger were responsible not only for listlessness and problems of concentration, but kept undernourished youth at permanent risk for severe illnesses.

Still, the school was not defenseless in its fight to draw and retain pupils in its classrooms. As its biggest foe, hunger turned out to be its biggest ally. The advent of the school lunch became the prime means of attracting young people to school throughout Germany. Arrangements for the lunch were made locally. The regional military administration often provided the food, but other countries like Sweden also donated items.[78] In the American zone pupils sometimes received chocolate, but children throughout Germany more often stood in line for a bowl of soup. They usually brought their own container and utensils, and many children ate from tin cans or their fathers' military helmets. The school lunch offered pupils a meager meal, and this served two, interrelated educational purposes. The prospect of food coaxed them into the school building, prodded by parents wanting to make more efficient use of limited supplies, so the teachers had the possibility to teach pupils. Nourishment also helped keep up young people's strength so they could pay attention during class. As a result, the school lunch became one of the dominant leitmotivs of the postwar school throughout all of Germany.

Pupils did not enjoy the look and feel of opening new schoolbooks on the first day of classes, or in subsequent years. Not even GDR historians would be able to frame the extreme lack of textbooks and other school materials in positive terms. Twenty years later, Gottfried Uhlig's otherwise laudatory account of the "beginning of the antifascist-democratic school reform" could do little more than impress upon readers just how little had been available to postwar classrooms. This situation was not only a result of postwar conditions, but also part of the typical administrative lag-time for producing educational materials. After 1933, new curriculum guidelines first had to be drafted and approved, and only then could new texts be written, approved, published, purchased, and distributed. In some Nazi-era classrooms, this slow process meant that some pupils had been reading Weimar-era texts into the first years of the war.[79] The bureaucratic machinery turned faster after the war, but only slightly. Although some school primers could be printed in the Soviet zone in the first school year, their availability remained limited through 1949. By the end of 1945, according to Uhlig, between 2.2 and 2.35 million pupils in the Soviet zone and 300,000 Berlin pupils shared 3,456,983 schoolbooks. The publishing house Volk und Wissen printed between four and six million by the end of the first 1945–1946 academic school year.[80] This figure of approximately one textbook per pupil must be evaluated within the context of how many schoolbooks a pupil actually needed: the eighth-grade class, for instance, included seven courses that required textbooks.[81] A parent association in Berlin-Adlershof wrote to the central school administration in October 1948, complaining that pupils had still not received books for school.[82] Reasons for such problems were legion, including paper shortages; delivery problems, especially to rural schools; and administrative incompetence at all levels.[83]

Historians and those Germans who once attended school in the Soviet zone often note that pupils used Weimar-era textbooks with any offending pages ripped out, pasted over, or with text marked through, but these anecdotes, collected after the Soviet zone period, do not accurately describe the materials being used.[84] Certainly, the judicious use of pre–1933 textbooks seemed logical to administrators and teachers alike, as evidenced by meeting of the Dresden "Provisional Committee of Antifascist Teachers," which described one of its main duties as evaluating Weimar-era textbooks.[85] Many of these works, however, contained too many militaristic passages to make this a viable long-term solution. Nor does this practice seem to have lasted long, although individual regional studies are lacking to confirm this inference. In any case, all of the occupying powers viewed the use of Weimar textbooks skeptically. In the British zone, the Textbook Section of the Education Branch approved only 8 textbooks of the 280 they examined that had been published before 1945.[86] This number included schoolbooks from all subjects, from German to mathematics. The texts accepted for German or history made up a small percentage of this total, since the commissions in all zones were particularly concerned with the previous militaristic quality of these subjects.[87] The SMAD quickly banned the use of pre–1945 textbooks with Order Number 40 of August 25, 1945, instructing schools to use those books published before 1933 only after removing all reactionary passages.[88] Soviet administrators made regular checks of schools to ensure that inappropriate texts were not being used, an offense for which school directors could be severely punished. Still, administrators reported instances of pupils using Nazi-era texts. In the case of one Berlin school, a director instructed a teacher to keep these books, because they would be useful once the "Siberian wind" stopped blowing, that is, once the SMAD turned its attentions away from the school (an event that did not come to pass).[89]

The shortage of paper and qualified textbook authors made it impossible to meet the demands for new textbooks, especially for the subjects German and history. The necessary approval by the SMAD, and for the case of Greater Berlin, the Allied Command as well, made the task even more difficult.[90] Moreover, in contrast to the Western zones, the Soviet zone did not cooperate with other zones and countries to collaborate in writing and publishing new textbooks.[91] To complicate matters even further, funds were not always available to pay the publishers. The school laws for Greater Berlin and the Soviet zone specified that pupils should not have to purchase instructional materials, yet finding money from the school's budget occasionally proved impossible. In some cases, schoolbooks sat at the bookstore if parents did not pay for them, as was the case at a Berlin school in 1948.[92] The more favorable conditions in the Western zones meant that the history series *Wege der Völker* could be published there as early as 1948.[93] Until the early 1950s, though, pupils throughout Germany often did not have textbooks for many subjects, and the first bound history textbooks appeared in 1951.[94] The slow process of textbook production and distribution is clear from SMAD Directive 150 of May 18, 1946, instructing Paul Wandel to have eight million schoolbooks ready by the beginning of the academic year, October 1, 1946, and a further seven million ready by December 15, 1946. The 120 teachers to be recruited as authors and assistants would receive extra food rations (*Arbeiterkarte Nr. 2*), officially recognizing the necessity of their contributions.[95] The quota could be met, and Volk und Wissen listed a delivery of 9.3 million schoolbooks in the second academic year 1946/1947, eleven million in 1947, and 13.6 million in 1948.[96]

At a conference in March 1947, educational ministers agreed that the publication of textbooks was a priority, but there were not enough qualified individuals to complete this task.[97] A frustrated Paul Wandel conceded that the situation was not ideal. Nevertheless, he warned, they should not explicitly recommend the publication of mediocre material. Yet they had little choice but to allow substandard texts into the classrooms. Teachers and pupils needed educational materials, and better texts could always be written later. The creation of "instructional and work pamphlets" (*Arbeits- und Lehrhefte*) alleviated the immediate need for some sort of materials. Presented in 1946 at a teachers' conference, these small pamphlets were to "replace" schoolbooks as well as "prepare their way."[98] The instructional pamphlets stayed in school for pupils' reference, while they could take the work pamphlets home in order to complete their homework. These work pamphlets needed to be updated more often than the instructional pamphlets, because pupils would see them on an almost daily basis.

The lessons in these pamphlets are full of forced and simplistic historical arguments. Authors reversed judgments on all achievements praised by the Nazis, while teleologically pushing workers to the forefront of all historical developments. The language was stilted and often unclear, and the structure of the argument did not provide an opportunity for pupils to form their own opinions on the material. The third volume of the secondary school instructional pamphlet series treated the "development of the Brandenburg-Prussian state until 1786." There, pupils learned that the Prussian state had abused its subjects: "This state only gave to each of its subjects in order to get more from them: to the bourgeoisie; in order to utilize the fruits of industry and trade for its financial policies; to the nobles, in order to lead them onto the slaughter fields; and in as far as it ever allowed the farmer to receive benefits, then only in order to use him as a tax payer as well as a soldier for the needs of the state based on power (*Machtstaat*)."[99] Within this historical determinist argument, pupils learned that the socialist state had progressed beyond exploiting its citizens.

A 1949 teachers' edition of a work pamphlet for the elementary school included only limited didactic guidelines for teachers. The vast majority of teachers were young and had been trained in intensive teaching courses, suggesting that their lessons would have been made easier by basic information about the texts. The pamphlet focused on England's industrial revolution, and included different kinds of documents to illustrate topics such as "England's colonial empire" and "the Chartists."[100] The section on "The Conditions of Workers and Their Children" devoted two pages to "The Poorhouse." For this topic, the pamphlet included an excerpt of the well-known scene in *Oliver Twist*: "Please, sir, I want some more." The use of Charles Dickens offered the well-read teacher a means of showing pupils how to conceptualize life in other times for other young people. Young Soviet zone people's own fight against starvation paralleled Oliver's plight: hunger gave him enough courage to ask the fat workhouse master for more food. The novel excerpt could even have allowed the teacher to begin a good discussion on the differences between fiction and nonfiction, or, depending on the level of the class, the use of different historical sources. None of this was likely to happen, however, if the teacher did not have previous knowledge of the novel. The text neither stated that the excerpt originated from a fictional source, nor did it give a synopsis of the plot. Nor were teachers likely to find adequate sources of professional support for these questions. Without any possibility of contextualizing the Dickens story within English society and historiography, a task otherwise within the grasp of teachers even in remote communities, both teachers and pupils would have walked away with nothing more than a vague sense of the terrible conditions in England for young people.

The Soviet zone and GDR history teachers' journal *Geschichte in der Schule* promised to offer history teachers the opportunity to dialogue about historiographical theories and lesson planning, which would have helped round out teachers' ability to present historical material.[101] First published in 1948, a year after history instruction officially became part of course offerings again, the journal fell short of addressing practical issues on a consistent basis. In spite of the editors' call to produce a journal helpful for classroom use, too many of the articles were written at a relatively sophisticated academic level. Given the types of submissions by teachers, these articles were of little help for their classrooms. Some authors offered practical advice for using historical documents, but others lost themselves in historiographical reflections.[102] Many articles contained musings of new teachers struggling with the concept of history. "Teaching history," a "young colleague" reverently explained, "means to learn from the past in order to understand the present and shape the future."[103] The first issue, devoted to the commemoration of 1848, contained some concrete didactic suggestions, including a class trip to Berlin to visit an 1848 exhibit and how to incorporate a film about 1848 into a lesson. But the extent of instructional support for history did not meet the needs of newly trained teachers in the way that, for instance, the monthly educational journal *die neue schule* did for Russian teachers, a subject with similar problems of underqualified instructors and absence of teaching aids. Teachers in rural schools or without a film projector did not benefit significantly from a majority of the suggestions.

Of the problems common to most school systems, the eternal disparity between rural and urban schools troubled educators in the Soviet zone the most. The lack of adequate school supplies or textbooks could be bridged with temporary measures. After all, pupils could write their essays on the back of discarded wartime documents, and the appearance of zonal educational journals provided teachers with lesson plans and tips for organizing the school day.[104] The "rural question" would take longer to solve. The Soviet zone and later GDR officially gave the impression—too often uncritically accepted even today[105]—of having eliminated one- and two-room schoolhouses from the country-side by the late 1940s. Much evidence from Berlin and elsewhere failed to reflect the actual state of affairs. Rural schools had many problems, as attentive educational administrators recognized. Pupils in distant smaller towns and villages received a day's instruction entirely removed from the controlling arm of educational administrators, and this realiza-tion caused bureaucrats sleepless nights throughout the Soviet zone. Moreover, rural inhabitants reacted suspiciously to reform measures handed down from Berlin, causing educational administrators to realize that the local residents were a major impediment to reform.[106] All of these factors contributed to the "Cinderella position" of the rural school, as one administrator of the heavily rural *Land* of Brandenburg com-plained. As the less fortunate stepdaughter in dire need of materials and personnel, the rural school found herself in a vicious cycle of not being able to improve her situation alone and not finding anyone to help.[107]

Attempts to eliminate real and perceived urban and rural educational differences or the programs designed to keep children coming to school show their commitment to this ideal. In this light, the stakes of Soviet zone educational policy become clearer. Educational administrators con-tinually emphasized two points: that the school was exclusively the pre-rogative of the state, and that the school should be the "cultural center of the city and countryside."[108] Soviet zone educational administrators strongly believed in the ability of education to offer common learning experiences to pupils, but only if the schools in cities and the countryside could be adequately controlled.

* * *

The "new school" required a significant amount of organization to even arrange for teachers to teach and pupils to learn. After identifying the objective of an antifascist, democratic, German school, educational reformers found themselves faced with the everyday structural difficulties of hiring politically acceptable teachers and providing materials for classrooms. Some of these obstacles could be surmounted by minor additions, such as school lunches. Others demonstrated the structural problems inherent to most school systems, for example urban and rural disparities, exacerbated by the postwar situation. Still others underlined the important role played by families and communities in the implementation of educational reforms. Where classes could not be held, or parents demanded that a teacher not be fired, educational administrators had to limit their own expectations of what the "new school" could accomplish. Postwar idealism about constructing a unified Germany by means of the new educational system gave way to pragmatic attempts to create an acceptable school system that, in spite of the difficulties involved in providing daily lessons, could positively influence young people in the Soviet zone. These conditions affected antifascist educators' programs to create a new unified, German, antifascist, democratic, national consciousness. The resulting *Einheitsschule* ultimately severed educational ties between the Soviet and Western zones, and furthermore called attention to the combination of diverse experiences within a single educational system.

CHAPTER 3

Rebuilding the School

The ideology and practice of antifascist democratic education had a clear setting: the school building. More than a place to hold classes, it was an integral part of the "new school," physically, pedagogically, and symbolically. Educational reformers frequently emphasized that the postwar school system needed an appropriate structure, one whose appearance would indicate the antifascist, democratic lessons being taught within its walls. Concern with the school building had several motivations. First, the idea that physical renewal could bring about intellectual and spiritual renewal was a metaphor so powerful that it permeated Soviet zone discussions and decisions about education. Allusions to architectural renewal of the school building provided a rhetorical device to argue in favor of educational reforms. As in other aspects of antifascist education, Soviet zone reformers relied heavily on previous educational structures instead of creating new ones; nevertheless, they believed that the appearance of school buildings and classrooms played a role in communicating antifascist democratic education to pupils. Second, the Soviet zone school building was the local point of access to the ambivalent task of reconstructing the German nation. The school building represented the physical site for chasing out old, undemocratic memories and creating new, antifascist ones. Also, it constituted a familiar, cherished, and positive landmark on the everyday landscape.[1] An intact, repaired, clean school building thus offered a physical space in which life had returned to "normal" for pupils, their families, and the German nation. Third, the school building was a site of memory and remembering. It resonated with traces of the past that educational administrators hoped to harness in the service of the nation. By decorating classrooms and bringing a sense of order into the building, antifascist educators believed that they could help faculty and teachers forget painful wartime memories and construct a new interpretation of their national and individual pasts. This chapter examines the complex functions of the school building as a metaphor of change, a local symbol of national stability and normalcy, and a site for creating new memories.

Structural Limits

"We cannot content ourselves with propping up a makeshift building from the rubble of our collapsed school," the school administrator Herr Viehweg called to his colleagues in October 1945. "The significance of the hour and of our task," he explained, "lies exactly in our erecting a totally new building, clear in its external construction and filled with a new, democratic, antifascist spirit on the inside."[2] Yet Viehweg did not actually offer concrete plans for major changes in any aspect of schooling or school architecture. Indeed, the curricular objectives he outlined implied more of a reactivation of an older heritage than a rupture with the past. For instance, he suggested using German instruction to win back pride in German culture, thereby teaching pupils the joy of singing folk songs.[3] Nowhere did he make radical proposals about how to teach antifascist democratic education. Two years later, Viehweg still worried about how the school building's actual appearance reflected its "spirit," but this time he did not exhort his colleagues to construct a new building. Instead, he advised teachers to decorate the school walls: "The teaching staff should on their own use pictures, etc. not only to make the school friendly, but rather to show that one is entering a democratic school."[4] In this instance, he emphasized physical improvements that could be accomplished easily. The financial and time constraints of setting up a school system had likely diminished his earlier expectations of reconstruction. He still believed that a school's appearance indicated its philosophy, and he also continued to demand external changes. But his suggestions now focused only on small adjustments in the extant physical and pedagogical structures.

Discussions about the school building were symptomatic of conflicting postwar longings for change and familiarity. Educational reformers called repeatedly for a new school building, and its condition and appearance surfaced throughout reports written by Soviet zone educational administrators, teachers, and pupils. But in spite of early references to redesigning the school building, antifascist educational plans did not actually include a new, national architectural design. Structurally, the antifascist school building was often a repaired and propped-up version of the earlier school building, the result of Imperial-, Weimar-, and Nazi-era educational systems. Not only did the same bricks find their old positions in the walls, but the school's function in the community retained its trusted character of educating children and organizing society by dictating the routines of a large number of young people, their teachers, and their families.

School buildings reflect not only educational programs and objectives, but also realities connected to city planning, including political, economic, and aesthetic debates.[5] An understanding of general architectural reconstruction in the Soviet zone thus highlights the situation in which school buildings were rebuilt, and about their role in the public consciousness.[6] The Soviet zone was a place with limited financial and material possibilities of physical reconstruction, in which residents at all levels of society were interested in how these resources would be used for the future appearances of their cities. The SMAD and Germans themselves had accorded education a key role in the antifascist democratic program, and they collected detailed information about the school building's appearance as part of reconstruction progress. By keeping the school building itself as a focus of discussions, education reformers successfully generated a conceptual framework that emphasized rebuilding rather than new construction.

Part of the reason for reconstruction rather than designing a new school building lay in the lack of dialogue between architects and educational reformers. This silence is at first glance surprising. Many first generation postwar architects wanted to draw on the unrealized plans of an earlier period, especially Weimar, analogous to the teaching philosophies of Soviet zone educational planners, who also looked backward for guidance.[7] Lively discussions among educators and architects about the perfect school building or classroom set-up had even been part of early twentieth-century pedagogical movements, such as Rudolf Steiner's Waldorf schools.[8] After the war, neither side took the next step of rethinking early twentieth-century movements for new architectural styles in schools. The architecture of Stalinist socialism that later became such a large part of the built environment of the GDR did not yet dominate plans in the immediate postwar period, although tensions between these architects and Stalinist visions of socialist cities already existed. Architects and educational planners even had similar wartime experiences: many Soviet zone architects active in postwar construction activities had returned from exile. Or they had remained in Germany during the Nazi period, practicing alternative "safe" careers that did not make them ideologically suspect after the war.[9] The apparent lack of postwar discussion between architects and educators about ideal building designs for schools is all the more striking. The ideological conditions existed to have made such a dialogue possible; the economic ones did not. The rubble metaphor of cleaning off and setting aside the still-usable pieces of foundations dominated the postwar discourse of physical and mental reconstruction. The school building was being hauled out and dusted off faster than it could have been reconceptualized.

The damaged postwar economy of Europe influenced the level of physical reconstruction that was possible.[10] Lack of funds and building materials limited the number of architectural projects that could have been carried out, and the crippled infrastructure made the implementation of even approved construction difficult. A despairing from 1949 report on the chaotic situation of school construction work in the Soviet zone contained numerous examples of misunderstandings and bad planning: "Saxony offers roof tiles to Thuringia, rejected with the reason that Thuringia has enough tiles. School building near Worbis doesn't have one single roof tile, but three tile manufacturers in the city deliver tiles regularly to Dresden."[11] The materials needed for school repairs represented a considerable amount of a city's budget, requiring politicians to make major decisions about finances and providing contractors and their suppliers with large contracts.[12] Schools relied upon city and *Land* budgets for construction projects, unlike the more financially liquid churches; nevertheless, the Soviet zone's architectural policy was not driven entirely by city planning designed to rescue the economy, nor was it based on rebuilding only the bare minimum of structures. Most school districts contracted several repair projects in the first period after the war.[13]

As evidenced by events such as the popular architectural competitions in postwar Dresden, the discussion for rebuilding public spaces involved a broad segment of Soviet zone society. Decisions that were made about the appearance of cities represented a consensus of residents and administrators.[14] The majority of the German population had been isolated from architecture and urban planning developments during the war, and they were ready to catch up with the international scene.[15] Yet the interest in architecture and urban planning differed somewhat at the everyday level of reconstruction. Postwar German families, busy satisfying their basic needs, were not ready to think about new architectural ideas for their homes or the interiors.[16] In an extension of this phenomenon, the sociologist Hilde Thurnwald noticed that those families able to remain in the ruins of their homes reported greater contentment than those forced to move into other quarters, even if the new residences were often newer and in better condition than their previous homes. In some instances, families returned from their assigned undamaged quarters back to their heavily damaged homes.[17] These Berlin families preferred the familiar to an unfamiliar environment, however desirable the newer accommodation might appear to an outsider.[18]

Schools, in the minds of public and policymakers alike, were not like other public buildings. Most telling in this regard are the kinds of buildings used in postwar representations of reconstructing the physical environment. Posters encouraging the efforts toward reconstruction or boasting of progress usually depicted recognizable, traditional, historical buildings: the Berlin city hall on Alexanderplatz, or the towers of the city halls in Leipzig and Dresden, often against the backdrop of the city's mascot or coat of arms.[19] Rarely did the more mundane aspects of architectural reconstruction appear in images of the city, such as hospitals, schools, or homes.[20] In the few examples picturing a school building, it occupied a different space in the public consciousness. The need for school buildings was on everyone's minds; its form, however, was not. When schools did appear on posters, it was to make a practical argument. One poster used a school building to symbolize peace, warning of the costs of war: Entitled "This Way or That Way?" (*So oder So?*), it listed a school building as costing the same as a canon; a hospital the same as a bomber.[21] Similarly, the 1946 poster *"Jugend baut auf!"* (Youth are [re-] building) mentioned an elementary school in Leipzig only as an eyesore, entreating local youth to volunteer to clean it up, along with a railway, and a hospital.[22]

Those posters that called for the rebuilding of schools did not focus on aesthetical concerns, but rather simply on the need for more schools. A Thuringian example, *"Wir bauen Schulen"* (We're building schools, see figure 2), showed two blonde children, a girl and a boy, walking hand in hand toward the image of a school building, sketched in chalk and superimposed on a map of the *Land*. The artist provided extensive information for most of the images on the poster: the boy is not wearing shoes—a typical problem—and all of the counties are listed that are to participate in the reconstruction program. Even the small buttons on the children's clothes are visible.[23] The school building, the actual subject of the poster, is void of such detail. Only the form of the walls, the windows, the turret, and the door are outlined; there is no doorknob, no window frames, no numbers for the empty circle that must be a clock on the turret. The local school building, although an integral part of the built environment, was never an icon in the way other monuments were. Although a major public building, it does not appear to have received mention in the aesthetical context of progress, appearing more as part of

Figure 2 "We're building schools. Everybody help out!"

Note: The map in the background is of the *Land* Thuringia. Note the boy's bare feet.

Source: "Wir bauen Schulen," Erfurt, [ca. 1945–1949], in *Plakate der SBZ/DDR* [CD-ROM], ed. Deutsches Historisches Museum (Munich: Sauer, 1999), Inv. P 90/4507.

the everyday, local built environment. Rudy Koshar has convincingly argued that the GDR tried to enlist "monumental history to build its version of the first socialist nation on German soil,"[24] but the discourse surrounding the reconstruction of the school building did not fit into this pattern. Adults treated the reconstruction of the school building somewhat differently from the work on other major public buildings. The first antifascist socialist school was to be housed in the hastily reconstructed and familiar school of earlier days.

Of course, the extent of bombing damage to Soviet zone cities varied greatly. The near complete destruction of Dresden and much of Berlin are well-known, including the phenomenon of the "empty desert" areas that resulted after builders demolished the ruins before constructing new buildings.[25] These cities presented the most extreme examples, though, and symbolized the widespread sense of despair at the destruction more than they epitomized the general situation. For the most part, the territory of the Soviet zone saw less overall damage than did the Western zones. In other ways, the Soviet zone was at a disadvantage as compared to its western neighbors. Its buildings were overall older than those in the West; that is, even those that had survived the war intact were nonetheless often in need of repair and maintenance. The Soviet zone also shouldered heavy reparation payments to the Soviet Union, which were often taken in material goods, making it even more difficult to finance and supply construction projects.[26] It is thus difficult to make broad generalizations about how the war affected educational structures in the Soviet zone, although educational administrators constantly emphasized the heavy damage suffered by school buildings as a general zonal problem.

The actual statistics collected by the educational administration on the extent of destruction suffered by Soviet zone school buildings did not present such a bleak scene. The first statistical overview of the Soviet zone school system, which appeared as an appendix to the reports from the 1949 Pedagogical Congress, offered the first major, published use of comprehensive figures for the school system.[27] The annual Pedagogical Congresses were zone-wide, and attended by large numbers of educators at every level of administration and by teachers. The information provided there, which detailed the state of affairs on November 15, 1948, a date late enough to have reflected several completed reconstruction projects, carried considerable weight and went a long way toward establishing educational policy and practice. The images of the school building, teachers, and pupils privileged some visual characteristics over others, and this communicated readable messages about how educational administrators perceived the school system as well as how they hoped to present it.[28]

The first diagram, "Schools—Pupils—Teachers," portrayed the ratio of school buildings, pupils, and teachers to each other and to the community. Thus, a sketch of a large group of male and female adult and child stick figures representing 10,000 residents boasted 6.29 school buildings, 1,530.3 pupils, and 36.5 teachers. Further, for every school building there were 243.2 pupils, one teacher for every 41.93 pupils, and 47.47 of every 100 teachers were female.[29] These ratios present teaching and learning situations that are far from ideal, but as average numbers they are not unusual for any public school system. According to the report, undamaged school buildings numbered 7,762 for the whole of the Soviet zone territory minus East Berlin.[30] This number was accompanied by a drawing of three large, one-story intact school buildings with six windows. The 2,741 slightly damaged school buildings are illustrated by a smaller building with the top right side crumbled away and only four windows visible, ostensibly the result of bombing. The 363 heavily damaged school buildings receive a smaller pictograph building with only three undamaged windows and most of the roof bombed away. A tiny pile of rubble with only the walls around two of the windows still remaining represents the 134 destroyed buildings. A second table at the bottom of the page, this time with five tiny piles of rubble, divided up the number of destroyed school buildings per *Land* (Brandenburg had the most at sixty-four, Saxony forty-two, Saxony-Anhalt thirteen, Thuringia nine, and Mecklenburg had the least at six).[31] The statistical table defined as "unusable" only those school buildings that were destroyed.

This means that in 1948 just over 1 percent of the 11,241 school buildings were officially not usable and that only 3.3 percent of the school buildings sustained heavy damage. However, these regional differences only hint at the range of conditions throughout the Soviet zone. A 1945 report from Leipzig described a more dismal scenario, likely representative of other heavily bombed cities. Of sixty-four elementary buildings, it listed twelve as destroyed and thirteen as heavily damaged, although the latter ones were still being used.[32] Berlin's schools were not included in these statistics, which would have dramatically increased the percentage of damaged and destroyed buildings, even though a much higher percentage of unusable school buildings would still remain a small overall percentage. It is also difficult to map out the level of damage that regions were repairing when they undertook reconstruction projects. In the school year 1947/1948, for example, Saxony reported having restored 84 school buildings and completed partial improvements on 528 others.[33] This impressive work record cannot be directly compared with the reconstruction experience of Berlin, which would have to wait until the 1950s to boast of such progress. The currency reform, the Berlin Blockade, and the ensuing division of the city made it impossible there to arrange for the finances and administrative overview necessary for approving building plans and hiring construction firms for large contracts.[34] Still, the detailed, worried attention given to the condition of school buildings cannot be justified by these numbers alone. The buildings themselves as described in these tables were, relatively speaking, not in terrible shape, and at least not in worse shape than other buildings.[35]

Lamenting the physical condition of the school building quickly developed into a powerful rhetorical tool because it drew on an existing value system that emphasized the importance of architectural progress. Construction and buildings had already firmly established themselves in Germany as positive signs of cultural progress and as representatives of governmental policy during the Nazi era.[36] The metaphor of rubble and reconstruction appeared immediately in the postwar months, and permanently structured the way Germans viewed the effects of wars on cities and the best way to rebuild them.[37] The symbol of reconstruction was particularly effective in the context of school policies because it could be anchored so concretely in the school building and then extended to other areas of antifascist education reform. "The Berlin school system, too, was buried under the rubble of the Nazi collapse," as the director of Berlin's central school administration Ernst Wildangel wrote in 1946.[38] Like other educational administrators, he blurred the line between symbolism and reality, alluding both to physical and ideological damage. Rubble and reconstruction provided vivid images to describe the horrors of the past and optimistic projects for the future. The school building provided educators with a ready object to make the metaphor come alive. The "school building" thus served a metonymical function for the whole of the Soviet zone school system, with the building's appearance standing for the entire condition of the Soviet zone educational system.[39] Calls to improve the school building's appearance used the physical site to represent an otherwise less tangible institution and organizational structure of "education."

The absence of a new architectural ideal for the school building was not only the result of needing to set up schools as quickly as possible. Education reformers were simply not ready for metaphorical discussions of entirely discarding the old school building and creating a new one. Moreover, actually constructing a new school would not have corresponded to their concept of antifascist, democratic education, which called upon past educational programs to structure the future school system. The educational tradition represented by the school building could not be invoked uncritically, but at the same time few educational reformers wished to abandon it entirely. Mentally, adults and children alike had first to chase away the unwelcome memories already threatening to establish themselves as a permanent part of the school building and then to select among happier ones as replacements. Physically, this entailed a precarious balancing act of calling for the construction of a new school that was not, in fact, entirely new.

Of course, the reconstruction of the school building served not only rhetorical functions. Pupils could not attend classes in significantly damaged school buildings. And if the school could not teach pupils, then the state had few means of reaching the younger generation. For instance, a school director trying to explain the potential dangers of pupils using slingshots during recess or before or after school told his teaching staff that a girl from a neighboring school had been hit in the eye and that some school windows had been broken. For lack of materials, the panes could not be replaced.[40] The prospect of a classroom lacking windows at the beginning of winter likely proved to be a stronger warning to teachers than did a girl with a black eye. Teachers understood that if they had to teach in a wintry classroom with only cardboard in the window, they could not expect to hold pupils' attention (see figure 3). Thus, monitoring pupils closely during recess to ensure that they did not damage the school building did more than keep the classroom a comfortable place to learn: a teacher thereby helped integrate pupils into the reconstruction of the nation by teaching them not to destroy, consciously or unconsciously, the progress that had already been accomplished within the school walls.

This insistence on a clean school building also stemmed from the convictions that pupils not only deserved unsoiled classrooms, but that they would learn antifascist, democratic behavior more efficiently in attractive school buildings. A letter sent out to parents by the Babelsberg (Brandenburg) chapter of the parent-teacher organization "Friends of the New School" (*Freunde der neuen Schule*) demonstrated this connection clearly. After explaining the dire need for the donation of funds and materials, the "Friends" listed eleven ways for parents to help speed up the reconstruction of their children's school. In the middle of suggestions to make a monetary or textbook donation and to volunteer work hours, point number seven asked, "Can you assist in approving the appearance of our classrooms by donating flowers, wall decorations, pictures?"[41] In an appeal to parental generosity, the "Friends" added, "Our

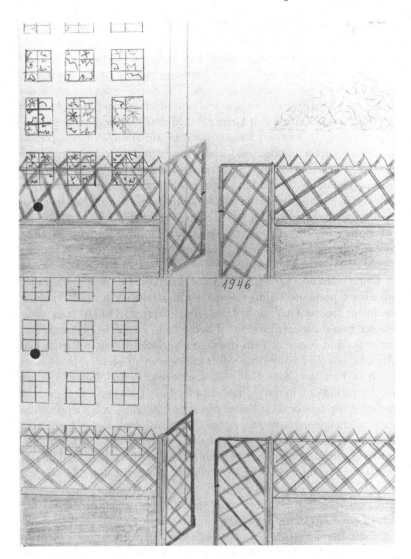

Figure 3 Pupil's Illustration of Building with Broken Windows and Piles of Rubble in 1945 (Top Half); New Panes in Window and Cleaned-Up Courtyard a Year Later in 1946.
Source: Anonymous, n.d. [1946], LAB/STA 134/13, 181/2, no. 423. p. 135.

children alone would be the beneficiaries" if everyone helped complete the school's reconstruction. The SMAD, however, had slightly different motivations for its interest in the appearance of school buildings. P. Zolotuchin, the director of the Soviet Education Department, complained to the German Educational Administration that unclean and poor conditions of the school buildings aroused discontent in teachers and pupils.[42] An unhappy school population would be more difficult to educate properly, they feared; worse, Germans might blame the Soviets for the unsatisfactory situation. The Soviet occupiers could not afford to exacerbate the already widespread anti-Soviet attitudes, and cleanliness was a virtue that Germans held dear.[43] A clean school building, the SMAD believed, would create an atmosphere of cooperation between occupier and occupied, and it would also help eliminate beliefs that the Soviets had unhygienic and thus barbaric social practices. Stories abounded of Soviet soldiers who used school hallways to defecate, admonishing German children to prove their higher degree of civilization by maintaining an unsoiled building.[44] These anecdotes called the authority of the SMAD into question: if the public perceived Soviet behavior as uncivilized, relationships between the two peoples would be even more strained.

Other references to the condition of the school building's appearance were based upon preexisting German political battles. In an illuminating comment from 1946, the city councilor (*Stadtrat*) Otto Winzer complained that the western regions of Berlin had "always" had better school buildings and materials than the poorer regions of eastern Berlin, and that the war had made these differences even greater.[45] The borders between East and West Berlin as described here by Winzer did not exactly correspond to those established by the Allied Command, since he included the poorer West Berlin regions of Wedding and Tiergarten as part of the poorer eastern regions.[46] The origins of this version of division expressed political conflicts that predated the war, when educators had already interpreted the conditions of the school building as manifestations of two different economic and therefore ideological camps.[47] To some degree, therefore, postwar East–West differences continued a tradition that had been established decades earlier. Reconstructing school buildings in the Soviet zone in this instance symbolized resolving long-standing perceived economic injustices between different German groups, ones that the school building had already been associated with. The repair of the antifascist democratic school building, like antifascist democratic education, was not an entirely new concern.

The school building's physical condition proved to be one of the biggest potential obstacles to realizing the ideas of antifascist democratic education. In fact, some school buildings were not even in good enough condition for pupils to attend classes. Where crumbling walls or cold rooms forced schools to close their doors or kept pupils from being able to concentrate on their lessons, educational reformers could not hope to use the educational system to enact major social change or influence political decisions. A speech by the Saxon education minister Wilhelm Schneller (KPD/SED) for a 1947 conference of county school inspectors in Dresden clearly identified the relationship between school buildings and their larger political role. Schneller began with the ominous warning that the 1948 London Conference would decide Germany's future as a unified or divided nation.[48] He went on to complain that the work of educating teachers politically had waned, so that an anti-Soviet Union attitude had developed among teaching staff.[49] With what at first seemed to be a sudden change of topic, Schneller then insisted that the suggestions to close schools as a means of dealing with the cold weather and inability to heat the buildings were not viable solutions to keep schools operating. In fact, Schneller's argumentative logic was in complete harmony with his fears about Germany's future. In educational reformers' minds, antifascist democratic education was the best hope for German unity. They could not control international political developments that affected their nation, but they did believe that, through antifascist education, they could guide the German population toward a course of unity. This program depended upon school being held, and thus on a functional school building.

The Local School Building and the Nation

Other important elements absent from official reports about school buildings were how the damaged school buildings made adults and children feel. A report from 1947 by the school county of Randow in Mecklenburg on the German-Polish border suggested that one possible reaction was a sense of panic. Randow lost all of its schools to bombing, and also lost part of its county to Polish administration.[50] Without school buildings, it was difficult to hold classes. To keep from losing all of its good pupils to other less desperate counties, Randow decided in July 1946 to open up an upper secondary school for gifted pupils in vacant barracks, in which pupils also slept. County administrators feared losing their upcoming best and brightest, sensing the area's demise if no one could be trained to take over the work of running the local government. This anxiety was certainly not unfounded, and was likely sharpened by the geographical losses sustained in the redrawing of postwar borders. Randow found itself in the border region's role of being the first and last topographical representative of its nation. The image of pupils sleeping in former military quarters to guard the survival of the region was difficult symbolically for a country recovering from a devastating war. The absence of a school building implied a threat to a community's existence, and this situation had to be remedied as quickly as possible.

Clearly, the school building occupied a unique place in the public consciousness. More than any other building in the Soviet zone, it represented the local application of a "national" institution—or at least an institution with national aspirations. Unlike the church, the mid-twentieth-century school building was not based on voluntary association and was a fully public institution. It nevertheless maintained a local character, based partially on the economic class of the surrounding neighborhood. Pupils in the Soviet zone attended a neighborhood school as defined by school district boundaries. In some cases this regulation did not always send them to the school nearest to their homes, especially if they lived in a district close to the Western zone's border.[51] The districting became more rigid, for many pupils as well as teachers, after the currency reform and the ensuing tightening of inter-German borders in the summer of 1948. The local character of the school was thus also defined by zonal regulations, which affected not only the curriculum structure but also composition of staff and pupils. Discussions surrounding the school building swayed between individual schools and the general idea of the German school and its building. The school building was clearly local, and clearly national, and as such provided Soviet zone educational planners with an ideal symbol for discussing their national aspirations at the local level.[52]

The joint character of education as a state and local institution meant that communities had a strong sense of ownership in the school building. Educational administrators in turn encouraged locals to see their school building as their community's participation in the reconstruction of the nation. A May 1948 regional educational meeting (*Landeskongreß*) in Schwerin, for instance, presented the local educational institution as a national cultural good. The opening concert set the tone of German culture with Beethoven's *Egmont Overture*, after which several speakers proclaimed the school as a key site for the struggle for a unified Germany. Such images effectively anchored the goal of German unity in the community and underlined cities' interactive relationship with the nation. A critical report on teaching methods then followed these speeches, given by the Mecklenburg educational minister Professor Ulrich Hoffmann (CDU/SED). Linking the appearance of the school building with its pedagogy, he first criticized the condition of school buildings. He then insisted, "The schoolhouse, as the center of culture of an area, should be its calling card, by which an area announces its will to be constructing culture."[53] In harsh terms, Hoffmann specified that every aspect of the school building's appearance should be presentable: "It is a cultural disgrace when schools do not have adequate or quality fuel for heating. It is equally a cultural disgrace when the cleanliness of schools is not exemplary."[54] This example illustrates all elements of the complex connection between the school building, the community, and the nation: the school building embodied local access to the nation, it helped create the nation, and it proved a city's worthiness to be part of the nation. The latter point of worthiness was particularly important in the face of impending German division. Educational administrators recognized the possibility that two separate German states might be created, and they were interested in demonstrating to themselves and the West that their half of Germany represented the nation better. A school building that looked antifascist democratic, that was clean, and that embodied national values of education and culture thus offered appropriate evidence of its right to be reeducating Germany's youth.

The school building's role as a local site of the nation accorded it a further symbolic status of a marker of Germany's well-being. After a devastating war and in the midst of uncertain Allied Command plans for Germany's future, Germans in the Soviet zone looked to the school building for local validation of their national progress. The Chemnitz city councilor (*Stadtrat*) Riesner expressed this dual function of the building in his own city at a 1945 meeting of school supervisors in Saxony. First focusing on his city's destruction after the war, he painted a grim picture of the work that had to be undertaken in order to begin school again. Almost all the school buildings in his city stood in complete or near ruin at the end of the war, potentially hindering schools from holding classes in the near future. But this tale of local despair, in Riesner's telling, ended happily for the nation. Residents labored hard to rebuild their city's school buildings, so that their children, too, could participate in the first academic year of the new nation. Riesner described heretofore unseen enthusiasm in October 1945 from educators, pupils, and parents upon his announcement that "the German schools have opened their gates again."[55] The sense of pride that he and other Chemnitz residents expressed in their schools emphasized the city's contribution to a larger cause, with the schools described only within the national context. Overcoming extreme conditions in Chemnitz represented an accomplishment primarily for the German nation.

If reconstruction of school buildings helped Soviet zone residents identify signs of their nation's progress, it is clear that progress often meant a return to an elusive normalcy. For this reason, educational administrators and the public alike insisted that the school building be returned to its primary function of an educational institution. Classroms had hosted a variety of noneducation activities during the war, from infirmary to bomb shelter to administrative office space. Where schools were being used by other institutions, antifascist education could not be taught, and the Soviet zone could not hope to realize its plans for transmitting a common culture through the schools. In many instances after the war, local communities and educational offices had to struggle to win back school buildings for their originally intended use. Adults and children in the Soviet zone pushed for a quick return to using their schools for "normal," educational purposes.

The reclaiming of educational spaces was sometimes accompanied by a return to the rigid adherence to guidelines for the use of the school, without attempting to make accommodations for the postwar situation. In the first volume of his 1992 autobiography *Zwischenbilanz*, the author Günter de Bruyn recounted a fateful conflict about the appropriate use of the school building in the village in which he had been assigned as a schoolteacher after the war.[56] The young unmarried de Bruyn, a native Berliner, landed in Westhavelland (Brandenburg), a village so small that even there was not a train station. This unfortunate teaching appointment followed his completion of a *Neulehrer* course, where his excellent grades still did not convince his superiors of his political reliability. De Bruyn claimed that their suspicions had made a more desirable Berlin position impossible. He noted that there were two school buildings in Westhavelland; that each also included rooms to house a teacher and his or her family, in which the other two teachers already lived; and that the school buildings, like the village itself, had not sustained damage during the war.[57] When one of the teachers moved to another village, the three rooms he had previously occupied became available and a heated battle erupted in the village about what should be done with them.[58] As de Bruyn recounted the story, one side wanted him to have all the rooms, in accordance with his position as schoolteacher. The other side protested the immorality of giving a young bachelor three rooms with a kitchen while the increasing flood of refugee women and their children slept in barns. Because the rooms belonged to the school, though, they could only be used for school-related purposes, making renting them out to other parties impossible. Both sides finally agreed upon a compromise, pressuring the twenty-year-old de Bruyn into marrying, so that he and his future wife could take over two rooms in good conscience. The community assumed the newlyweds would soon need the second room for children, and decided to convert the third room into an extra classroom to accommodate the many refugee pupils. De Bruyn divided the village residents into two groups. The first group, those who were in favor of him moving into the school apartment were unwilling to prolong the exceptional circumstances that resulted from the postwar situation. They insisted that rules of propriety should continue to be followed, and the teacher should be given the accommodations that belonged to his station. These feelings were doubtless stronger because the village had been physically untouched by the war, with the teacher whom de Bruyn replaced even having had the luxury of residing as a single man in the school's three rooms. The war had not forced residents to live differently, so they saw no reason to subject themselves to new rules after the war. The second group could not justify continuing to adhere to regulations about the use of the school building in the chaotic postwar period, and thought that these rules should be bent to reflect the changing physical landscape brought about by refugees. The compromise acknowledged the influx of refugees, but it worked within the realm of school practices. Child refugees could be accommodated as pupils; their families' need for shelter was not the responsibility of the school. Any other solution would have highlighted the unstable social network by redefining the school building's purpose, whereas residents were more interested in searching for signs of social normalcy.

Hilde Thurnwald described a similar sociological phenomenon on the part of educational authorities. When the unusually cold winter of 1946/ 1947 and fuel shortages made heating difficult or impossible in Berlin, many schools closed for extended periods.[59] Teachers and pupils requested to hold classes in other available and heated rooms, at least to keep the children occupied. They were forbidden to do so, although it is not clear if this decision came from the Allied Command or the German educational authorities. The refusal to allow improvised instruction in nonauthorized schoolrooms might be interpreted solely as inflexibility and an overzealous attachment to regulations, or even a simple refusal to share resources. But perhaps the motivations in this Berlin case paralleled those of the villagers among whom de Bruyn lived. A conviction, felt but not articulated, existed among the administration that school buildings must be reserved for school activities, and that school activities must be reserved for school buildings. As in other areas of everyday life, antifascism in the Soviet zone tended toward bourgeois sensibilities of reconstructing the present exactly as residents (or Germans) remembered the past—and this desire precluded using buildings for unusual purposes. Individuals who attempted to hold classes in other places or to permit nonschool activities in school buildings generally met with stubborn resistance throughout the Soviet zone.

Certain instances of offering the school building for other activities were not as problem-ridden for educational administrators or the public, especially those uses that had an educational aspect. Exhibits on local culture held in the school auditorium, for instance, reinforced the school building as a guardian of regional culture, thus reinforcing the local character of the national educational system. Victor Klemperer thus wrote approvingly of an exhibit in the schoolhouse of *Erzgebirgs-Spielwaren*, local wooden toys and crafts for which the region is well-known.[60] Numerous celebrations in schools for the memorial year 1948, anniversary of the failed 1848 Paulskirche Revolution, had local expressions as well. A speech by the *Ministerialrat* Viehweg for a Dresden school inspectors' conference noted that 1848 stood for three things: the Paulskirche bourgeois-liberal revolution "under the sign of unity and freedom"; the birth of the workers' movement with the Communist Manifesto; and the first meeting in August 1848 of the *Allgemeine Deutsche Lehrerversammlung* (All-German Teachers' Meeting) in Dresden, of which Dresden teachers should be particularly proud.[61] An emphasis on the local aspect of national memories did not undermine their national character, but made them more accessible to the community by infusing them with regional meaning.

The school building also hosted a limited number of events that were not strictly educational, but responded to pupil's physical needs, thus improving the health of the nation. The "half-liter of school meals [*Schulspeisung*],"[62] in most cases a thin soup, helped combat near starvation levels and furthermore encouraged school attendance. The advent of the school lunch, a relative novelty in Germany, enabled the school to continue in its function of education.[63] The former *Neulehrerin* Käte Agerth remembered solving tardiness and attendance problems with food, giving a hard roll in the morning only to those who arrived on time. " 'You catch mice with lard,' " she quipped, "[T]his time it was really lean lard, but it still attracted them."[64] The school building also functioned as a collection site for regular *Schuhaktionen* (shoe donations and clothing exchanges), also designed to make it easier for pupils to walk to school, especially in cold weather.[65]

School lunches and shoes were not a prerequisite for attendance. Photographs from the period show barefoot children in school, and school lunches were not always available at all schools.[66] An October 1945 questionnaire filled out by an elementary school stated that about half the children did not have winter shoes or coats. Many came to school without breakfast. These conditions worried the teaching staff and the school doctor, because they resulted in children who got sick, were listless and inattentive in lessons and at playtime, and had a decreased joy of learning.[67] More generally, the heartbreaking sight of shoeless and hungry children, who wolfed down all their small rations of food early in the morning and then cried hours later from hunger, was a painful reminder of the war and of the chaotic situations still present throughout much of Germany. Clean windows and tables did not suffice for the school building to truly symbolize a return to normalcy. The school building's occupants were part of the physical landscape as well, and in too many cases, barefoot and starving young people offered visual proof of the ongoing consequences of the war.

Unusual uses of the school building could also reverse the emphasis on the local aspect of recreating a nation, putting the nation in the role of asking for local help. A zonal memorandum in the fall of 1946 on collecting apple and pear seeds went out to the various regional administrations, turning school buildings into a site of hope for the continued health of the national population.[68] The memorandum was sent out by the DVV, but it likely had been the brainchild of German administrators or the SMAD—possibly the health department. According to the memorandum, the harsh winters of 1939/1940 and 1941/1942 had severely damaged the fruit trees throughout the Soviet zone. Pupils therefore were to be enlisted in a seed-collection project through their schools, with all seeds being sent to a central collection site in Magdeburg, a prize of 100 Reichsmark going to the school with the highest seed collection.[69] There is unfortunately no follow-up memorandum on the success of this project, and plenty of teachers and pupils were likely more concerned with their own immediate survival than projects for planting fruit trees. However, this suggestion, like others proposed by administrators, would have found its way into some classrooms, perhaps incorporated into part of a science lesson, offering pupils the chance to see the importance of their local school to the reconstruction of the German nation.

The uses of any space are influenced by its relationship to other sites.[70] The Soviet zone school building existed in response to community life, and structured residents' everyday routines.[71] Other local institutions in the Soviet zone assigned the school building the role of cooperating in social projects. One such request came from the Berlin police chief, addressed to an elementary school in Prenzlauer Berg. He asked the school to help teach children the danger of playing around vehicles.[72] The school minutes noted suggestions of dealing with this problem in an essay or a dictation, possibly using a mathematics lesson to discuss statistics of traffic and pedestrian accidents. Such willingness of other public organizations to view the school as a partner in finding solutions emphasized a belief in school administrators' capabilities as well as a degree of trust in the school's effectiveness socially. A visit paid by a police chief or a letter sent to the school director reaffirmed the school building as integral to the local community and placed it at the center of the community's intellectual and spiritual reconstruction.

This constellation of a community's institutions brought about some exceptions to schools being used for activities not related to educational objectives of educators or the state. These instances usually resulted from the ambivalent relationship between schools and churches. In an extreme example, Klemperer reported that his wife had heard an acquaintance of theirs preach in the Dresden-Dölzschen schoolhouse. But such an event was rare in Soviet zone schools.[73] More frequent was a sharing of classroom space between churches and schools, although that was a situation that school administrators tolerated only after having successfully eliminated religious instruction in the Soviet zone. They could thus be generous as the victors, and their cooperation with the churches demonstrated their willingness to work as equals with other local institutions. The relationship between the school and the church was nonetheless tense, and the forced mutual sharing of facilities regularly led to boundary disputes. The number of classrooms listed as not being in school buildings—3,330 of 58,850 total elementary school classrooms, for instance—would have comprised bunkers and other buildings that could no longer be used for their original purposes, but it almost certainly included a significant number of church rooms. The school building, as an integral part of the social fabric, thus found itself even more solidly tied into the activities of other institutions.

Adults were not the only ones to determine how the school building should be used. Pupils had specific ideas about the need to return educational functions back to the school building. Their school essays expressed a sense of ownership in their classrooms and schools as well as impatience with the obstacles to getting the school back in shape, such as the seemingly impossible task of heating the classrooms adequately.[74] Citing positive examples of reconstruction in the first year after the war, fourteen-year-old Margot Rosenthal noted with satisfaction that the district's administration had moved back to their former buildings, after having been housed in school buildings.[75] Describing how his school looked after the war, the dismay and indignation were still present in twelve-year-old Wolfgang Banach's essay that people had taken the school benches and window frames to use for fuel. Worse, "They stole the equipment that they had absolutely no use for out of the physics room."[76] Other, unavoidable impediments emerged that prevented pupils from using their school at all. The pupils of one elementary school in Prenzlauer Berg attended another elementary school during the winter because the former had neither windows nor heaters to keep out the cold. When they were allowed to return to their own school in April, the pupils breathed a sigh of relief, "Because it's always nicer in your own school," as young Eva Schmude noted.[77]

These feelings extended to pupils' desires to have the school look attractive as well. A school building should not only be used for school; it should also look the part. Young people were determined to be among those people who helped achieve this objective. An eighth-grade girl explained that her teachers wanted to bring back the former *Gemütlichkeit* (comfortableness) and *Sauberkeit* (cleanliness) into their school, but that the teaching staff could not do all that work on their own. Therefore several mothers and all the pupils assisted in the clean-up, with the girls even bringing in flowerpots and the mothers washing the curtains, as the eighth-grader Vera Rietz stated, "because you learn all the better in a clean classroom. No one feels at home where paper lies around, the tables are dirty and the windowpanes peer at us unhappily."[78] Another pupil described how the schoolchildren and teachers removed and cleaned the offensive piles of roof tiles that had been thrown in the schoolyard from bombed roofs.[79] Young people gladly took part in turning their schools back into orderly places of learning.

Essays by pupils in Berlin about reconstruction indicate that the repairing and clean-up of school buildings and school grounds took on a more concrete meaning for them than did mythical stories of the *Trümmerfrauen*, the "rubble women" who sorted through the building material of destroyed buildings. Pupils wrote with a distanced and heroic tone of women who cleaned up rubble throughout their city, praising "the woman" as the main person of reconstruction,[80] explaining how "housewives and those persons who don't have a permanent job" tore down old houses for their usable stones.[81] In essays about cleaning up the school, these anonymous women took on a personality, becoming mothers of pupils and, instead of sorting through rubble, washed curtains and windows.[82] More often, pupils and teachers noted their own involvement in the reconstruction of the school. Here, as through-out so many other areas of postwar life, the world of children was interdependent with that of adults, but it was not a miniature version of it. Young people sang the praises of *Trümmerfrauen* as did the adults around them, but pupils saw themselves as responsible for the reconstruction of the school building, the sphere of national and local culture that was arguably most theirs. They generally reported with obvious pride on their contributions to the reconstruction of Germany through their maintenance and repair work in the school buildings. Occasionally, a pupil was less sure of the reasons for having to help with the work. The upper secondary school sixth-grader Ruth Frahn seemed to struggle with this question when she wrote that the calls for even more participation in construction projects should not be understood as "propaganda" (which she placed in quotation marks), but as "the only possibility to show the world that we want to fight and work for our right to the existence of our German people."[83] But most pupils' essays related a satisfaction in the group work in which they participated.

The use of pupils in such tasks not only made practical sense, but also mimicked the desire from the Allied Command to give Germans the main responsibility to rebuild their nation. It also fit in with educational administrators' plans to educate pupils toward self-government in schools.[84] This practice corresponded to other programs making use of young people in the labor force. Training boys in handwork and crafts (*Werkarbeit*) was not only important for their future careers, but provided skilled help in the necessary construction work in the school buildings.[85] Pupils proved that they were up to this task, understanding quite well that getting things done in postwar Germany sometimes involved not only hard work but also improvisation. In one essay, the seventh-grader Margot described the painstaking work of each class finding enough nails, wood, and glass panes for windows to winterize their classrooms. Some of the materials could be found at home, but much was found in seldom-used storage areas of the school. Pupils ran back to their classrooms "in feverish haste," so that other classes would not steal their spoils.[86] Her essay reads like a mystery when her class (apparently all girls) discovers the next morning that another class has stolen their windowpanes and nails, and they decide to steal them back. The culprits turned out to be from a boys' class. The girls showed them more consideration than they themselves had been given, afterward carefully covering up the others' windows with cardboard. In another class, a girl reported that their teacher divided her class one morning into groups and then had them count the number of panes still missing from their school and measure how much glass would be needed altogether: "With the measuring and calculating of the panes we had a nice, informative, practical geometry lesson."[87] In all these examples, the school building provided a space for young people to practice functioning as a community, whether they worked together to steal each other's materials, or to calculate measurements for a reconstruction project. The school was not just a microcosm of society; it was children's access to society. By placing the school building in the center of young people's lives, educational administrators and teachers encouraged pupils to find a sense of membership in the school community, and thereby in the German nation.

Memory and Forgetting

The school building also had a special role as an explicit "site of memory";[88] German and Allied Command education administrators alike targeted the school as a key site for teaching young Germans a new way of remembering their individual and collective pasts. The school's development illustrates the creation of a site of memory and forgetting, in which a collective consciousness as well as private memories were anchored, discovered, and elaborated in a process of defining a Soviet zone-specific version of the German past. This function in turn enabled the school building to become a touchstone for orientation toward the present and future. Education became a key element in the reconstruction of the German nation, and the school building itself occupied the most important physical and imaginary space in this program, for administrators, teachers, parents, and pupils. The school building represented a tangible and familiar space in which educational administrators hoped to construct a local sense of membership in a Soviet zone collective.

Yet the school building was more than a factory for constructing memories, even if this role alone would have granted it a special status. It symbolized in a physical and safely secular form a long tradition of German education from which could be drawn a structure for postwar educational plans. Discussions about physical damage were a convenient means of addressing multiple concerns about school buildings. But damaged foundations and broken windows did more than render a school unusable. They reminded the local population of a time when school buildings throughout the Soviet zone had witnessed the war in other capacities. School spaces had been key sites for wartime events. Classrooms and auditoriums had served as shelter during air raids, an infirmary for the wounded, or office space for the occupying armies. The idea of reconstructing a school building brought with it the desire to triumph over the past, or when this goal proved difficult, to at least wallpaper over painful memories of chaotic times.

School administrators felt threatened by the memories and attitudes presented in and by the school building in its dilapidated state.

Participants at a 1948 meeting in Dresden-Wachwitz reflected on the first days of the reopening of school after the war in October 1945, leading county school administrator Herr Erler to exclaim, "Think of the schoolhouses, classrooms, teaching materials, libraries, and school furniture that you saw, and then perhaps you will understand the thoughts that I want to explain here briefly. The children ran wild. They weren't used to discipline at all, and parents had lost interest in the school."[89] The chaotic state of the school building became more than a metaphor for emotional and intellectual confusion. Educational administrators claimed that it prevented young people's successful transitions to antifascist, democratic thought.

The work of eliminating unpleasant or unacceptable memories from the school building actually began before the war had ended. Lieselotte Walter, an elementary school teacher, described a physical example of this memory cleaning in an essay about how her school experienced the battle for Berlin.[90] Walter began with the ironic sentence, "We wrote April 20, 1945 [on the blackboard]—the 'Führer's' birthday!"[91] She described the school courtyard as lit up red and yellow from fire, and a group of frightened people running across it to the air raid shelter in the school's basement. Some of those in the cellar soon learned that their houses had been destroyed, and they were told to stay in the shelter that night. "This was nothing new for us, since we had had such guests quite often at that time," she noted.[92] At one point, others in the bunker began discussing who was actually responsible for the bombing. Frau Walter assumed that the fighting had not yet turned serious, since she had been secretly instructed that in the event that enemy tanks or planes attacked the capital, a five-minute siren would signal, providing time to destroy important files.[93] Soon, however, the Russians were in the city and street-fighting broke out. Frau Walter noted that "our [Nazi] party members" became uneasy, finally leaving the cellar in order to take down anything in the school or their apartments that would identify them as sympathetic to the National Socialists. In the process, they physically began to change the school building from a site full of reminders of the Nazi past to one ready for new memories.

Educational administrators' attempts to help young people create new memories in the "new school" meant helping pupils actively suppress old ones. Thus the school director of an elementary school in Prenzlauer Berg reported taking school files to his house at war's end because of looting problems, thereby deciding that he would be in charge of physically removing evidence of the past from his school.[94] Of course, he might have received orders to destroy sensitive material from his superiors. Regardless, his resolve to take charge of getting pupils into classrooms again was unmistakable, and protecting school files and the memories therein functioned as his first step in guarding his domain. He then decided, apparently on his own, that as soon as the first shock of the destruction of the school buildings had passed, to find temporary accommodations (*Notunterkunft*) to supervise children. Having located "private rooms" to use as classrooms, he then received instructions from a school administrator to begin registering pupils on May 19, 1945. He proudly reported that, on May 25, they began provisional instruction (*Notunterricht*) of approximately 1,000 children, divided into twenty-five instructional groups with thirteen teachers.[95] In his justification for the time and energy thus invested, he claimed that the enthusiasm and willingness to work manifested by his teaching staff filled him with pride and joy, and he cited all these accomplishments as his "own, happy success." With an attempt at modesty, he explained that this was all his "reward in the service of the reconstruction of our so badly beaten *Volk* and the so necessary educational work on our German children."[96] The other, potentially less flattering interpretation of his combined actions points to a determination to keep his earlier role in the school undiscovered by the administration and forgotten by his colleagues.

Not all old memories could be so easily spirited out of the building. Many of them continued to pervade the school building in many instances, attesting to the difficulty of creating a new, antifascist atmosphere. Teachers at a Leipzig school printed a satirical newspaper for Christmas 1947, "without the permission of the military government,"[97] they noted, revealing a tense work environment among teachers there. The various fake announcements "written" by different public organizations and administrators leave the reader with an impression of frustrated *Neulehrer* surrounded by human and ideological remnants of the Nazi period. One notice criticized both a zealous denazification program and its targets with the statement that the SMAD had forbidden all wearing of any NSDAP medals, "even as sock garters." A rather biting comparison also appeared under "Political Mosaic in the School" between the earlier treatment of German soldiers and the attitude toward *Neulehrer*. The first paragraph read, "Earlier: The German soldier is always and without interruption to be kept occupied. Even when no fruitful occupation is available, it is to be ensured that no boredom appears, which bring about lapses of discipline and signifies the undermining of the military forces." It was signed "OKW (Operations Staff)," *Oberkommando der Wehrmacht* (Supreme Command of the Armed Forces). The next paragraph followed with "Today: It is to be ensured that the *Neulehrer* be kept fully occupied, even when the lessons cannot be held as planned because of insufficient coal. The school directors are requested to monitor this. Department of Education."[98] An editorial note at the beginning of the newspaper asked readers to have tolerance for the sometimes harsh comments and to understand that criticism can be constructive. Without further documents from their school, it is impossible to know whether their colleagues appreciated the obvious parallel between the *Wehrmacht* and the Department of Education as constructive criticism. Still, the paper's existence implies that these *Neulehrer* did not content themselves with passively accepting the situation as it was. They did not see themselves as separate or even excluded from the German nation they wanted to rebuild, and they actively protested against the reminders of earlier Nazi power structures and harsh new ones.

In a very real sense, the memories being made of terrible postwar conditions in classrooms were an improvement over others made in school buildings during the war. Aside from the central role of teaching Nazi ideology, the school buildings had been used for a variety of unpleasant purposes by the end of the war. After losing their house in the bombing, Hans Joachim and his family had to seek accommodations in the Sonnenburger school, which served as temporary shelter for those bombed out of their homes: "We were assigned a room with six plank beds where we lived. That was the worst time of my life. We were there for about a week."[99] Hans Joachim had experienced the worst time of his life already; anything after that he evaluated as progress. Another pupil, Horst, remembered the Greifenhagener school in two roles: not just as shelter, but also as infirmary. He and his family abandoned their bombed and burning house to run to the school. "Our burning house, like a burning torch, showed us during this time the way."[100] Before joining his neighbors in a dark basement of the school, Horst delivered his wounded father to a room in the school building being used as an infirmary.

The school building as the symbol of having lost a home, or as a place in which friends and family members lay hurt or dying: these were the memories that children and adults attempted to erase by scrubbing tabletops and hanging pictures. Yet even before the situation had become so critical, pupils had begun to resent the intrusion of the war on their daily lives. Fritz, who apparently fled the city when the bombing got severe, complained that the bombing had kept him and his peers from attending school, "because barely after you arrived at school, the siren started wailing yet again."[101] Even the war's end brought unpleasant memories to the school building. The sixth-grader Hannelore wrote of the grim scene that greeted pupils when they could finally enter their school again: "After the capitulation, the Russians occupied our school and lived there. We found filth and refuse, droppings and feces in the halls and on the stairs."[102] It is not difficult to believe that, once young people finally entered their reconstructed and cleaned-up schools, many really did participate in lessons with "new energy."[103]

Educators and parents also saw the school building as a place that, if nothing else, kept children off the streets, out from underfoot, and out of trouble.[104] The school's ability to be a caretaker was limited, however, and depended on pupils' own attitudes toward school and their free time. Thurnwald established, for example, that older pupils saw the need for school, and resenting it being closed, whereas younger children saw school as drudgery. Younger boys especially preferred interruptions in their daily routines for the possibility of new experiences outside of school.[105] Attending school supposedly offered an alternative to working in the black market, but it was easier to preach about the black market's evils than to persuade young people not to participate in it. Equally optimistic was the educational administration's hope that cultural events at the school could draw pupils away from the dance bars.[106] Black market trips on the way to and from school and related activities in the schoolyard, such as bartering, in fact indicate that some young people integrated illicit diversions into their school routines. Pupils in the Soviet zone invested the school building and its grounds with new traditions, weaving new routines into established ones.

It is impossible to know precisely how well pupils incorporated and made sense of their own frightening and unpleasant memories when faced with learning new ones. The temptation to claim that pupils were expected to forget their wartime and postwar memories dims when one realizes how often pupils and their teachers were asked to remember their horrifying pasts. Nor did all young people feel prepared to leave behind their memories of the war and take up new ones in the school building. A sixth-grader at an upper secondary school distanced herself entirely from the children she observed at her school: "Happy children's voices ring out again on one side of the playground. Good that the war spared at least you carefree playing children such deep wounds in your souls!"[107] Another girl expressed weary resignation at the Soviet decision to blow up the Friedrichshain bunker instead of using it for a hospital. One could almost hear her sigh with the sentence, "Well, there is nothing to be changed in our fate."[108] These sorts of mournful and pre-cocious comments remained in the minority, however, and the young authors almost never made them in conjunction with remembering specific traumatic events. Children recounted past or present difficulties with sadness, but generally not with despair. When the elementary pupil Hannelore wrote of the extreme cold in her classroom, which even "ate through" the blankets they brought to wrap themselves in, she nevertheless finished with the challenge, "So everyone help to give us warm and watertight classrooms!"[109] She was talking to a broader public, from the police chief to the minister for health, certain that they were as concerned with this situation as she was.

* * *

The school building loomed quite large in the minds of residents in the Soviet zone trying to come to terms with their pasts and presents. It was not a bureaucratic representative of a faceless government. The school building was part of the community landscape, in which parents sent their children to school and took an interest in what went on there. Directly connected to the same buildings under National Socialism, and yet invested with different if related functions, school buildings represented neither rupture nor continuity, but rather, as the whole of antifascist democratic education, an uneasy mixture of each. School buildings presented a physical sign of cultural progress in the Soviet zone to the world, legitimating its ability to function under the sign of the German nation. Continuing to work within the framework of the Soviet zone program of national education, regions nevertheless saw school buildings as part of the local community. Cities and towns pulled the school building firmly into the social fabric, forcing it, in cases such as the church, to retain its relationship to other institutions. School buildings also offered young people their own space to remember both individually and collectively. Here, pupils established themselves physically and mentally in the reconstruction of their nation, actively participating in the primary site of memory that they could claim as their own.

Teachers, pupils, families, and administrators spent a considerable amount of both financial and labor resources in the cleaning up of the school building and reconstructing educational spaces. How, then, can recurring concerns about school buildings' numbers and appearance be reconciled with the lack of interest in their actual form? At least one response would be that not discussing its form implied a decision to keep its form, much as no one seriously discussed changing the appearance of other damaged national cultural treasures. The very familiar nature of the school building lent it its hallowed status. Despite rhetoric to the contrary, the administrators and the population did not really want to begin anew. Sweeping out a few of the memories and washing the curtains became the easier solution. In short, the school building represented the physical space and limits of efforts toward renormalization in Soviet zone life, one that allowed for an acceptable fusion of the often-contradictory tendencies of historical continuity and rupture. This was not an organic process. The school building, as a symbolic ideal, assumed these roles while educational administrators promoted the Soviet zone, rather than the Western zones, as the worthier successor to the German nation, and while teachers and pupils attempted to normalize their daily lives.

CHAPTER 4

Rubble Children and the Construction of Gender Roles

Soviet zone educators and policymakers were uncertain what role should be assigned to children in the rebuilding of the nation. While adults' idealizations of childhood and youth promoted the children of post–World War II as an untainted generation, one that would reconstruct an antifascist, democratic Germany, other, competing notions conceptualized youth as a phase of emerging sexuality, and as the end of innocence. Adults used categories of "female" or "male" and "old" or "young" to describe their hopes and fears about Germany's future. They ultimately burdened youth with the task of regenerating the diseased nation. At the same time, young people in the Soviet zone became objects of adults' suspicion and, occasionally, their scorn.

The literature that examines the complexity of differing national gender experiences acknowledges that women and men do not function as monolithic subgroups of the nation, but seldom do the categories "boys" and "girls" enter as elaborations of these identities.[1] Many national and social policies target younger generations, but these groups have not been adequately investigated as participants in the nation-building process. Young people's multiple social roles in the Soviet zone ultimately affected the kind of antifascist democrats, and thus Germans, that they became. The practice of antifascist democratic education maintained many of the social differences that educators believed it would eliminate. This chapter considers the relationship of Soviet zone youth to the new nation, focusing on issues of age, gender, and sexuality. First, an analysis of girls and boys in the Soviet zone situates young people in the context of the rest of the population. The chapter then shows how adults used the concept of youth to express both hope for and fear of the future. A further section analyzes the role of youth in the construction of a national consciousness, focusing especially on the coeducational aspect of the *Einheitsschule*, which promised to enfranchise girls and boys equally in the nation. With the antifascist democratic school reform, adults looked to the younger generation as a possibility for a new start for Soviet zone society.

Images of Young People

Educators in the Soviet zone generally discussed school policies in terms of two age categories of young people. The larger group comprised children old enough to think creatively and independently. They were approximately ten–fourteen years of age and usually in the upper grades of an elementary school or the lower grades of a secondary school. Older adolescents and university students loosely constituted a second, much smaller group. Young elementary school children remained hidden in these educational policies, partly because their undeveloped ability to write or reason made it difficult to evaluate the effect of teaching methodologies on them. Adults used terms such as "children," "youth," and "young people" fluidly. When writing about young people for statistical and analytical purposes, educational administrators and other researchers sometimes employed terms such as "schoolchildren," indicating youth no older than eighteen years of age, or, at other time, referred to a more general category of young people from fourteen–twenty-six years old.[2] To further complicate an analysis of the effect of antifascist democratic education on young people, demographic statistics for the Soviet zone were incomplete and inconsistently collected, and in constant fluctuation, due primarily to massive refugee movement from the East.[3] Furthermore, statistics occasionally excluded East Berlin, or included both eastern and western halves of the city ("Greater Berlin").

The war had changed gender ratios but for the very youngest and very oldest age groups. Adult women now outnumbered men by 15 percent, as compared to their 1 percent majority of 1939.[4] In 1945, the population of the Soviet zone, including East Berlin, numbered 16,194,626, with 6,581,979 males (40.6 percent) and 9,612,647 females (59.4 percent). Returning soldiers and refugees from the East caused sharp and steady increases in the first months and years after the war, so that by 1946, the population had already swollen to 18,488,316, of whom 7,859,545 (42.5 percent) were male and 10,628,771 (57.5 percent) were female.[5] Of these, excluding East Berlin, 2,450,854 were pupils in elementary schools in 1945, by far the largest component of the school-going population, comprising more than 90 percent of all school-goers.[6] This number, again excluding East Berlin, reached 2,660,926 in 1948, comprising 1,355,658 boys (50.9 percent) and 1,305,268 (49.1 percent) girls.[7] In secondary schools for the same period, excluding East Berlin, there were 43,768 boys (60 percent) and 29,494 girls (40 percent). Greater Berlin presented a similar picture. For the school year 1945–1946, there were 128,739 pupils in Berlin's lower elementary schools, divided almost equally among boys and girls at 61,987 and 61,163, respectively. The middle schools had 2,445 boys (43.7 percent) and 3,144 girls (56.3 percent), and there were 7,076 boys (59.9 percent) and 4,743 girls (40.1 percent) in the upper secondary schools.[8] Educators estimated that over 90 percent of school-aged children, approximately 15 percent of the entire population attended school.[9]

The ratio of elementary school-aged girls to boys (through the eighth grade) in the Soviet zone was near parity, which differed from the entire population's ratio of more women. This younger generation had not suffered as many casualties from military service, reflecting a more normal demographic composition of females and males. The numbers vary somewhat for fourteen- to twenty-six-year-olds, a statistical category that included fewer school-goers as well as more of the generation who fought in the war. In 1945, young women fourteen–twenty-six years of age made up 17 percent of the entire population; young men in this age group were only 13 percent.[10] Of the small group of young people who attended secondary school, the majority were boys, even though there were demographically far fewer boys in this age group. This unusual situation was the result of fewer girls than boys having attended school before and during the war. Although boys still outnumbered girls in secondary school classrooms after the war, the latter were there in larger numbers than ever before, although secondary school attendance was not compulsory. Clearly, Soviet zone educational administrators had succeeded in making upper secondary school classrooms more attractive to female pupils. The percentage of pupils older than the normal age for their school grade hovered around 20 percent for boys and 18 percent for girls throughout elementary and secondary school, but in the older grades the differences were significantly higher. There, 21 percent of boys were older than the normal age range, compared to 15.8 percent of girls for the grades nine through twelve.[11]

The picture of young people in the "new school" is thus one that reflected the demographic changes inflicted by the war. Typically, many young people had experienced events that did not normally correspond to typical incidents for their age group. Classrooms had a significant number of female and male pupils older than the normal age range, with boys in the higher grades constituting the largest percentage of older than average pupils. Yet adults continued to use age as a convenient marker, referring to concepts such as youth metaphorically. In this context, definitions of rites of passage became unstable. For example, school boards usually placed young men who had returned from the war back in their last attended grade, an attempt to reconstruct normal educational structures.[12] By reassigning ex-soldiers the role of pupil, however, educational administrators negated young men's military service as a passage into adulthood and created classrooms that singled out an overwhelmingly male experience by highlighting age differences. Similarly, even those young men and women between approximately twenty and forty years of age trained in short-term courses to become *Neulehrer* were not always considered fully responsible adults.[13] At times, administrators portrayed them as young people, fresh and ready to replace the "extremely overaged faculty" (*stark überalterte[r] Lehrerstand*) and take on the task of national renewal.[14] In other instances, *Neulehrer* were judged incompetent because of their inexperience. Educational administrators occasionally substituted the more positive term used to describe these teachers, *neu* (new), with *jung* (young), especially in criticisms of their abilities: "Even with the best intentions, briefly trained or even untrained *Junglehrer* [another term for *Neulehrer*] cannot match the capabilities of an experienced *Altlehrer* ("old teacher")," as one administrator insisted in 1947.[15]

The *Neulehrer* age issue overlapped with gender categories as well. Educational administrators viewed young men as strong authority figures for the classroom; young women were seen as their necessary maternal counterpart. Thus a newspaper article hoping to attract *Neulehrer* candidates noted that, "This call goes out especially to women and girls, who should see their foremost duty in the education of youth."[16] A report from 1945, on the state of teacher training and entitled "Women who educate our children," observed that 70 percent of *Neulehrer* were women.[17] Overall, female teachers comprised just under half the number of all teachers; however, they made up less than a third of upper secondary school teachers. Public images of new teachers, though, most often portrayed a young, enthusiastic man. Posters and newspaper photographs showed a male teacher helping an elementary school boy and girl with their assignments, or else an omniscient man teaching geography to a group of enthusiastic boys.[18] Female teachers did not appear in such authoritative poses. References to female teachers focused exclusively on the need for a maternal approach to pedagogy, since the National Socialist years, "in every regard a period focused only on male characteristics," had ignored the role of women in society, as the CDU newspaper *Neue Zeit* noted in December 1945.[19] Educational administrators did not promote or even view women as academically rigorous instructors; correspondence about female teachers instead focused on the special contributions of women to the antifascist classroom. Women, administrators explained, brought a gentler, feminine element to the schools. In this manner, female teachers did not threaten men's status in the school hierarchy. Administrators recommended women as an antidote to the overly male pedagogy of the past twelve years, but the public and policymakers regarded male teachers as the bearers of antifascist democratic education 9. As in other western nations, the stigma of the teaching profession as a feminized one kept female teachers in the classrooms and out of public myths.[20]

A slightly younger social group, made up of both males and females, symbolized most clearly to adults the possibility of Germany's antifascist future. Older adolescents made appearances on posters as *Trümmerjugend* ("rubble youth," a variation on the "rubble women" who cleared away the physical debris of the city) or able-bodied workers. At first glance, boys and girls seemed to have equal status in these labors. The posters suggested that the physical task of national reconstruction necessitated hard labor and vitality, and was the responsibility of both sexes. Numerous posters appealed to girls to engage in physical labor. One example called for female bricklayers, showing a muscular young woman in front of a brick wall, obviously capable of what had previously been men's work.[21] In spite of such messages of equality, some jobs remained in the male domain. A poster advertising for miners declared: "A career for real guys."[22] Posters often showed boys and girls working side by side to emphasize the sexes' equal participation in the rebuilding of Germany, but the division of labor tended to slip toward traditional constellations. In one poster, a trio of adolescents performed various chores to help reconstruct their nation. One boy carried what might be pipes or wood, another boy laid bricks, and the girl carried a basket of laundry.[23] Society was not ready for equality of the sexes to extend to young men taking on typically female vocations.

The fate of younger children in the difficult postwar years also concerned adults, and contemporary images reflected these fears. Girls in particular were portrayed as vulnerable and guileless and in need of adult protection. Thus an educational administrator acknowledged that a mother might be worried that "wild, robust lads" would possibly terrorize her "tender and shy girl" the first day of school.[24] More symbolically, visual appeals to improve the nation for the benefit of the next generation used very young children, either girls or boys standing alone or with a family, to communicate a promising future.[25] Children, by virtue of not having reached an age of accountability, represented much-needed proof that parts of society had been uncorrupted by National Socialism. Youth bestowed innocence, according to this interpretation, so that young people were not responsible for the politics of their elders. Children embodied the future of the nation both metaphorically and concretely, making images of children a convenient means of promoting ideas and policies that had "our children" at heart. The other tenet to this interpretation was that, as innocent beings, young people had been most wronged by adults' actions under Nazism. Children thus provided a physical reminder of their parents' guilt.

A 1946 Christmas poster captured these ambiguous feelings that young children elicited in adults (see figure 4). The poster, designed to laud the improved consumer situation, portrayed a rosy-cheeked young girl walking through a snowy night, carrying a basket full of wine, boughs, and presents in one hand and a brightly shining candle extended before her in the other.[26] The scene, for all its associations of peace and joy, calls to mind more disturbing messages from the fairy tales of Grimm's "Little Red Riding Hood" and Hans Christian Andersen's "The Little Match Girl." What had happened that this young girl was the sole person in her family responsible for acquiring Christmas gifts and provisions? She is not old enough to have earned her own money for the items, and as the Brothers Grimm have taught us, too young to be sent out on an adult mission. A host of wolves lurked around the corner, threatening to rob her of her innocence. Abandoned or betrayed by adults around her, she made an easy victim. The match girl allusion told an equally distressing tale of a desperate girl's hopeless last fantasies, the candle in one hand illuminating the unreachable dreams held in the other. Implicit in the young girl's youthful innocence and joyful anticipation were a recent unhappier past, as well as the dark threat of future perils. Adults saw an accusation of their own guilt for the past and responsibility for the future reflected in their children's seemingly trusting gaze.

Against this contradictory background of symbolic innocence and power, young people often found themselves faced with an overwhelming responsibility to right their elders' wrongs. Pupils such as those attending graduation in Leipzig in 1948 learned that their choice of professions would affect more than their personal lives: "Both the decision about your careers and the decision about Germany's fate will determine your personal lives as well as the life of our *Volk* . . . Our future and happiness lies in your hands."[27] Similarly, the committee *Rettet die Kinder*

Figure 4 Christmas, Communism, and Consumerism.

Note: This advertisement reminds its audience of the joys of worldly goods, while suggesting a disturbing image of a young girl (the Little Match Girl) in peril.

Source: "Weihnachten 1946 Arbeit und Aufstieg 1947 Konsum," Dresden, [1946–1947] *Das politische Plakat der DDR* [CD-ROM], Inv. P 94/619.

("Save the Children"), formed to improve the situation of children, claimed that Germany's future lay "in the hands of those [children] whose fresh spirit, in the radiance of the tenderest age, is the fertile ground for a blessed flourishing of our new body of thought, [and] whose hands are clean of the blood of those peoples subjugated by Nazism."[28] Such optimism in youth's abilities to generate renewal in Germany was unbridled, but had a parallel in ironic fears that children would not prove themselves up to the challenge of showing Germany's ability to change. Similarly, some references to young people pointed to their malignance and waywardness, both "natural" and learned. Alongside ideals of children's purity existed the possibility that youth corrupted. In this view, young people knew nothing but the teachings of twelve years of Nazi rule. They could not now be trusted to distinguish right from wrong, so they easily let themselves—and their nation—be led astray. "It must unfortunately be recognized that through the exigency of time, through bad examples, through the excessive length of the war, moral and discipline, the feeling for justice and injustice have suffered extremely," as elementary school teachers in Prenzlauer Berg worried at a staff meeting in 1946. They offered antifascist democratic school as an antidote: "This is where the new democratic school must fulfill its educational duty. It must re-educate youth in their basic convictions and bring them to the point that they freely obey the community and do things that are just even when it is disadvantageous."[29] Unfortunately, children were caught in a vicious circle of imitating adults' examples and receiving adults' harsh judgment for this. Worse, in mirroring their observed world, children pulled adults into a complicated process of self-reflection and self-hate that interfered with a much-needed saintly, even heroic, vision of the innocence and redeeming qualities of youth.[30]

In a typical example of this phenomenon, a campaign poster that appeared shortly after the war featured a very young girl in braids, holding a bouquet of poppies in her hand at which she gazed sweetly. The caption called to mothers to ensure the future of their children by voting for the KPD.[31] This message of wanting to keep young children away from the horrors of the previous twelve years also hinted at the tragedy this little girl might already have survived. For whom had she picked the flowers: a missing or fallen father or brother? To brighten the table of her single working mother? The choice of flowers—poppies, instead of the traditional red carnation of the socialists and communists—elicited the memory of another postwar era. World War I soldiers had circulated the legend that the massive numbers of red poppies all over the graves of those slain in battle sprang from the blood of the fallen soldiers.[32] Where would postwar Germans believe the girl had picked her poppies: a soldier's grave, or perhaps from a wheat field, both fresh and fertile sites for the many seeds of the poppy? In the first instance, the quickly wilting poppies became an appropriate bouquet for a girl mourning the loss of a young family member. In the second, poppies suggested a future of fecundity and harvest. The connection between death and fertility is made by the young girl. Her poppies will wilt within minutes, she will toss them aside, and their seeds will be strewn about. New poppies will comfort the survivors, offering sleep to still pain, serving as memorials to the dead, and promising a brighter future.

Such images of young girls completing the circle of life and death contrasted starkly to the rhetoric of equal education for girls. Portrayed amidst flowers and idyllic natural scenes, girls embodied immature, but healthy sexuality. Adults depicted young girls as a symbol of necessary and desirable national reproduction, but sending young women to school meant encouraging them to partake of the fruit of knowledge. This development elicited contradictory feelings in adults. As sexual beings, girls were not seen as entirely innocent, necessitating in one instance a teacher's admonishment to her class to respond politely to Russian soldiers when spoken to, but not to chase after them: "You mustn't forget that you are German girls and shouldn't run after the Russians."[33] The teacher connected her pupils' sexuality directly to appropriate behavior for German girls. Their emerging sexuality was accompanied by certain obligations to the German nation that superseded individual biological or emotional desires. Boys' and girls' sexuality, believed many adults, had been perverted by Nazi teachings and the immorality of wartime, and now threatened the well-being of the nation. At a conference in Potsdam that took place just before the official beginning of the first school year, a speaker explained the extent of damage that youth had experienced under Nazi rule: "Through its [Nazi education] prolongation and emphasis, this [immoral] condition became a normal one for youth—also in sexual relationships. Thus homosexuality and sadism developed into epidemics, and the girls became the victims of the SS."[34] School administration meetings repeatedly included reports on the potential damage that sexually transmitted diseases (STDs) among young people could wreak on Germany. With horror, educators reported that 30 percent of patients with STDs in Berlin were less than twenty years old.[35] Educators and doctors alike described sexual diseases as serious *Volkskrankheiten*, national diseases.[36] Educational administrators' proposed solutions centered around early sexual education, beginning with explaining biological reproduction and the dangers of STDs to ten- to eleven-year-olds, providing thirteen- to fourteen-year-olds with information on sexual problems, and informing fifteen- to sixteen-year-olds of the importance of using condoms, and not switching partners.[37]

Sexual education was a sensitive yet essential aspect of antifascist education. Discussions of STDs among young people also referred to a problem in the Soviet zone that was less of an issue in the Western zones: rapes by Soviet soldiers. The hushed-up issue of massive Soviet rapes of German women and girls did not have a parallel in the American or British zones, which instead had the problem of prostitution and semiprostitution (sex in exchange for favors such as providing cigarettes or chocolate).[38] The magazine *Benjamin* for young people in the British zone engaged in open and heated discussions about fraternization between occupation troops and German women.[39] Authors and letters to the editors warned of the unkind nicknames for young women who went out with soldiers: *Veronika Dankeschön* in the American zone, "Veronica Thank-you very much," abbreviated VD, and thus a direct reference to venereal disease; and *Schokoladenmädchen* in the British zone, "Chocolate Girl," for the sweets given them by the soldiers.[40] In both cases, the warnings about STDs surfaced frequently in public discussions. Young German men accused those German women intimately involved with Allied soldiers of national treason and betrayal. The women pleaded true love and a need for romance. Regardless of the speaker, relations between western soldiers and German women focused on women's agency. Allied anti-STD propaganda portrayed soldiers there as the unwitting prey of exploitative German women, continuing the wartime language of aggressor and victim. Young German men saw their female compatriots as conniving, opportunistic back-stabbers. Young western German women described themselves as following their hearts and transcending petty national quarrels, rejecting allusions of prostitution.

Romantic relationships also existed between Soviet soldiers and German women, as did ambivalent administrative and public feelings about them. These instances could not be discussed openly. Anger, fear, and humiliation as a consequence of retaliatory rapes by Soviet soldiers of German women found few sustainable avenues of public expression, yet overshadowed silently many discussions about nonviolent Soviet-German relations. Atina Grossmann has estimated that 110,000 or more Berlin women were raped by occupation soldiers, mostly Soviets, resulting in as many as 10,000 of the women's deaths.[41] Women turned to doctors in large numbers both to terminate the incurred pregnancies and to be treated for STDs, so that rapes were treated primarily as a "social hygienic" problem.[42] After initial open discussions about rapes and appropriate medical solutions, in the first weeks after the war, the subject disappeared explicitly from public discourse.

Nonviolent, consensual relationships between young German women and Soviet soldiers were therefore a delicate topic for antifascist educators. Overt criticism in classrooms of Soviet-German couples would have jeopardized Soviet-German relations, but many Germans clearly believed that women in the Soviet zone were lowering themselves in social as well as cultural status by dating Soviets. The essay part of an examination, given in January 1946 to candidates of an eight-month teacher training program in Saxony, left little room for doubt about which sentiments toward the Soviets were common, and which ones were acceptable to educational administrators. One question treated the potentially dangerous juxtaposition of love and patriotism: "A German girl falls in love on December 20, 1945, with a Russian soldier. Her girlfriend, who is engaged to a German, accuses her of offending the national honor. What position do you take in this argument? How do you justify your position?"[43]

The question typified the delicate situation between eastern Germans and their occupiers. It could not be addressed through nonfraternization laws, as the SMAD did not pass these restrictions until the summer of 1946, a half-year after the examination was given.[44] Although repeated aggressive incidents between Soviets and Germans made the need for nonfraternization laws increasingly obvious, teacher candidates in January 1946 had no legal arguments against a Soviet-German couple. Candidates hoping to make a good impression on German and potential Soviet readers would have been wise to conclude that the murderer was a racist, anti-Soviet German who had not internalized the teachings of antifascism, and then express outrage at such senseless violence against someone who had sacrificed so much for his own country and Germany. A response that spoke to the need for increased education of the public about racism and crime would have brought attention back to the candidate's role of future teacher, promoting schools as constructive solutions to social problems. Answers in this vein proved that the candidate had learned well the unconscious lesson of keeping issues like rape alive, but not in explicit, public terms. Having been pushed to maintain the issue of rape in the German collective conscience, but without endangering oneself to Soviet authorities, a candidate focusing on the need for more understanding between Germans and Soviets would thus, in the eyes of the DVV, make a perfect antifascist teacher for the "new school." With the hostile atmosphere toward acknowledging the problems of Soviet rapes in the Soviet zone, the continual discussions of STDs by school officials leave the question of their opinion about the rapes open. Perhaps the officials perpetuated the problem by refusing to consider publicly a connection between young people's STDs and rapes. Nowhere, for instance, did anyone ask if these STD strains were the same or not—which would have suggested either that the rapes by Soviet soldiers extended broadly to young people or identified larger consensual and semiconsensual sexual relationships between Soviets and young Germans than assumed. On the other hand, the prevalence of venereal disease and the problem of rape forced German educational administrators to address sexual health. In thematisizing STDs and racist sentiments, educators appear to have found a safe arena in which at least the physical consequences of rapes could be addressed and, through increased sexual education, perhaps avoided.

Educational administrators also worried about pupils' behavior that suggested continued Nazi practices, such as young people's violence against their peers. Teachers most often reported boys as creating the more serious disciplinary problems, keeping boys at the center of this discussion. The discipliner tended to be male, as well, particularly if brutal punishment was meted out. In late May of 1945, a Dresden school director filed a report of four teenage boys, approximately fifteen years old, beating a mentally and physically handicapped classmate to keep him from attending their class.[45] They beat two other classmates in the process, using the "battle cry" of "communist pig." The director of the school, noting that the ringleader of the pack had been a former Hitler Youth leader, identified this behavior as part of persistent Nazi ideology of dealing unfairly with lesser-abled persons in order to maintain racial purity. He slapped the boys and gave them fourteen days of detention from 7 a.m. to 7 p.m., with an hour for lunch and *Arbeitsdienst*, manual labor service. One of the four boys who had not used the "political insult" received only ten days of detention, and the pupils not directly involved in the fighting served three days of detention. Later, another pupil who heard that the director had slapped the boys accused them of having let themselves be beaten by a "communist pig." The child then punched a boy in the face who reported this insult, and had to go to the director's office where his mother came to defend him for being a "good boy." After the mother left, the boy became verbally abusive. The director slapped him twice, and the boy attacked him physically. Security came in, hooked the boy's arm behind his back, and he confessed to everything. He then spent four days under police arrest with decreased food allowances, followed by ten days of hard labor and night arrest. The director finished his report with the comment, "The clearly political incident necessitated the most strict punishment."[46] It is doubtful whether the director intended any ironic allusion to his own behavior.

Educational reformers had launched a campaign right after the end of the war to end corporal punishment, which they identified with Prussian and Nazi methods of disciplining children that resulted in militaristic and fascist attitudes.[47] Antifascist educators deemed its continued practice unacceptable, but were particularly upset when girls were the targets. A thirteen-year-old girl from a girls' class wrote an essay on her thoughts about her school, which was read by members of the Berlin central school administration. She complained that the director hit the pupils. This section in her essay was marked in red by an administrator with the surprised question, "Corporal punishment—does he hit the girls as well?"[48] The alternatives to spanking or hitting children that classroom teachers developed did not always conform to an antifascist democratic vision of eliminating drastic and violent punishment. Possibilities ranged from rewarding positive behavior to withholding food as punishment, and seemed to be employed against boys and girls. Two faculty meeting entries from a coeducational elementary school in Prenzlauer Berg (Berlin) illustrated the shift from corporal punishment to other equally severe disciplinary measures. The first, written in October 1945, admonished teachers, "Don't hit the children!" Nine months later, an entry explained that pupils who skipped school repeatedly would have their food ration cards taken away, since this method effectively kept pupils in school: "Whatever could not be achieved with instructive and education measures was immediately solvable with the 'stomach question.' "[49] Yet extreme examples of punishment continued to be directed only at boys. The city councilor of Meißen, for instance, explained that his city had introduced youth arrest as a disciplinary measure. Offending boys had their heads shaved and were then assigned to "work duty," a sentence of hard manual labor and a humiliation all the more remarkable for its associations with typical punishment of women who had fraternized with enemy troops, or the delousing of war prisoners.[50]

The search for a viable means of punishing young people grew out of a perception, more felt than observed, that children threatened the social order through their lack of discipline and a disregard for the law and moral codes. German educational administrators realized that they were not the only ones looking to their young people as a measure of Germany's progress toward antifascist democratization. Educational administrators constantly worried that children's actions would further damage Germany's national image. The school director of an elementary school in Berlin-Köpenick even warned his teachers in 1945 that their pupils misbehaved so badly so as to horrify Allied sensibilities during school observations: "Foreigners, the English, Americans don't understand such lack of discipline. There, a respectful calm and quiet reigns."[51] In another example, some thirteen-year-old girls' use of Nazi propaganda in 1947 created a real dilemma for the school board in Berlin. Unable to decide if the teenagers understood the context of their actions or if it had been a thoughtless prank, the school board members did not inform the Allied Command of the incident. Fearing that this case could be used to strengthen Allied doubts about the schools' effectiveness in creating antifascist, democratic citizens, educational administrators ignored standard reporting procedures.[52]

From all sides, faith in young people had its limits, and sometimes extended to outright mistrust. The irony of the generation responsible for the Nazi regime judging children was seldom a topic of discussion. German educators complained that young people could not be trusted to withstand the mixed messages their daily experiences gave them. Occasionally, however, classroom teachers and school directors treated pupils' struggles to come to terms with their new social situation with sympathy and understanding. A 1948 report from Berlin-Köpenick describing the character of the graduating class seemed to reflect upon the emotional and intellectual work involved in being an adolescent in postwar Germany, describing young people as still caught between Nazi and antifascist teachings: "It could be seen in many [of the pupils] how great the rubble left behind by this war in the intellectual area is. Some of the graduates are uncertain and innerly torn. They struggle to cope with the demands of life."[53] More often, teachers and administrators concerned themselves with directing pupils' thinking toward acceptable forms of expression. Educational administrators discussed in June 1947 whether children should even be allowed to write and act out their own plays, or whether they might internalize unacceptable lessons if given too much creative liberty. The argument against youth's plays focused on one child's piece that reenacted a robbery, which some administrators feared might lead young people to see theft as acceptable under certain circumstances. In this case, the prevailing opinion noted that teachers could use such plays as learning opportunities, and that children already read accounts of robberies every day in the newspapers.[54] At that date, antifascist democratic education had been in place for two academic years. Many administrators apparently did not believe that pupils had learned their lessons well enough and thus did not entirely trust young people to draw their own conclusions about democratic values.

Antifascist educators' concern with their pupils' behavior extended beyond the school grounds. Children's presence in illegal and otherwise questionable activities outside the school caused educators considerable worry. Participation in the black market, for instance, disrupted the struggling economy, broke social mores, and kept children from attending school, in turn preventing them from internalizing the antifascist lessons taught in class.[55] To combat this tendency, adults forced to young people spend considerable time during class learning about the dangers of illegal practices. Pupils wrote essays about the economic damage inflicted by the black market: "The biggest enemy of reconstruction is the Black Market, because it makes the people poor," as twelve-year-old Manfred dutifully stated.[56] Given the prevalence of young people active in the black market, it is likely that many of them, after completing their schoolwork, then went out to hawk wares illegally. Black market dealers used children to transport goods, and young people often sold items themselves for their families, often at the request of their parents. For instance, the communist-oriented family Q, interviewed by Hilde Thurnwald in 1946, expressed support for the "new school," but sent their twelve-year-old daughter on black market errands.[57] Although teachers criticized these situations, they did not usually intervene when they observed children exchanging goods on the black market, to the chagrin of those parents hoping to keep their children away from these activities.[58] Ultimately, young people's unlawful behavior in black market activities implicated adults, who ultimately supported this underground economy and whose actions thus led children astray. On a less criminal level, children even mimicked the bartering society of their parents, down to the bickering and hard feelings such transactions evoked. According to one teacher, bartering had taken the place of games: "Bartering brings the children together again and again in spite of all the disputes. Whenever you see a group of children tightly huddled together, they are definitely trading."[59] In this case, the teacher then indicated a certain degree of understanding for the allure of bartering, especially when money gained in such transactions could be used to go to the movies, where it was both warm and entertaining.[60] The theories of antifascist democratic education did not provide guidelines for these everyday needs and desires of young people, leaving children and their teachers to decide for themselves the extent of wrongdoing that an antifascist society could accept. In spite of classroom lessons to the contrary, the margins of tolerance were generous.

When adults looked at young people, they saw a manifestation of Germany's past and future. Discussions about educational policies thus demonstrated adults' attempts to come to terms with what the younger generation represented to them. Young people comprised a social group that reflected the consequences of the war, from demographic changes to ideological teachings. Youth symbolized innocence, but it also represented the recent years of National Socialism. Many images of girls and young women portrayed them as helpless victims, while others alluded to their sexuality as a threat to the nation, particularly if they associated with Soviet soldiers. Although adults saw boys and young men as powerful forces that would rebuild society, they also viewed this group as particularly unruly and disobedient, and at times still in need of corporal punishment. These tensions in adults' perceptions of young people constantly worked against antifascist educators' objectives of unifying society. Young people remained a separate, distinct group in the Soviet zone; within this group were further divisions of age and gender. Antifascist democratic ideals of entirely erasing social divisions proved impossible.

Antifascist Coeducation

One aspect of the antifascist democratic classroom best illustrated the difficulty of using youth to construct a new, unified society: the education of girls. Like other aspects of antifascist democratic education, the discussions here drew on ideas of early centuries. The specific education of girls was problematized publicly in Europe with François Fénelon's 1687 work "De l'éducation des filles," translated and widely read in Germany in 1698.[61] Fénelon, like Jean-Jacques Rousseau after him, favored a separate girls' education that would correspond to their character and social roles.[62] New ideas about citizenship a century later paved the way for female members becoming enfranchised in the nation, but proved difficult to implement.[63] In 1786, the German educational philosopher Joachim Heinrich Campe commented with considerable frustration that, when it came to the female sex, the state seemed not to care if it was educating "girls or meerkats."[64]

Europe in the late eighteenth century resonated with new ideas about citizenship, the state, and women, and the school and the concept of national education played leading roles in these discussions. Accused of "irrational" thought processes and subservience to the church, undereducated and religiously observant women were seen by men and other women as a key impediment to a nation's fulfillment of Enlightenment goals.[65] Three solutions emerged from these debates: exclude women from the political nation; find a gender-specific, national role for them; or, the increasingly selected option, secularize and rationalize them in schools.[66] Girls' schooling has thus become the most obvious symbol for demarcating between "modern" western societies and "backward" traditional societies.[67] Denying girls equal education meant that they would not receive the same training for membership in the nation as boys.[68] With the French Revolution, the fountainhead of national educational systems and compulsory military service, states increasingly enlisted male youth in the service of the nation.[69] The gradual implementation of mandatory universal education over the course of the next century, in contrast to the exclusively male army, extended this nation-building project to girls.

Coeducation had a long history in Germany. Gender-segregated classrooms have been the rule throughout most of history in the western world, but examples of limited coeducation have existed in many areas. Girls found their way into boys' classrooms, particularly at the elementary school level, when no other schools were available, or when girls needed specific skills associated with boys' education, such as simple accounting.[70] The necessity of coeducation in smaller rural schools, which had a one-room schoolhouse or at most two classrooms, had been accepted without much debate in the nineteenth century. Marianne Horstkemper described the practical approach to this problem by educational philosophers and policymakers: "Separation when possible, coeducation when necessary."[71] Demographic shifts brought about a rethinking of this situation. Although the one-room schoolhouse was the most common means of schooling for boys and girls toward the end of the nineteenth century, this practice changed in the early twentieth century as more schools were built, making separate schools possible even in the countryside.[72] Coeducation debates increased as segregating the sexes became a matter of choice. For the first time, scientific studies about gender differences accompanied the debates about a separate girls' education, including whether girls should be admitted to secondary schools. Many studies concluded that girls and women were physically incapable of advanced learning or directing schools, because females had smaller and more delicate brains, hearts, and lungs.[73]

In the mid-twentieth century, military service would still separate "Mars and Venus."[74] This continued to be the case throughout the GDR once an initially voluntary *Volksarmee* (people's army) was founded in 1955.[75] Schools in western Europe, though, were beginning to make at least topical concessions to educating their *citoyens* and *citoyennes* similarly, and United States schools had been largely coeducational since the nineteenth century.[76] Girls had been granted the right to a state-supported education, even if the law did not guarantee them an equal education. In the immediate postwar period, a military organization did not yet exist for Germans in the Soviet zone or GDR. Not only was the school the only institution to matriculate the entire population, but also remained, until 1962 and the introduction of compulsory military service for men, the only site where the state had the right to demand the presence of any of its members.[77] A coeducational school system therefore took on an even more symbolic and concrete role as the representative of the state in the Soviet zone.

The defeat of Nazi Germany reopened the discussion on girls' and boys' education, with different solutions for East and West. Educational administrators in the Soviet zone chose mandatory coeducation—the schooling of girls and boys in the same classroom; the Western zones retained gender-segregated classrooms after elementary school for another two decades. Antifascist educators wished to eliminate the consequences of unequal access to education for girls and its implications for the future of an antifascist, democratic society, which adhered to the socialist interpretation of gender discrimination as the consequence of capitalist exploitation, a theory elaborated most clearly by the nineteenth-century writings of August Bebel.[78] The existence and perpetuation of gender-segregated classrooms, differing career expectations for girls and boys, and gender-specific courses all threatened to interrupt the progress toward a unified German nation that would offer all of its citizens the same education. The concept of coeducation for boys and girls thus became a cornerstone in the discussion of how to rebuild Germany and a convenient metaphor for enlisting the school in the struggle for women's equality. As DVV director Paul Wandel insisted in 1947, "We are still of the opinion that women and men must receive exactly the same opportunities based upon school education. That is, there should be nothing in the school that from the beginning says that the woman should actually be in the home. Instead the entire education in the elementary school and in the secondary school must proceed in a manner that assures the same education for women and men."[79] Although he spoke in the name of all Soviet zone educators, he did not represent a unified front. Eliminating the gender bias from schooling proved a daunting task. Coeducation threatened the traditional school order, called the authority of the church into question, and, because it was practiced in only one half of postwar Germany, separated the Soviet zone from the Western zones. To the chagrin of social democratic and communist educators, furthermore, coeducation did not resolve the issues of girls' education or participation in the nation. It created new questions about gender, sexuality, and age, all tied to the realization that a single, revolutionary policy could not change society by itself. Contemporary observers of Soviet zone educational reforms expressed as much frustration with the slow progress of girls' educational reforms as Campe had.

Although girls' education became one of the most significant issues in rebuilding society in the first German antifascist state, there was little uniquely antifascist or socialist about coeducation programs. Nonetheless, historical discussions about girls' education found an echo in the Soviet zone, making coeducation a key tenet of antifascist democratic education. However, subtle biological postulations about girls' and women's natural roles as future mothers continued to be part of Soviet zone decisions about coeducation. Advocating equal education for girls and boys existed alongside the question of whether to offer certain courses to girls that were considered necessary for their futures as wives and mothers. Educators at a 1945 conference in Saxony voiced resistance to having the same education for boys and girls, if it was at the expense of ignoring girls' domestic and maternal instincts: "Instruction in home economics has the duty to transmit some basic home economic knowledge and a modest amount of practical skills to the adolescent girl, thereby attending to the housewife disposition of girls."[80] Similarly, a women's committee for the DVV that met in 1947 to discuss girls' education complained that the school curriculum lacked mandatory domestic skills courses for girls.[81] One member noted that she could not imagine that women would find this situation acceptable.[82] The educators present finally suggested that domestic skills should be offered to girls, parallel to boys' courses in handicrafts, and that biology should be taught separately to boys and girls, at least in the seventh and eighth grades. The only dissenting voice was Käte Agerth, who later went on to become a respected teacher in the GDR—indicative, perhaps, of her early commitment to antifascist democratic objectives, such as eliminating unequal access to education.[83] Even as late as December 1948, an educational administrator in Berlin suggested that the absence of infant care courses in the secondary school would create real problems for girls later in life, and proposed the establishment of four- to six-week courses that would cover the "necessities."[84] Many antifascist educators, although they accepted the premise of coeducating pupils for most courses, were not willing to abandon traditional female courses involving housework and childcare and, in some instances, biology.

Even arguments in favor of coeducation alluded to psycho-biological differences in women and men that needed to be overcome. Discussions about girls' education often positioned girls and women as having been unable to resist Nazi propaganda because of a lack of appropriate reasoning skills. A conference in 1947 on girls' education concluded that girls needed the same education as boys in order to acquire the same logical reasoning skills and political-historical orientation. Only such an education could equip girls and young women to resist mass suggestions as during National Socialism. "Today's school must educate a new generation of women who are immune to fascism and who will later raise their own sons and daughters antimilitaristically, for peace and the reconciliation of peoples, and who will confidently take their places in public life, in the economy, and in politics," as a 1947 report on the DVV's position on girls' education insisted.[85] A 1945 meeting of six male elementary school teachers in Dresden described the perception that women were more psychologically labile in even more concrete terms. This self-appointed "Provisional Committee of Antifascist Teachers" raised the concern that, of those teachers who had not belonged to the NSDAP, some were politically reliable, while others had nevertheless participated in some Nazi activities. The group suspected that these individuals "were likely to be primarily young female faculty."[86] One point of these "scientific" arguments was clear. If Germany was to become truly reeducated, it was not enough to reeducate only half the population. Its other message was equally clear. Mothers had a special ability to raise antifascist, democratic children, regardless of women's otherwise equal place in society. That reeducation programs especially targeted mothers and girls as future mothers indicates the extent to which postwar life circled around a very uneven gender distribution of labor, in which girls and women carried the larger responsibility.[87]

Proponents of coeducation also championed its economic, political, scientific, and pedagogical benefits. On a practical level, maintaining separate institutions for girls and boys was not financially feasible.[88] Teachers, school buildings, windowpanes, and coal were already in short supply, making judicious resource sharing a priority. When the economic savings of coeducation did not significantly impress its opponents, educational administrators added ideological considerations. Usually referring to Prussian and Nazi schools, they claimed that previous educational systems' mistakes would be remedied by coeducation. An internal report from the department for schools in the Department of Education pointed to Nazi policies that incorrectly justified gender-specific education with biological differences, resulting negatively in a more "feeling-oriented" education and increased course-work for girls.[89] Pointing to the need to educate boys in household skills as well, the author further noted that as soldiers German men had previously learned how to clean their bedrooms and cook.[90] Since armies would not initially be playing a role in postwar Germany, it was obvious that men would have to learn these lessons elsewhere.

Some teachers who had experience in coeducation classrooms claimed that the learning atmosphere was better, since boys and girls challenged each other positively.[91] For those adults who worried about girls and boys' interactions, the author of an article for the newsletter for the teachers' branch of the trade union (*Freie Deutsche Gewerkschaftsbund*) assured his readers that sexual tensions diminished between boys and girls who were educated together.[92] Keeping up with international pedagogical developments became another powerful argument for coeducation, with its implicit comment that Germany's educational system had not progressed under the Nazi regime, while education reforms had moved forward elsewhere. This also drew on discussions from the turn of the century that looked to the United States as a progressive role for its policy of coeducation.[93] The early twentieth-century arguments that "coeducation" was a foreign word and therefore an inappropriate foreign concept did not exist as such in the Soviet zone.[94] Nonetheless, references to foreign educational systems in lecture evenings and journals often noted pointedly that other countries' educational practices did not necessarily speak to German needs.[95] After tensions between western and eastern Europe and the subsequent emerging cold war, Soviet zone educators no longer cited the United States as a modern nation that practiced coeducation.[96]

Educators in the *Einheitsschule* were not all convinced that girls had the same abilities and interests as boys, and some of them continued to promote some gender-segregated courses. Still, neither educational reformers nor the public made explicit arguments to reserve an elite education for boys, and even proponents of coeducation did not conceptualize gender-segregated classrooms in terms of boys' advantages. The needs of the labor market were also absent from these discussions, even though it would be dominated for a generation by women. Girls' education continued to be addressed in the same terms as it had been since the beginning of the century. Politically, changing girls' education was seen as necessary for the well-being of the nation. Scientifically and morally, arguments had disappeared about girls being less capable, making their separate and inferior education unjustifiable. In spite of differing levels of opposition to identical education for boys and girls, the state and a majority of social democratic and communist educators, nevertheless, wanted to see girls receiving the same education as boys to pave the way for a more active participation of women in the "new Germany."

Although educators succeeded in pushing coeducation through as an official policy in the Soviet zone 1946, it did not immediately become reality. As in so many other areas of educational policy in the Soviet zone, the optimistic goals set in the early days of school reform for true coeducation faced practical obstacles, making it impossible to implement fully. Nor were regions actually required to provide coeducation. Although the 1947 "School Law for Greater Berlin" specified a coeducational system for Berlin, the 1946 "Law for the Democratization of the German School" did not actually name coeducation as a specific condition of Soviet zone schools.[97] Its preamble guaranteed a school system that gave "all youth, girls and boys, city and rural children, without regard to the wealth of their parents, the same right to education." Separate boys' and girls' schools, which continued their existence for a time after the war, did not therefore violate the letter of the law.[98] Some districts practiced coeducation because of practical, rather than ideological considerations. Many separate sex schools throughout postwar Germany were not actually entirely gender segregated, regardless of official policies. Furthermore, shortages of classrooms and teachers made coeducation the only solution in many areas. However, other tangible impediments to implementing coeducation represented more complex issues, such as the lack of separate bathroom facilities for girls and boys, a major concern among educational administrators.[99] Assigning boys the bathrooms on one floor, and girls the bathrooms on another floor usually solved this problem. Such concerns highlight more emotional opposition to coeducation, suggesting adults' fears of overlapping male and female spheres.[100] As the most intimate domain of the school building, parents', teachers', and educational policymakers' constant allusions to the bathroom conveyed discomfort with children's emerging sexuality, couched in arguments about practicality and hygiene.

The uneven implementation of coeducation might also have resulted from a resistance to changing previous school attendance habits, especially in those cities in which the original buildings remained at least partially intact. New policies such as coeducation would have been easier to enforce with the creation of entirely new school communities, instead of redividing school districts. The Western zones' general stance against wide-scale coeducation did not have major consequences for the majority of the Soviet zone, although it did create some problems for

Berlin, which functioned under slightly different administrative authorities. None of these problems posed obstacles to complete coeducation, and yet in the summer of 1949 a report in the Berlin central school administration could do little more than modestly report that coeducation was being increasingly implemented throughout the Soviet zone, with Berlin having introduced coeducation in the first and fifth grades (i.e., at the beginning of elementary and secondary school, respectively) to prepare its "organic development."[101] In the end, coeducation belonged to those policies whose simple provisions couched a host of demands about restructuring society for which much of the Soviet zone population was not ready to commit itself fully. Nevertheless, the official decision to eliminate segregated classrooms from both the elementary school and the secondary school represented a significant departure from previous conceptions of gender-appropriate education and women's role in society.

Arguments about gender-specific coursework had two sides, though. The language and examples educational reformers used in discussing coeducation focused primarily on the question of how equal girls' education should be. Boys' education, however, also became an explicit subject of debate. For example, separate woodworking and metal classes for boys fulfilled not only future needs for these skills, but offered a practical solution to providing repairs for badly damaged schools.[102] Such examples demonstrate boys' presence in arguments about equal education, even if they were seldom explicitly mentioned. Educators most often conceptualized gender-specific coursework in terms of its relationship to women's rights, but coeducation also implied changes in boys' education as well as their later status in society. Coeducation was the harbinger of a reorganization of gender relations, and debates about girls' education were also debates about boys' education.

Yet, the teaching of girls and boys in the same classrooms did not bring about educators' utopian visions of a truly equal education for all. Coeducation classrooms continued to manifest gender differences, especially in areas of pupils' interests and knowledge base. Girls wrote about decorating the classroom with flowers, while boys recounted military battle details in their essays. Obligatory coeducation represented one of the most radical and physical changes in the antifascist classroom, but it was not met with "broad consensus," either publicly or in schools regardless of what official histories of East German coeducation have led historians to believe.[103] Pupils especially had ideas of their own, and they sometimes worked against the reforms in ways that surprised educators. Young people did not always act in accordance with the spirit of new policies, as the Thuringian education minister Walter Wolf (KPD/SED) complained in a report on his region's schools in 1948 and 1949.[104] Regarding girls' education, Wolf concluded with some chagrin that the system of allowing all pupils to choose one of three subject tracks— math/science, classical languages, or modern languages—did not result in an even distribution of girls and boys throughout these fields. Girls preferred modern language courses to science or classical languages. He was certain that this was the result of social pressures, and not a reflection of different gender abilities. But getting girls to reject social norms was as difficult as changing the messages.[105] Coeducation represented yet another program for unity that continued to be plagued by elements of division. Coeducation and its larger context of girls' and boys' interactions represented another layer of policymakers' and educators' attempts to construct a unified nation and push the idea of the *Einheitsschule* to its limits.

Underlying all of these justifications for coeducation, finally, lay the constant moral reference to girls' right to the same (*gleiche*) education as boys.[106] A 1948 report on girls' education stated decisively, "The school reform of the East zone fundamentally recognizes and guarantees a woman's human right to the same education, to work, and to the free choice of a career."[107] Soviet zone educators did not stop at the "same education," which did not explicitly call for the identical (*selbe*) education. Separate education could not be considered equal. Still, the educational reformer and teacher trainer Heinrich Deiters (SPD/SED) told Austrian educators in the summer of 1948 that coeducation in Germany was not yet achieved. "The question of coeducation has not yet been definitively decided upon in the Soviet Occupation Zone"; he explained, "but developments are leading to coeducation. In Berlin the division of sexes generally still exists. In the other parts of Germany," that is, the Western zones, "coeducation is fundamentally rejected and only allowed in exceptional cases."[108] Coeducation did not become a policy in the Western zones, and attempts to implement it there in any educational arena generally failed. Churches continued to play an important role in educational policy in the Western zones, and coeducation was associated historically with policies antithetical to Christian beliefs: in 1929, Pope Pius XI issued the papal encyclical *divini illius magistri* against coeducation as opposing the doctrine of original sin.[109] Policies of coeducation thus highlighted the confessional/political divides between the Soviet and Western zones, as illustrated by a 1947 incident in the Western zones reported in the Hamburg young people's journal *Benjamin*.[110] The authors of the article, members of the *Falken*, the socialist youth organization in the Western zones, responded to accusations in CDU newsletters that girls and boys had slept in the same tent on *Falken* outings. The CDU critics had called such policies immoral and un-Christian. This incident had in fact occurred, the authors explained, but it was an accident. The *Falken* did allow younger children to sleep in the same tent, but older boys and girls slept separately. Furthermore, the authors noted that coeducation was a foundation of their work, and that they were a democratic organization and did not take sides on religious issues. Coeducation could not have succeeded as a policy in the Western zones because conservative parties there associated it with socialist policies of the East. The division of Germany made itself felt in such everyday considerations as the composition of classrooms.

Even after implementing coeducation as fully as possible, though, educators in the Soviet zone had a difficult time evaluating its success. Some administrators worried that they had gone too far in encouraging girls to take on traditional male professions. When Ernst Wildangel, chair of the Berlin school board, visited a girls' twelfth-grade vocation class in 1948, he found that all of them wanted to be doctors or architects. His comment that no one wished to become a seamstress, or enter educational careers like preschool and kindergarten teachers, was met with some ambivalence by his fellow school board members.[111] On the other hand, a newspaper reporter interviewing six female pupils "of marriageable age" at a girls' school in Thuringia walked away dismayed at how little these girls valued education. They laughed at the reporter's frustration that they had little knowledge of geography and regarded French literature as degenerate. As they explained, a woman's place is at home with her family, so that she need not know very much. In a telling moment, one girl stated, "If I were ugly, I wouldn't have any other choice but to work towards a doctorate."[112] As significant a step as coeducation was toward women's equality, it did not change persistent gender dynamics either in or outside the classroom.

Based on their experiences and biases, educators presented certain characteristics of boys and girls that both reflected and reinforced their gender roles. Beliefs in the equality of women notwithstanding, many educators agreed that girls did have a certain kind of nature that differed from boys', one that might need to be addressed as such even within the structure of coeducation. This approach took various forms, ranging from uncommented observations of girls as less unruly, to more explicit suggestions for how to meet girls' special needs. The *Kreisschulrat* (county schoolboard administrator) Lehmann of Leipzig noted that the *FDJ* (*Freie Deutsche Jugend*, Free German Youth, the state-sponsored socialist youth group) had a difficult time attracting female members. He described the "female sector" as having "fallen into the sleep of Sleeping Beauty." He did not propose a Prince Charming, though, but rather suggested that more women were needed in the upper secondary schools to change education's traditional male-dominated structure.[113]

Not all of these characterizations of girls found favor with all Soviet zone educators. They accepted certain stereotypes uncritically. A man interviewing for the position of school inspector in June 1947, for instance, claimed that girls were much easier to control than boys. No one on the committee protested this observation.[114] Negative comments about girls' political behavior, however, were met with skepticism. Frau Päslack, a candidate for a school inspector position in April 1947, had made an unfavorable impression when she claimed that girls were very unpolitical. "[E]ven the question of women's equality is difficult to convey to them," she explained.[115] When confronted with such statements, which questioned the school's ability to influence girls' antifascist, democratic views, educational administrators were quick to name counterexamples. Ernst Wildangel thus responded that he had heard that girls held more radical views than boys, whether they were fascist or antifascist opinions. After Frau Päslack left the interview, Wildangel insisted that she was wrong about girls, because they expressed themselves far more decisively than boys.[116] The woman did not receive the position, apparently because the educational administrators believed that she did not adequately understand girls. Both of these examples drew on anecdotes and impressions, with educational administrators drawing on evidence that satisfied their idealistic aspirations for girls' potential. Individual experiences with girls that did not correspond to these images were identified as misinformed.

Certainly, young girls and boys demonstrated many similar interests and behavioral patterns. One common issue discussed by all young people was the future of Germany. In multiple instances children looked with hope to the new democratic Germany;[117] praised the progress made in areas such as water availability, the street cars, and the theater; and mourned the damage to their country by National Socialism.[118] Others expressed uncertainty about the Soviets, including surprise at how friendly they were as well as discontentment with how Russians took many goods back home.[119] In general, boys and girls expressed a fervent desire to live once again in a Germany they could be proud of, one that had returned to "normal," one where there would again be "peace, instant powdered drink mix, and a flag."[120] There was no doubt in young people's minds: they were German, and were part of the new Germany.

Other children's sources show some fundamental differences in the habits and activities of girls and boys. Small autograph books collected by girls, containing short verses and notes from girlfriends and family, remained a very female pastime. Typical entries used phrases clearly remembered from a different era, such as "*Frisch, fromm, frei/das deutsche Mädel sei*" (Fresh, pious, free/be ever the German girl) or "*Im Glücke niemals stolz/Im Unglück edelmütig/Den Freunden stets getreu/Und gegen Feinde gütig*" (Never proud in fortune/Noble in misfortune/Always true to friends/And generous to enemies).[121] Practices such as these were learned outside the school, from older female family members, establishing a continuity in gender roles that worked against antifascist democratic principles of eliminating social differences. Pupils also demonstrated gender-specific behavior in classroom assignments, such as the boy fascinated with the English ten-ton bomb and the last military battles of World War II,[122] or boys' excited accounts of watching a bunker in East Berlin demolished.[123]

Several pupils' essays written in the Berlin district Prenzlauer Berg on the topic "*Frauen räumen auf*" (Women clean up) reflect young people's struggles to reconcile traditional gender roles with new lessons about women's role in society (see figures 4 and 5). For instance, many female and male pupils praised the value of women's role in cleaning up the rubble of the city, but occasionally noted that this, task entailed women having to perform "men's work." Pupils themselves thus implicitly recognized "rubble work" as a prime example of the larger societal question of whether women's equality actually meant assigning them separate spheres of action or whether equality could or should imply shared spheres of work.[124] One pupil—most likely a boy—went so far as to note that women could take their places alongside their men when called to do so.[125] His teacher underlined the sentence with the comment "awkward," although if this referred to the wording or the content would have been difficult for even the teacher to say with certainty. Other essays described women in less flattering terms, using familiar stereotypes of talkative or domineering women. Thus, Helga Katz's essay on the air raid bunkers described nervous mothers and the gossips in the building, then exclaimed that the bunker was the perfect place for their gabbing.[126] Another girl bitterly remembered a mean landlady who had made her family's unpleasant life in the first postwar period.[127] In contrast, pupils' descriptions of male figures did not focus on men's virtues or vices. Men more often surfaced as missing fathers, absent figures who would be able to help the family set up the household again.[128] Such essays reflected a changing society in which young people, like the adults

14.

XX, 30

Frauen räumen auf.

Jeden Tag sieht man Frauen auf
den Straßen den Schutt wegräu-
men. Sie haben Karren und
Schippen, damit laden sie den
Schutt auf und fahren ihn weg.
Die Frauen räumen die Steine
auf, weil wieder neue Häuser ge-
baut werden müssen. Wenn
der Schutt aus den Straßen weg
ist, dann sehen die Straßen
viel schöner aus. Die Frauen
klopfen die holprigen Steine grade, weil sie dann besser zum Bau-
en verwertet werden können.

Christa Klatt
Wörtherstr. 20
Kls 5 b

Figure 5 A Pupil's Essay and Illustration of "Rubble Women."
Source: Christa Klatt, Fifth Grade, n.d. [1946], LAB/STA 134/13, 182/1, no. 37 and 38.

around them, attempted to make sense of new roles and past beliefs. Additionally, these young people had learned the lessons the new Germany had to offer them: their nation would be cleaned up again someday, and they could take pride in the German women who contributed to this task, and men were still needed to do some of the heavier work. Regardless of educational policies that promoted gender equality, girls and boys in the Soviet zone continued to see their world as divided between women's and men's spheres. Gender roles in the Soviet zone were in flux, caught between traditional male–female divisions and new ideas about men and women.

The idea of unity in antifascist democratic education came up against a difficult obstacle with the introduction of the gender issue in educational policies of the Soviet zone. Too often studies on nations and gender and sexuality focus on the metaphorical side of gender—the nation as mother, or the state as father. The Soviet zone presents a picture of the concrete issues involved in analyzing differing gender and generational experiences of a population. Competing notions of gender equality interacted with previously held concepts of gender and generational roles that could not be quickly eliminated. Gender is not a monolithic category, as demonstrated by the need to treat girls' and boys' experiences separately from women's and men's. The extent to which young women and men experienced the creation of their nation differently, though not quantifiable, is tangible. The increasing coeducational classrooms represented a partial success for antifascist democratic education, which sought to unify girls' and boys' educational experiences. The antifascist classroom did not eliminate age or gender differences, however. Girls and boys brought gender-specific experiences with them into the classroom, created new gender-specific experiences there, and left school as young German women and young German men.

CHAPTER 5

The Antifascist Narrative

When Soviet zone young people were not gathering potatoes in the countryside, or dealing on the black market, or tucked in bed under a few thin blankets against the cold, they were in school. There, when they were not scrubbing desktops, crying for hunger, or huddled around the classroom stove for heat, they were learning the lessons of antifascist democracy. In essays for German and history classes, pupils lauded national heroes, recounted their memories of the recent years, and speculated about Germany's future. Their real and imagined stories demonstrate a struggle to make sense of the antifascist historical narrative offered them by their teachers. This narrative positioned the Soviet zone as largely innocent of National Socialist crimes and as the worthier successor to the German nation. Using what I identify as "memory lessons," antifascist educators helped young people interpret and narrate their individual pasts and their collective (national) past. These lessons did not comprise an explicit program, but were rather a function of the communication between school and society. Writing essays, young people gradually learned to narrate their own and their nation's experiences within an interpretative framework that encouraged them to remember certain elements, while suppressing or deemphasizing others. The extent to which this project was and was not possible determines not only how to evaluate the postwar curriculum strategies of the Soviet zone, but also the effectiveness of the school in the political and historical education of its pupils. Ultimately, this antifascist narrative helped pupils make sense of their recent experiences, a precondition for accepting the "new school's" and the new nation's legitimacy.

This chapter examines the role of remembering in the denazification and antifascist democratic education programs in the Soviet zone, focusing not on the content of history and German lessons but on the actual antifascist narrative framework that informed them. Considering the multiple influences on young people's memory formation, such as the school and the family, this chapter then demonstrates how pupils themselves used the antifascist narrative to shape their memories and histories. Examining how pupils wrote about their history and memories in school essays, the chapter asks in a final section how young people incorporated the lessons of antifascist education into their worldviews. Pupils consciously struggled with the difficulties of being presented with a new version of world history and making sense of their memories of the recent years. Young people's ability to connect their individual memories with the new antifascist memory of their nation's past enabled them to envision themselves as part of a new, antifascist Germany.

Learning to Remember: Memory and Young People

The way in which individuals remember and narrate individual and collective memories reflects a myriad of socialization and personal interpretative strategies, none of which is easily controllable by the state or other agencies. Many external considerations influence the content of memory as well as the structure of its narrative. The school is the most far-reaching state-sponsored institution capable of aiding in memory work and public history construction on a national basis, but children begin the school year with a considerable amount of memory development and practice in remembering already in place.[1] Reeducation and thus memory lessons in occupied Germany were to teach young Germans new modes of thinking and offer them new schemas for interpreting their memories. In the case of the Soviet zone, this program involved replacing Nazi ideology with an antifascist understanding of Germany's past and future. This was not an easy undertaking. Like their parents and teachers, young people sifted and weighed their memories critically against the new narrative structures they were offered.

Niklas Luhmann differentiated between the individual level of memory (*Erinnerung*) and a society's collective consciousness (*Gedächtinis*). He insisted that collective consciousness is a function of a social system that enables communication and is therefore far removed from the physical neurological process of remembering.[2] This separation of social from individual memory is a useful distinction, especially because it avoids the pitfall of projecting actual cognitive functions onto a metaphorical way of looking at collective memory (or consciousness). His claim that there are in fact multiple systems for collective consciousness opens up the possibility of recognizing the many roles and manifestations of memory. At the same time, this strict division between social systems and individuals risks overlooking their interdependent and interactive relationship.[3] Personal and collective memory work together, affecting a nation's public history construction.

Classrooms themselves contribute to this process of individual and collective remembering. Literature on teacher and pupil interactions has demonstrated that the dominant culture in a school environment dictates how pupils narrate experiences—that is, how they remember.[4] Certainly, the confused postwar situation in Soviet zone classrooms makes it difficult to even specify such concepts as a dominant culture, but such initial findings call into the question the possibility of a complete "Sovietization" of the school system in classrooms of German teachers and pupils. Pupils are active participants in their school experiences, regardless of the political system. Educational planners understood this premise of interaction, as stated in a 1949 draft for the "Guidelines for the Didactics and Methodology of the German Democratic School": "<u>Lessons are the joint *work* of teacher and pupil</u>."[5] Young people themselves create meaning for their school experiences, and these attitudes are not easily predictable or controllable.[6] Male and female pupils in the Soviet zone participated with their families, teachers, and peers in the process of learning how to interpret their individual and collective memories as well as how to recount them.

The reopening of schools had occurred with very little guidance or coordination in the first weeks after the war. Nevertheless, even in this time period before the official zone-wide beginning of school in October 1945, educational administrators and teachers agreed that German and history lessons should play the key role in the elimination of Nazi ideology and reeducation toward an antifascist, democratic mindset. In this spirit, the curriculum goal for German lessons stated, "German, more than any other subject, has the possibility of affecting the entire person and to instruct pupils in the true spirit of humanity. It thus has a decisive role to play in the education of the new generation."[7] Further on, the goal of German for the first eight grades was to enable pupils to express their rights as citizens in a democratic form.[8] Likewise, the guidelines of the Allied Educational Commission for history instruction in Berlin schools stated that the most important goals here were to educate pupils to think independently as active state and world citizens who would be able to evaluate Germany's present situation in light of its historical experience.[9] Decisions about curriculum content also occupied educational administrators' time. At one point in the fall of 1947, the historical commission of the education department in Berlin worried that if they allowed a subjective political line to shine through their curriculum drafts, the Allied Command would most likely not approve them.[10] However, the more interesting aspect of these stated curricular goals was their continual emphasis on getting pupils to look critically at their history. This aspiration, more than any other question of which heroes to praise, represented the most significant break from the past as a new way of remembering, thinking, and learning. As becomes evident in pupils' essays, the formulation of "critical thinker" implied a definite type and style of antifascist criticism and thinking, even if it also allowed for more questioning and doubts than had been the case in Nazi classrooms.

It was not an easy task to rewrite curricula for subjects like German and history, however. One significant challenge was bureaucracy. The actual decision of what type of history should be taught and how this should ensue was not made by one single, homogenous administrative organization. Consensus in official policies was reached through significant compromises, accounting not only for noticeable empirical differences in the execution of directives and the presence of confusing multiple regulations, but highlighting the numerous parties participating in the construction of the "new school." In many cases, teachers took matters into their own hands when their administrators failed them. As late as July 1949, for example, the *Neulehrer* Helmut Otzen from Meiningen wrote in an exam that he was attempting to start a history study group for the pupils in his school since this subject was not yet being taught there.[11] Toward the end of his essay he complained that *Neulehrer* did not have enough time to adequately prepare themselves for teaching as well as do the assigned reading for their own teacher training. This announced intention to voluntarily take on more work suggests that Otzen agreed with educational administrators and the Allied Command that history instruction was key to the reeducation of young Germans.

One of the key aspects of antifascist democratic education included assigning teachers a critical role in teaching pupils how to remember their national and individual pasts. This practice drew on a larger tradition of believing that schools and their authority were among the most important aspects in the formation of young people's world views. As one educator at a 1948 regional pedagogical conference insisted, the teacher "is an unbelievably important social factor, even a factor in the construction of society. Because without the school there can be no social life."[12] Those postwar teachers not certain of how accountable society would hold them for their presumed sway over pupils had only to read one of the many accusations being written against teachers who had taught during the Nazi period. For example, a 1945 Saxon newspaper article assigned "double guilt" to any teacher who had been a member of the Nazi Party: "He was a propagandist of the Nazi Party in the school and a political activist of the Nazi regime at the same time."[13] The public and the educational administration viewed teachers as accountable representatives of the state.

In spite of the general agreement that teachers communicated National Socialist ideology to pupils during the previous twelve years, there was no question in the postwar period of minimizing teachers' potential authority. Indeed, this role was seen as a fundamental aspect of modern society that could be harnessed for the greater good. "70,000 teachers who teach two million children daily make up an impressive pedagogical and political power," as the parent-teacher association "Friends of the New School" claimed.[14] A 1946 essay by the *Neulehrer* "new teacher" candidate Erik Braune demonstrated the prevalent sense of reverence for teachers' responsibilities: "I am aware of the seriousness of my task, and want with all my energy to attempt to learn all of the necessary knowledge in order to become a teacher; but my joy, my love, that I already now have, belongs to the children of my first class, because I want to be a good pedagogue."[15] Braune's fervent desire to be a good teacher mirrored a general conviction that antifascist teachers and pupils maintained an almost symbiotic relationship, with the teacher as the stronger party.

The Leipzig teacher trainer Theodor Litt remained one of the few critics of this belief in the power of teachers. His response to an article that idealized the teaching profession is one of the few documents suggesting that the very conceptualization of teachers' power in structuring an antifascist democratic society was overstated. "I think much more modestly about the extent of power and possibilities that a teacher has. If you were right, then the teacher would be the creator of the future, as far as the inner life of humans is concerned."[16] For the most part, many of Litt's colleagues believed strongly in the teacher as the bearer of Germany's antifascist, democratic future. Or at least they believed this scenario could happen with the right teachers in place. Administrators' reports on the quality of teaching complained about poorly trained teachers or teachers with Nazi or fascist attitudes: "Parts of our rural population and large parts of the not yet adequately trained teaching faculty, who are supposed to construct the democratic *Einheitsschule*, find themselves in the position of someone who is supposed to build a new house out of rubble but doesn't know how to build a proper, good house because he has never known one," as one observer complained in 1947.[17] But administrators' awareness of these conditions did not change the role they assigned teachers. The tasks remained in place. No one seemed interested in questioning the wisdom of a system that invested so much trust in one authority figure—if, in fact, this one person did have so much authority.

The SMAD and German educational administrators believed that they only needed to create the right kind of antifascist materials, including teachers, to reeducate pupils; any problems of reception, they assumed, arose primarily because of structural problems in the plans. This belief in the power of adults to influence children's understanding of their pasts led the Allied Command to forbid the teaching of history in the first year of school in postwar Germany, as if pupils' and teachers' historical thoughts could be put on hold until the authorities were ready with new material.[18] The desire to prevent children from thinking about the past while school administrators went about the business of retraining teachers and writing new textbooks might have seemed to be the most logical step under the circumstances. But young people continued to gain knowledge of their country's past in other arenas, including other school subjects. In one attempt to keep all aspects of history out of classrooms after the war, a Soviet evaluator of a proposed curriculum plan for English approved its treatment of grammar, but protested that it also contained an historical focus. According to this SMAD education inspector Uljanow, history and language should be kept separate, especially since it would be otherwise impossible to control the teachers' personal opinions on this subject.[19]

Such instances demonstrated the impossible task of preventing history from being discussed in or out of schools. Pupils confronted reminders of the past throughout their days, such as the many teachers who stood before them from earlier times. Joked one school alumnus to a British reporter in 1947, "It was just like listening to ghosts from the past seeing my old teachers, all of them over seventy, trying to do their jobs again."[20] The past was also a topic at home. Whether or not pupils received history instruction in school, their families played a large role in teaching children about their pasts and how to remember them.[21] Children and their parents discussed the "new school" regularly, comparing their new teachers and schools with those during the war. For instance, two Berlin brothers from an antifascist family, sixteen and seventeen years of age, complained to an interviewer that many of their present teachers had not belonged to the Nazi Party but believed strongly in its tenets. Better teachers, who had joined the Nazi Party, had lost their jobs, they claimed. Their mother worried that this situation had caused the boys to look fondly upon the Nazi period, when they remembered learning more.[22] In some families, such discussions led parents and children to agree that "everything [had] worked" in the Nazi school; that is, there had been enough teachers and books. Other families believed the "new school" to be better; that is, more academically rigorous.[23] When teachers finally began the work of officially teaching about the past in history classes, they had to work around and often against children's perceptions of the past developed elsewhere, whether at home or in other school subjects.

When skeptics did step forward to insist that young people could not simply be expected to take on whole new ways of thinking about Germany's past or future, they most often went to the other extreme. They combined criticisms of the persistence of Nazi ideology among young people with a sense that this group had been incapable both before and after the war of resisting Nazism. A report on the 1946 school year in the region Mecklenburg-Vorpommern complained that Nazi ideology had "coursed" (*durchblutet*) through the veins of the "young and especially very young *Neulehrer*" for too long to expect democratic behavior from them: "Only someone who is politically blind can believe that a young person who breathed Nazi ideology in school, on the street, in society and in public life could remain untouched by this ideology only because that person was not in the [Nazi] party."[24] The point was certainly correct, but the author failed to consider the difference between "touched" by an ideology and convinced of it. The combined ideas that young people had been exposed to Nazi ideology during their "crucial formative years" (*die entscheidenden Jahre*), that those who had not been in the Nazi Party or Hitler Youth were nevertheless affected by Nazi ideology, and that those who entered democratic parties after the war often did so without having the necessary democratic attitude painted a very opportunistic and labile picture of the younger generation in the Soviet zone. Educational administrators seldom suggested that young people had been capable of critical thought during the Nazi regime or after the war.

Regardless of young people's political attitudes, the state (through the school) did not and could not have offered its pupils an entirely new set of memories of their own and their nation's pasts. Only rarely do adults or children succeed in exchanging one conceptual memory system for another. Even when individuals develop entirely new modes of viewing a subject, they are still working with a system of knowledge for processing further information within that same schema, relating new information to old information and vice versa.[25] Instead, the "new school" offered a means for young people to reinterpret their personal and collective memories and pasts, even adding in new elements. All this could only occur within a framework of previously existing knowledge. The new antifascist understanding of history was not a "new" history, but rather one that reinterpreted both (personal) lived and (national) imagined experiences, whether in the recent or distant past. The postwar school offered a "second" history that was in fact an edited version of the first.[26]

Children Recount Their Memories

Young people quickly incorporated the version of an antifascist narrative into their assignments when discussing historical events from other epochs. It was not difficult for them to accept new or rehabilitated distant figures or episodes in German history as part of their cultural heritage, such as the 1848 Revolution. When asked to express specific political viewpoints, they understood and accepted the new ideas about their nation. Pupils admired Germany's cultural icons and wrote critical essays for their history lessons, such as the tenth grader from the Saxony-Anhalt city of Güsten who wrote a report for German class in 1948 on the medieval minnesingers, referring to them as "Germany's glory." A year later, she completed a critical essay on nineteenth-century German chancellor Bismarck's absolutist foreign policy, demonstrating the long-term damage that such militaristic and class-based behavior inflicted upon German society.[27] It was in the elaboration of how the new democratic Germany was supposed to look, historically, contemporarily, and in the future, that pupils' essays showed uncertainty and a struggle for orientation. Particularly when pupils wrote about their own lives, their essays took on a different, more emotional, and even questioning quality. Over the course of the postwar years, young people learned to use the antifascist strategy of placing Soviet zone residents in the role of victim to denounce Germany's fascist past, without looking critically at their families' participation in the Nazi regime. In this process, they both contributed to and reflected the ongoing process of reconstructing and identifying Germany, critically evaluating how the contents of their second history should look.

Young people in the immediate postwar period had directly experienced little else but war and its trying aftermath. In an instructional environment that encouraged them to learn from and work on their private and public memories, pupils learned to incorporate memories that they had not always personally participated in into their approaches to their individual and collective pasts, presents, and futures. Equipped with fewer experiences than adults that could have served as desirable scenarios of political and private life, pupils relied heavily on adults and peer groups to make sense of their world. This relationship was not entirely one-sided, however; children and adolescents questioned new ways of remembering and thinking that could not easily be assimilated into their hitherto existing cognitive patterns. As such, pupils were historical actors who are easily ignored in evaluations of the success of the postwar school in the creation of new Germans. Reading pupil-generated sources entails separating out young people's voices from those of their parents and teachers wherever this is possible, which can be a daunting task.

The fifth-grader Otto Dieß wrote essays that vividly illustrate these different aspects of memory and remembering in the Soviet zone. In December 1944, during the last months of the war, Otto wrote an essay for his German class entitled "My Friend." Here he proudly described a powerful, gray-eyed, blonde boy with his hair parted on the left and good skin color, a strong nose, good muscles, and a clear enunciation, who had earned a leadership position in the Hitler Youth.[28] There is then a silence of several months in the school notebook when, presumably, the final months of fighting kept Otto and his classmates at home. During the first month of the reopening of schools in the Soviet zone less than a year later, Otto completed one assignment on his visit to a dentist, followed by the assignment "The Rebuilding of the Countryside and City."[29] Here, Otto began with the standard judgment that the criminal deeds of the Nazis led to misery throughout the entire world. It becomes clear, however, that this essay lacked the earlier tone of self-confidence. This change of tone was due in part to Otto's misunder-standing that the Potsdam Accords intended for Germany to become an agrarian state: "Germany is to be placed into a central European standard of living, that is, such a life as is the case in the Balkan coun-tries, thus living in huts."[30] Otto believed that it would be hard for Germans to get used to life in this new Germany. He added that the reparation payments "demanded by those peoples that we attacked" made him, like others around him, doubt the newspapers' rosy visions of quickly reconstructed cities. Yet after expressing frustration at having been lied to by the Nazis, he finished with a positive closing tone, proclaiming that the future remained open for Germany.

Otto's specific situation demonstrates the limits on the school's ability to dictate a new understanding of his past to him. He lived in

Oschersleben in Saxony-Anhalt, a region that switched hands from the Americans to the Soviets. The Morgenthau Plan, which had called for an agrarization of Germany, was a topic of conversation in the Western zones that would have reached Otto's ears, and could easily have been confused with the Potsdam Accords, which focused on the "4 Ds": demilitarization, denazification, decentralization, and democratization.[31] Perhaps Otto overheard outraged and even sarcastic discussions among his parents and their friends as they bemoaned Germany's plight, during which he won the impression that he might have to start living in a hut. Yet he did not uncritically accept all that he heard. He was aware of optimistic reports in the newspapers, but these were not enough to convince him that everything would proceed smoothly in the rebuilding of his city. Nevertheless, he joined the ranks of determined Germans ready to write Germany a better future.

It is a confused and searching essay, written by a young boy who understood what was being asked of him. Even as they established themselves as partial victims, young people like Otto did not see themselves as helpless receptacles of knowledge or memories. Otto could not have entirely refused the new version of history presented to him, but this did not mean he or his peers could be given new memories upon which they had no influence. Assignments such as these point to an internal tug-of-war in Otto's sense of his past and present and between him and the adult world. Memory is elastic, but it is not endlessly flexible. It often balks at accommodating new, conflicting memories, and cannot solely be dictated from an external source.

Essays from other pupils a year later demonstrate a clearer internalization of the new "memory lessons," which implicitly encouraged young people to use personal experiences to explain the Soviet zone's collective history. When thirteen-year-old Vera Müller wrote an essay for the subject of the last days of the war, she wrote about her family.[32] This might seem an obvious point of departure, and it was certainly often selected by school children. But not all pupils used the narration form of the subjective family experience to complete similar essay assignments. Kiaz Kiewicz' essay "*Kämpfe um Berlin*" (Battle for Berlin) stands out as a counterexample with its table "*die letzten wichtigen Ereignisse*" (the last important events) of the final military battles of the war and the newspaper clipping of an illustration of an English ten-ton bomb.[33] More often, though, the theme of the family's traumatic experiences during or after the war served as the frame of reference in many pupils' essays. This tendency fits in with the specific outlines of the Soviet zone German curriculum for what types of writing pupils were to learn in the fifth through eighth grades, in which narration (*Erzählung*), description (*Beschreibung*), and reports (*Berichte*) were considered to be the most important types of written expression, with a particular emphasis on fantasy and reflection.[34] The personal memories and experiences of pupils lent an untouchable, authentic quality to them. On closer inspection, these accounts fit a clear style of antifascist narration that elicited this perception. This style is evident in Vera's essay, in which she recounted a bombing during which she and her mother were attempting to run to their house: "I told my mommy that she should wait for a second so that we didn't all run across together, that might get noticed. Outside it was deathly quiet, only the dust of exploded grenades lay in the air. I ran as fast as I could, but too late. Just as I reached the sidewalk, I was hit by an exploding piece of grenade."[35] Vera did not know at the time what had actually happened to her mother. "They told me everything but the truth. Only after I had been in the hospital for four weeks with no news from home did I find out that my parents had been victims of the Nazi regime."[36] Her mother had apparently died during the bombing; her father's experience is subsumed without further explanation in this victim narrative.

There is a curious tension between the personal trauma recounted by Vera and the more distanced, formulaic description of her parents' status as "victims of the Nazi regime." In fact, it is unclear from her essay whether both parents died during this particular episode. It would seem that only the mother and she were present, although the phrase "so that we didn't all run across together" might indicate other persons. Another possibility is that Vera's father died in another context: in a concentration camp, perhaps, or as a soldier in combat. This interpretation would explain why Vera, after being hurt and brought into the air raid cellar, only asked about her mother.[37] Pupils like Vera learned quite early to adjust the degree of subjectivity and emotion with which they related experiences depending on the function of the narrative. The essays portrayed the suffering and injustice felt by the young authors, but within a clear antifascist interpretive framework. Thus, the retelling of even a painful memory about the war almost always further added an evaluative comment about the senselessness of the war or the fault of the Nazis for disrupting everyday life.

It was not always easy for young people to remain within this narrative framework of identifying the victims. Trying to reconcile too many interpretations of the same situation, pupils found themselves caught in contradictory narrative strategies that clearly elicited discomfort in the young authors. Without commenting on its implications, Vera added the statement that she was the one who decided that she and her mother should cross the street separately. From the entire scene, this decision remained important enough in her memory to mention it in a school essay. Rather than pose the heartwrenching question of what might have happened had her mother run with her, Vera learned to give responsibility to the Nazis—the elusive others. And yet, her own role in this scene, which did not neatly fit into an analysis of Nazi responsibility for postwar sufferings, could not be forgotten, any less than the quiet protest that no one would tell her the truth about her mother. In a time when school children struggled with the realization that the lessons taught to them during the war were lies, such small perceived injustices of being told "all sorts of stories" did not serve to smooth the transition into accepting the validity of the new system for pupils like Vera, even as she began to accept a new way of understanding Germany's, and with this, her own, history.

An in-class essay by the middle-school pupil Edgar Günther, "An Old Banknote Tells its Life Story," provides an interesting juxtaposition to Vera's essay. Although Edgar does not note anywhere if this topic was entirely his own idea or his teacher assigned it, other sources suggest that teachers assigned this theme of a "traveling banknote" throughout the Soviet zone, perhaps following a suggestion in an educational journal.[38] One can assume that Edgar had limited or even no choice in the actual topic of the assignment but at least some freedom in how he completed it. Against this background, the main similarity in his and Vera's essays is also the biggest difference. In both cases was a voluntary decision about how to complete the assignment. Namely, Edgar also employed a family motif to narrate his story. His was, however, a radically different use of the family. In contrast to Vera's clearly subjective tone, Edgar's more imagined family memory became a tool that appropriated the framework of subjective experience and then shifted the narrator's voice to that of an "objective" observer.

Edgar told the history of the Nazi Party from the point of view of a banknote that started off in the hands of a worker misled by Nazi promises.[39] After being given to the Nazi Party by the worker, the banknote landed in Hitler's office with other money, where it heard the "treacherous plans concocted by the Nazis."[40] From there it found itself in the hands of soldiers, "the very same thugs who called themselves SS [the *Schutzstaffel*, the Nazis' paramilitary organization] and murdered the thousands of humans in the concentration camps."[41] After the war, the banknote appeared in the pay envelope of Edgar's father: "Now it was among workers again, but they were different ones than before."[42] One can ignore the obvious unlikelihood that those workers before the war switched positions with different workers after the war. It is equally improbable that Edgar meant the change in workers to be an intellectual transformation rather than a physical change of personnel. More important is the essay's sense of otherness, of removal from culpability, which extends even to Edgar's description of his father, or to the teacher's portrayal of how children should conceptualize their fathers. Even if the form of the essay would have been largely voluntary, such antifascist, socialist narrative structures had been taught both directly and indirectly, both in and out of official history class. Here, too, the family served as a reference point for pupils' evaluations of the Nazi period and became a key arena for a socialization of innocence that dominated official school lessons. Appealing to the privileged status of the family as a near inviolable sphere of personal experience, the essays from Vera, Edgar, and others like them allowed young people to position themselves and their immediate circle of intimates on the side of antifascist victims and victors. The groundwork was thus laid for an individual self-image that mirrored the state-sponsored program of creating distance between the Soviet zone and the Nazi era.

Other elements can be drawn out of the essays about gender context.[43] Despite policies about coeducation, the likelihood that Vera attended a predominantly girls' school and Edgar a predominantly boys' school is good. Vera's essay was most probably part of the larger Berlin school administration project under which teachers were asked to assign a variety of themes to their pupils.[44] Without knowing more about the conditions under which Edgar received his essay assignment, one can guess that the themes of workers and capital were introduced by his teacher within the context of teaching the history of the Nazi Party. Although both Edgar and Vera employed the family in their essays of their own recent pasts, the structures of their narratives differed greatly, pointing to a gender difference in pupils' use of the antifascist narrative. This structure was not coincidental. Their modes of narration reflected typical male and female approaches to relating stories, evident as well in other essays, and these differing types of essays were assigned by teachers who were teaching either boys or girls. It is not inconceivable that a girl would have written an essay more in the style of Edgar's, or vice versa— gender boundaries are never absolute—but neither is it surprising that the essays follow a certain recognizable type of gender-specific narration. These sorts of learned gender differences have numerous factors, of which the school environment plays a key role.[45]

Vera's essay reflects a typically female form of elaborative description in which her feelings impart a sense of validity to her story. Her narrative draws its strength from the sense of a painful memory that, even though it is personal and individual, achieved a status of universality through its structure. Vera recounted the scene in such a way that it invited contemporary readers to recognize the familiar sense of tragic fate that then segued into the moral of the story: they were unfortunate victims. In fact, Vera's device of mentioning her suggestion to cross the street separately, without explicitly questioning how this decision affected her mother's death, allowed her and her readers to make this connection implicitly. Her audience and she were then led to immediately proclaim her innocence. In this manner, her readers could also exculpate themselves from their own similar, unarticulated fears of guilt, in private as well as public matters.

The use of memory in Edgar's essay is less obvious. Like other boys, Edgar narrated his "memory" through a seemingly neutral individual or object. The banknote served as an omnipresent, seemingly disinterested party that, because it was not clearly directly Edgar's voice, assumed the voice of an objective narrator with a privileged view of the family's structure. Although this memory is clearly not a "true" memory in the way that Vera's most likely was, it has a certain sense of authority to it that contrasts with Vera's more personal style. The banknote judges the Nazis as treasonous and refers to the SS as thugs who murdered "the thousands" in the concentration camps, again positioning the Nazis as an external group, separate from those Germans—workers—who possessed the banknote after the war. It is in fact Edgar's father who, in receiving the banknote after the war, embodies this "different" worker. If the teacher prompted the essay style, all the pupils could similarly count their fathers among this new style of German. With this essay, Edgar created and then accessed a memory that would help contribute to his understanding of how his country was both perpetrator and victim, and specifically offered him a means of locating his father in this past. Equally relevant, Edgar learned to narrate memories in a convincing manner, including imagining himself in a fictitious setting and drawing conclusions from this that would shape how he viewed his past and thus his future.

Girls, too, used imagination and fantasy to present a moral or relate a metaphorical event, but here emotions helped legitimate these narratives instead of a sense of objectivity. A group of seven girls from a Berlin-Prenzlauer Berg middle school wrote a short theater piece for the assignment to illustrate the progress made in Berlin in the areas of water, gas, and electricity availability.[46] The main figures in the first scene are the girls themselves and one of the girl's families. The girls relate scenes they have most probably experienced themselves, or at any rate, scenes the reader believes the girls could have experienced directly. The sense of authenticity comes from the vivid descriptions of the family in the kitchen, the long line at the water pump, the discussion on the street about whether progress is being made in the reconstruction of Berlin (the answer turns out to be yes) (see figure 6). The narration retains an emotional side, even in the scenes that are presented without the physical presence of the girls. The "hundreds of people" standing in line for water are all "tired, [and] longing for quiet and a meal";[47] even the new gas flame that finally burns again in Berlin-Neukölln, though weak, shows "its good intention."[48] By creating as subjective a narrative as possible, such assignments placed girls in the role of informed participant, or at least observer. Girls tended to create authenticity for themselves by including as much emotion as possible in their memories; boys tended to find legitimacy in a sense of factual, distanced recounting. Both structures were learned.

Pupils' essays that did not conform to this model of experience narratives still drew on elements of memory and remembering. In particular, the ambiguous desire for things to be the way they were "earlier" was an oft-repeated wish by pupils. Alongside essays that proclaimed the need for a new Germany, pupils recounted memories of an unspecified better past that became their goal for orientation toward the future. This apparent contradiction was possible because pupils tended to separate memories of a physical space from memories of political history. Thus ten-year-old Gerhard Krüger proudly predicted that if everyone kept up the good work, Berlin would once again be an attractive city: "Today our school looks almost like before, and if the works keeps us, then it won't be long until Berlin is once again just as pretty and clean a city as before."[49] In a similar vein, the eighth-grader Vera Rietz wistfully described how her school building had looked earlier. "Our school used to be a true model school. Every classroom had curtains, there were flowerpots on all the windowsills, pictures decorated the walls, and clean tables and benches peered out at us. Everything was there that belongs to cleanliness and a sense of well-being."[50] Locating the source and motivations of such memories presents an interesting puzzle. As an eighth grader, Vera Rietz would have had most of her schooling during the war. Any physical damage done to the school building would have likely occurred only toward the end of the war, even if supplies might have run out earlier, so that she could have experienced a clean, bright school during at least her early school years. Gerhard might also have attended a year or two of school in a time before the war had drained away supplies and staff and kept children at home to help out their families. Were these Nazi-era memories of school and city

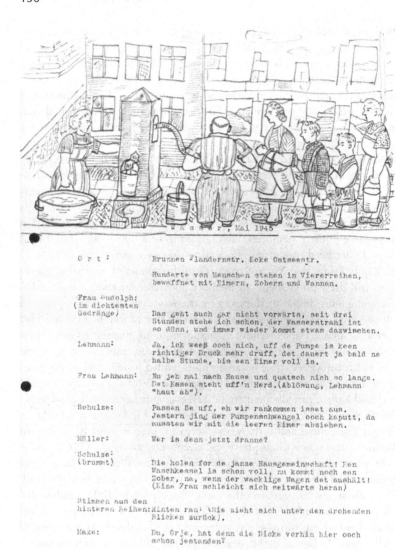

Figure 6 Pupils' Illustration and Play about the Long Lines at the Water Pump after the End of the War.

Source: Christel Novak et al., n.d. [May 1946] "Wiederaufbau Prenzlauer Berg," Girls' Middle School, LAB/STA 134/13, 181/1, no. 112.

really the conditions to which young people hoped to return? What did teachers and educational administrators hope to accomplish in assigning essays about the damage pupils' school buildings incurred?

The subject of bombed and burned-out school buildings took place within the context of a vast disruption of the earlier landscape. Many pupils lost their entire neighborhoods, including their homes, in the bombing (see figure 7). The physical condition of the school thus offered pupils the chance to discuss the destruction of their physical space in a more distanced, less immediately painful manner than the absence of a roof over their heads. The school became a tangible element around which pupils could organize a chronology of memory that included the personal experience, without focusing on the more anguishing aspects of postwar private life. This technique provided an important contrast to the function of memories about family experiences. As a central element in young people's lives that was by definition a public institution, the school building also required young people to think in collective terms, rather than allowing them to isolate themselves in the private worries of family life. The school building pulled them into the work of defining the new antifascist nation together, of what it used to look like and should look like. In this framework of remembering a better time, pupils most often wrote about reconstructing a previous Germany rather than constructing a new one, true to antifascism's tendency of correcting the past instead of developing new concepts for the nation.

Such school exercises did not require pupils to define when this earlier, better time occurred. They did not address such sticky issues as the fact that Vera's "true model school," if she meant the one of a very recent past, did not only consist of clean, decorated rooms but also of Nazi ideology. If pressed for a periodization, the adolescent authors might equally have remembered a time before 1933, or even earlier, dovetailing onto memories they received from parents or teachers through narration or photographs—a time before many had been born. The nostalgia for better times was so prevalent that neither pupils nor their teachers saw any contradictions in lessons that on the one hand called for a break with the Nazi past, while on the other hand sought both mentally and spatially to recreate the physical contours of this same past. The fifth-grader Helga Hoch wrote an essay indicating the broad popularity this past enjoyed, noting that "people" wanted things to return to the way they were: "All of the rubble that used to cover our streets is mostly gone, and people are working on the rebuilding of or *Heimat*, and are happy that Berlin can again be in the next few years the way it once was many years

138

g. Maßper
23. Wolke/Köln
Kaggolallen 41/42

Berlin, den 25.1.46.

I, 68

Erlebnisbericht.

Bomber gingen um.

Es war der 23. November 1943. Wir ... in der ... Es war im gefähr ½ 8 Uhr.
Plötzlich erlösten ... in unmer ...
Wir ... und ...
gen in den Luftschutz-Keller. An diesem
Abend verloren viele Menschen ihr Heim
und viele wurden ein Opfer der Bomben.
Als wir eine ½ Stunde im Keller saßen,
fing es schon mit Flakgeschütz an, das
immer ... empfindvoll, bis plötzlich ein
... der ... uns zu Haus
erschüttern ließ. Man hörte ein geräusch-
volles ... Bomber ... Bomben
schlug ein. Die Pfeiler des Kellers bebten
... den ... ein-
schlagenden Bomben und Minen. Es war
... als ... die Hölle los war. Als es wieder
... wurde, ging ich mit noch anderen
... hinaus, um zu sehen was ...
... Überall brannte es. Im Hinterhaus
... brannte es auch. Ich holf beim
... Es war nicht ...
... durch glimmender und ...
... Zur ...
Hier und da erklangen ...

Figure 7 Pupil's (Probably Boy's) Essay and Illustration of Bombing of Neighborhood in 1943.
G. Massner, January 25, 1946, LAB/STA, 134/13, 180/1, no. 524/1 and 524/r.

ago."[51] In this aspect of socialization through the school, as was the case in almost all facets of school life in the Soviet zone, pupils reflected surrounding attitudes as well as contributed to their formation. No less eager than the adults around them to live in an environment that did not constantly remind them of their uncomfortable and painful past, young people used their own memories and learned new ones to imagine a future for themselves and their nation.

* * *

The cognitive nature of memory and remembering means that it can be impossible in many instances to verify if we really experienced events of which we later have memories. In the process of being offered a second history, however, this distinction is not meaningful. The "fake" memory of Edgar and his banknote guided him toward an antifascist memory that connected his personal history to his nation's history, a function also played by the possibly "real" memory of Vera Reitz and her school building. When pupils had the liberty to reflect on their new and old memories, they showed an awareness of the work this entailed, and consistently employed a form of memory narration that coincided with larger social expectations of how to remember. The school played an important role here, but it did not work in isolation. Its influence should not be overestimated, just as pupils' influence on the school and their own school experience should not be underestimated.

The language and ideas used by pupils in the postwar period to describe their war experiences, or how they survived the first bitter winter after the war, point to an antifascist narrative structure and form influenced by parents and teachers. In those instances when pupils expressed uncertainty and doubt in their essays, they did so within a framework that had been learned in school and family environments, including stopping short of harsh criticism that educational administrators discussed in other sources.[52] Aside from introducing pupils to a common body of knowledge, the school contributed to training pupils to express their memories within a certain antifascist framework that used real and imagined experience as a means to establish authenticity. Pupils accepted this strategy and employed it to begin the work of mastering their private memories, establishing themselves as producers and receivers of memory. When Otto, who confused the Potsdam Accords and the Morgenthau Plan, finished his essay with a sense of determined hope for his nation, it was not an empty sentiment. It was an acceptance of a proffered role of helping to reconstruct the nation, and a challenge to the adult world that the second history they envisioned for him would not go unchanged as he struggled to assimilate it.

CHAPTER 6

"Vati's home!": From Defeated Nazi to Antifascist Hero

S chools and the emerging state viewed the family as a key partner with schools in the antifascist socialization of young people. Families were all the more important as an institution, since churches in the Soviet zone played almost no role in the overtly secular program of antifascist democratic education. With the elimination of religious instruction from the school curriculum, in fact, educational reformers sought to limit, even eliminate, the influence of the churches upon young people's education. But was the family, so recently the object of a broad National Socialist policy designed to permanently fix parents and children in the service of that state, adequately prepared to assume suddenly its role as coantifascist educator? Popular representations of the family suggested that the answer was a worrisome no. At the heart of this concern was the rehabilitation of the father, whose war experience had often left him dejected and confused—when he was even lucky enough to have survived.

Throughout post–World War II Germany, children waited for their fathers to return from the war. Some of these young people waited in vain, only to learn that their fathers had fallen during combat. Others were left to wonder what had happened to them. At times children never received definitive answers about the fate of their fathers; other families might be surprised by the arrival of a father years after all hope of his return had evaporated. Fortunate children experienced their father's joyous homecoming in the first months and years after the war's end, and the tearful scene of embracing families that played itself out in front hallways throughout the country became the bittersweet stuff of film and literature for years to come. This happy return of the father-soldier became the core of a Soviet zone program to replace Nazi ideology with an antifascist, socialist worldview. In an odd continuity of roles, the same men who had fought for Nazi Germany and now returned as the defeated enemy were now responsible for the country's victorious antifascist reconstruction and reentry into the community of nations. Fathers were to remain heads of the household, literally and figuratively—an interesting proposition in a society that had lost large numbers of its adult male population.

The widespread desire to keep the father at the head of the family was in part an attempt to maintain a traditional family structure that seemed at risk of disappearing. Germans in East and West feared that World War II had damaged the family as a viable institution.[1] Research on postwar Germany has generally corroborated this impression. In this interpretation, the demographic changes resulting from the war, particularly regarding the father's role, significantly weakened the family as a social unit.[2] Certainly, families had suffered considerable damage after World War II. In August 1950, for instance, of the 6,190,217 Soviet zone households (not including those in East Berlin), 33 percent were run by women.[3] Approximately 18 percent of these women-led households comprised two or more persons. This situation meant, in the majority of cases, that the woman was widowed, divorced, or waiting for her husband to return from internment, and that she lived with individuals dependent on her, usually children.[4] Where the father lived with the family, the mother often continued to work outside the home. An accurate depiction of a typical Soviet zone family could thus never assume a working father and housewife mother.

Yet more general research on the family offered a different viewpoint, one that takes a longer view of demographic changes: this "new" type of family had appeared long before 1945. After the end of World War I, sociologists in Europe and the United States questioned how major social ruptures such as war had affected various institutions throughout history, particularly the family.[5] The end of World War II and the perceived need for reeducation renewed discussions in Germany about which institutions should be entrusted with the education of the nation's young, and how the family might play a role. In 1948, the German sociologist Hilde Thurnwald published a study of 498 Berlin families interviewed in 1946 and 1947. Her research situated the postwar German family historically and socially, and finished with in-depth portraits of twenty-five families. Thurnwald concluded that the traditional family unit of two parents and one or more children had slowly been undergoing changes that were neither unique to Germany nor solely the result of two world wars. She rejected the thesis that extreme economic need necessarily had a negative impact on family relationships, and in fact uncovered many examples of families functioning stronger than ever in the face of a seemingly hopeless economic situation. She also noted that many families had long ceased to conform to popular images of the family unit. The patriarchal two-parent family, the definition in the sociological and statistical literature of an "intact" family, still dominated the demographic landscape, but it had been joined by many variations of this structure.[6]

Despite such findings and an obvious physical landscape dominated by women, the traditional family structure of a strong and stern father as head of the household and a housewife mother remained the ideal of postwar Soviet zone Germans. It mattered little to anyone that this fantasy did not correspond to most families' situations. The dream of an intact family prevailed throughout popular and administrative culture, nurtured by the emerging antifascist democratic narrative of national redemption through physical reconstruction of the nation. This vision created a tension regarding gender roles in the new Germany and undermined the antifascist socialist promise of creating equality for women and men.[7] All fathers were now antifascist heroes come home to rebuild the nation, regardless of their wartime activities. Whether at home or in a POW camp, whether healthy, deceased, or disabled, Soviet zone fathers remained present in the minds of their families. Women faded into the background. Unfortunate heroes of a defeated nation, families and the state looked to men as fathers to return Germany to a position of power. The construction of antifascism was to be a male undertaking.

Public Myths

The most obvious statement of men's role in turning the Soviet zone away from its National Socialist past could be seen on the big screen. The SMAD viewed the film industry as an important vehicle of cultural reeducation—the cinema department was even part of the German Educational Administration in the Soviet zone.[8] It thus comes as no surprise that films by the official Soviet zone (and later GDR) film company DEFA (*Deutsche Film Aktiengesellschaft*) and educational reforms reflected similar goals of rehabilitating institutions that had been abused during the National Socialist regime. Although most directors and actors in DEFA could not claim the resistance past that antifascist educators boasted, the so-called rubble films of the immediate postwar era advised the same solutions that the "new school's" teachers taught: reconstruction of the nation entailed physically rebuilding destroyed landscapes and reeducating the people.[9] Accordingly, the Soviet zone's first family film made the connection between strong fathers and a healthy nation explicit. Gerhard Lamprecht's *Somewhere in Berlin* (1946) was the third production of the newly established DEFA, and the first film intended for a young audience.[10] Often cited as a neorealist "bridge" between the expressionism of the Weimar Republic and the later "Berlin Film" genre, *Somewhere in Berlin* is stunning for its realistic, grim assessment of children's life in postwar Germany.[11]

Filmed amidst actual ruins of the city, *Somewhere in Berlin* offers an idealistic vision of the near future and an embrace of the antifascist democratic lesson of redemption through reconstruction. The central plot revolves around young Gustav Iller and his mother, who struggle to get by as they wait hopefully for the return of Gustav's father from a POW camp. Gustav's best friend, Willi, has few illusions of a brighter future for himself. Orphaned during the war, he has been taken in by the well-meaning but weak-willed owner of a stationery store, Frau Schelp, and has become an assistant to the black market dealer Herr Birke, one of Schelp's boarders. Gustav's and Willi's very different lives and future prospects intertwine to tell the story of life for young people in the postwar Soviet zone. While Gustav's family struggles to reintegrate Herr Iller back into the household, the fatherless Willi wanders from one misadventure to another, helpless against negative societal forces like the black market. Feeling abandoned and alone, Willi falls to his death from the wall of a ruined building, an accident with undertones of suicidal intent. Gustav's "Uncle Kalle" (Karl) refuses to allow young Gustav to give in to his grief, though, telling him that Willi's death was a necessary sacrifice for the greater good. Armed with the knowledge that only hard work can make sense of such tragedies, Gustav recruits his friends and Uncle Kalle to push Herr Iller to reassume his role as breadwinner of the family. In the antifascist leitmotiv of assigning children partial responsibility for the future of their nation, the film invests Gustav with the power to push his father toward his antifascist duties in the new Germany. In a curious fluidity of roles, the state—through the school—has educated young people to educate their parents, so that the latter can in turn become antifascist educators of young people. The movie ends on a positive, heroic note, with Gustav and dozens of other boys helping Herr Iller reconstruct their small part of the new antifascist nation: the Iller family garages.

Audiences approved of the world presented in *Somewhere in Berlin*, in which the key to redemption for the Iller family was the father's successful return home. This idea was so important to the plot that the scene of Herr Iller embracing his family was used to advertise the film on posters, as well (see figure 8).[12] The film's critics complained of a story-line that was not entirely cohesive; they praised, however, the attention to authentic details of everyday life in postwar Germany. The reviewer Peter Kast, writing for the socialist newspaper *Vorwärts*, noted with satisfaction the director's optimistic message that the postwar youth, "formed by Hitler barbarianism and a war or bombs," was not a lost generation.[13] Granted, Kast found the closing scene of men and boys working joyfully together to rebuild the Iller family garages "too sym-bolic," and he wondered if the lengthy shots of ruined landscape were too depressing. But ultimately, he claimed, it was impossible to deny the film's "nice ethical value," or its complex treatment of life amidst rubble. *Somewhere in Berlin*, he insisted, was not just about the war-torn city of Berlin; it was emblematic of the postwar situation throughout Germany. He closed with the hope that audiences, especially younger ones, would recognize the similarity between the film and their own situations. If he found the film too sentimental in its praise of a "new beginning" for a Germany that lay in ruins, he did not argue with the formula of fathers leading their sons to reconstruct the nation as a solution to the mental and physical rubble left by the Hitler years.

Kast believed that *Somewhere in Berlin* presented a "poetic truth" in its portrayal of the daily lives of postwar Berliners, but he too readily accepted the film's problematic representation of fathers as the key figure in both Germany's rubble problem and children's moral education. Following this logic, the orphan Willi plunged to his death from the remaining wall of a ruined building because no father guided him safely through childhood. If this scenario could be expected from other, simi-lar situations, then many more of Willi's peers would end their lives trag-ically. Even Gustav, despite the temporary father figure of Uncle Kalle and the eventual return home of Herr Iller, is largely without a true father until the young boy can force Herr Iller back into his paternal role. More children in the Soviet zone suffered the loss or incapacitation of their fathers than were able to welcome them home, alive and well, even if popular conceptions of a traditional and healthy nuclear family persisted throughout the postwar years.[14] The father dominated the household and disciplined the children in the state's and society's imagi-nation, and neither educators nor the public believed that a different model could result in a happy ending. Despite all evidence to the con-trary, postwar Soviet zone society continued to wait for the return of *Vati* to make everything right. Later GDR myths would position women

Figure 8 DEFA Film "Irgendwo in Berlin" Poster.
Note: Notice the homecoming of the father and son against the background of rubble.
Source: 90/6647 SBZ/GDR CD-ROM. *Plakate der SBZ/DDR* Inv. Nr. 90/6647 SBZ/GDR [CD-ROM].

as the heroes of reconstruction, idealizing in particular the "rubble women" who cleared away the destruction of the war. Postwar society, though, was not interested in praising women for completing men's work.[15] The sight of women running businesses or hauling away piles of bricks was a painful accusation of an emasculation of the Soviet zone. It was also a painful reminder to women that they might not be able to ever abandon the workplace. After all, even if many women found their wartime and postwar employment liberating, surely others longed to return to their prewar lives as full-time *Hausfrau* (housewife) and *Mutter* (mother). Reconstruction necessitated every free hand in the Soviet zone, regardless of gender, but the physical labor involved in it was coded male. A KPD poster in the Soviet zone called out, "All Hands for the New Construction!" but the three arms pictured—laying bricks, cutting with a sickle, and writing with a fountain pen—were clearly masculine.[16] Female workers could be seen throughout the Soviet zone, but popular depictions of reconstruction did not acknowledge that women had permanently left their traditional domestic sphere.

By viewing these women as placeholders who would soon return to their kitchens, postwar Soviet zone society could believe itself merely temporarily impotent, a condition easily rectified.[17] Herr Iller reflects this situation in *Somewhere in Berlin* after returning home to find his wife and son alive and healthy. He thanks his friend Kalle for helping them survive the war years, but does not once congratulate his wife and son for their own contributions to their happy state of affairs. Instead, everyone in Herr Iller's world immediately works to put him back in charge. We soon learn that the family's plans to reinstall the patriarch actually began during the war, with the wife carefully mending his suit for his return ("quality material from before the war," notes the tailor who takes in the waistline for Herr Iller) and the son keeping the suit clean. Herr Iller does not protest that the suit's precious wool might better have been used to keep them warm during harsh winters, just as the tailor does not question that the suit is still intact. Continuing this leitmotiv of communal sacrifice for the father, Gustav's friend Willi steals food from Herr Birke for Herr Iller in order to still the hunger of his best friend's father. Willi gets in trouble for stealing, but not for his intention. Family, friends, and the emerging state mobilized all available resources to get men back on their feet. Official and public rhetoric thus ignored and repressed any acknowledgment of women's movement out of the domestic sphere.

It was not that women went unnoticed. Numerous postwar documents from adults and young people demonstrate that they recognized how hard the women around them worked. Under the broadly assigned topic of "reconstruction," for instance, several pupils in Berlin—Prenzlauer Berg focused on women's contributions during the first months following the war. Two boys from different schools wrote about the women and young people who formed a chain to haul away the rubble from demolished buildings, and one of the boys noted that it was "mostly women" who did this work.[18] Still, in keeping with the impression that this work was not actually appropriate for women, Ingrid Höll commented that, "Many a woman even performs men's work, she manages to get a wagon filled with stones to its place, that is definitely not easy, because a stone weighs 7 to 8 pounds."[19] Children and adults commented on women's unpaid labors, as well. The eighth-grader Vera Rietz, for instance, suspected that her teachers would not have been able to clean up the school without those mothers who volunteered their help.[20] Politicians, too, spoke publicly about how the war affected women's lives: the new East Berlin city councilwoman for the health department, Frau Schirmer-Pröscher, received a journalist's praise in 1948 for her remark that "women are the pack animals of every war."[21] But Schirmer-Pröscher was not referring to "rubble women," or even women taking over family businesses. Her main concern was that the city should provide women, as the family caretakers, with enough coal to keep their children from freezing, and fresh milk to keep their children healthy. She then mentioned briefly that women now working in factories made socialized healthcare an important issue, thus highlighting the tension between women's working realities and society's desire to ignore these women.

Thurnwald described such contradictions as a "problem of female employment": women had to work outside the home, but their responsibilities in the home did not diminish.[22] As if to offer an example of this phenomenon, she explained in the introduction to her book that the research took longer than expected because almost all of her investigators were women overwhelmed with their many responsibilities between home and workplace. A few pages later she defended her decision to spend more time interviewing the women in the families than the other family members. "As housewives and mothers," she claimed, women typically were better informed about their families' situations than the men. The excerpts from her interviews verify her statement. Given how extensively they discussed their families, surely not one of her respondents, and none of the pupils who wrote about women, would have disagreed with her.

Yet elsewhere in pupils' writings and in family-oriented films of the Soviet zone, women all but disappear from sight. Where they are present, they are either caricatures of helpless females or merely substitutes for the men who will return to their jobs as soon as they are able. In the 1949 family film *Die Kuckucks* (The Cuckoo Clock), for example, Inge Kuckert capably takes care of her siblings until a man enters her life. At that moment, she turns into a fragile child in need of his help.[23] Girls and women in the popular imagination waited eagerly for the moment when they could hand off their responsibilities to the first available man. This world that adults painted for children and that children learned to dream of had nothing in common with official socialist plans for an egalitarian society that would resolve the "women's question." Nor did it reflect the society that postwar Germans actually lived in.

Somewhere in Berlin left no doubt that women should play only marginal roles in reconstruction. The two main female characters in the film, Frau Iller and Frau Schelp, both have maternal responsibilities. But neither is a strong matriarch, and leave the men to educate the boys. In an early scene, Frau Iller has informed Uncle Kalle that Gustav has skipped school. It falls to Uncle Kalle to discipline the young boy in the absence of the father, as Kalle tells a colleague. Only a man could exercise the necessary authority over a misbehaving child. Frau Schelp, who functions as a surrogate mother, holds even less power. When police officers discover that both Gustav and Willi have been stealing food from their homes to trade for black market fireworks, Frau Iller explains that Frau Schelp is just too weak to raise him properly. Frau Schelp's shortcomings stem from two attributes she had no control over: she is female and elderly, a combination that suggested moral lability and ineffectuality to viewers. In case the audience misses this point, after the police come to investigate Herr Birke's black market dealings, Frau Schelp tries to throw out her criminal boarder. Herr Birke laughs, writing a ninth notch on the wall to mark the number of times she has powerlessly threatened to evict him. Despite her claims to moral superiority, he mocks, she was only too glad to partake in the black market food he brought back for dinner every night. As Frau Schelp's plight makes clear, a woman could be an effective mother only alongside a father. In any other scenario she was incapable of raising her sons.

Indeed, the important children in these depictions were always sons. Positive images of girls were almost nonexistent. More often, girls appeared in film and other popular media as trivial creatures, noteworthy only for their powerlessness or, occasionally, deviancy. A sexual health campaign poster from the Soviet zone warned against "uninhibited living," encouraging anyone with syphilis or gonorrhea to see a doctor. An unhappy, unshaven, drunken, and ostensibly infected man in the foreground is smoking, ready to exchange all the family's ration cards for illicit pleasures.[24] Two sad women look down, probably his wife and daughter, abandoned by their caretaker. The moral and physical downfall of the father meant the destruction of the entirely family. Elsewhere, women and girls appear equally in need of male protection. *Somewhere in Berlin* presents two young girls in only a few scenes, their youth and smaller statures underlining their physical and even emotional weakness. Their speaking parts comprise only a few phrases that characterize the girls as unpleasant, shrill nuisances. In one instance, young Lotte threatens to tell on a young boy who has lied; the scene serves only to portray her as an annoying tattle tale. Later in the film, the camera draws the viewer's attention to a chalk stick-figure drawing of a girl on the ruins of a building with the observation, "Lotte is dumb." There is no dialogue during the shot, leaving the audience with the sense that the young artist's judgment needed no debate. As is clear from the film, if boys had little time for girls, adults showed even less interest in them. When the neighborhood's children go to say good-bye to the dying Willi, his nursemaid turns Lotte and a younger boy away, telling them that they should come back at a different time. Heads bowed, the children agree and leave. Lotte has no better luck in the final scene of the movie either. While all the boys, Herr Iller and Uncle Kalle joyously begin the important work of reconstruction, the girl can only watch, entrusted with the easy task of holding Herr Iller's jacket. The reconstruction of Germany was for fathers and sons only.

A year following the production of *Somewhere in Berlin*, another film that addressed children and fathers in postwar Germany appeared in European cinemas. Roberto Rossellini's 1947 *Germania anno zero* [Germany Year Zero] was an Italian-French-German production, filmed partly in Berlin with the cooperation of DEFA.[25] In this classic example of Italian neorealism, Rossellini presents a sober look at the devastation wrought by the Nazi era. Many scenes recall *Somewhere in Berlin*'s view of society and its cinematography, in which rubble landscapes become dangerous playgrounds for innocent children who assume difficult and even dangerous adult roles. Despite similar story lines, however, the mood of *Germania anno zero* differed significantly from the lecturing yet optimistic tone of *Somewhere in Berlin*.

Not interested in DEFA's positive spin on the antifascist message of redemption through reconstruction, *Germania anno zero* ends in despair. Most significant, father figures in this western interpretation of postwar life do not achieve heroic status. Indeed, according to Rossellini, they are to blame for Germany's physical and emotional ruins. The central character is eleven-year-old Edmund Köhler, who must struggle for his and his family's pitiful existence. The older male figures have all failed him in their fatherly duties: the actual father is present only as a bedridden, ineffectual old man; the older brother is a returned Wehrmacht soldier hiding in their apartment from the authorities. There are no female figures in Edmund's life to function as authority figures—his mother is absent, and the neighbor women are shrill, judgmental meddlers. Not even Edmund's older sister, Eva, is of help: a symbol of female moral decadence, she has turned to semiprostitution with Allied soldiers in her own fight for survival, leaving the boy with no family members to instruct him. In this vacuum of authority, Edmund turns to a former schoolteacher, Herr Enning, a pedophile who still preaches a National Socialist message that weaker beings must make room for the strong. Acting upon Herr Enning's philosophy, Edmund fatally poisons his ill father. The former teacher expresses shock rather than pride at Edmund's confession. Edmund realizes his mistake, and throws himself from the top of a building under construction while his father's funeral procession passes underneath. There is no redemption in Rossellini's view of postwar society: whereas DEFA used the orphan Willi's accident/suicide to push fathers to shoulder their responsibilities, Edmund's suicide is only one more installment in the German tragedy that resulted from the country's National Socialist past. With this credible and compelling plot, Rossellini had succeeded in producing a film that was almost a documentary, but this feat did not attract large audiences in Germany.[26] Unable to resurrect a father-hero, *Germania anno zero* was a depressing reminder of too many families' feelings of hopelessness, and offered no assurance that missing fathers might someday return to make everything better. Although DEFA's postwar films find order and justice in the chaos, this example of the neorealist film movement begins and ends on a note of despair.

Despite numerous examples of fatherless families, postwar Soviet zone Germans seemed unable to imagine a child without a father, or a nation without fathers. Children who behaved when the father was absent seemed unthinkable, warranting special mention, such as the four teenage children in Family M interviewed by Thurnwald.[27] Granted, the M children occasionally snitched food from the already bare pantry, but they were otherwise a help to their mother. Such positive family relations in a fatherless household were almost inconceivable in the public mind. In popular images and contemporary reports, families and friends generally sought a substitute father for children when none was present. The replacement might be another male family member or friend, such as

Gustav's Uncle Kalle, or perhaps a male teacher. The widowed Frau T., for example, another of Thurnwald's interviewees, noted with relief that her twelve-year-old son was very close to an older male teacher at his school, even bringing him home for meals.[28] In the Family L, the nineteen-year-old son felt responsible for helping his mother rear the younger siblings, since the father had died during the war.[29] Such a sense of responsibility was logical as regarded his thirteen-year-old brother, but his paternal role also included watching after his eighteen-year-old sister. His mother "discussed everything with him," according to Thurnwald, thus strengthening the sense that he and his family had pushed him to take the place of the deceased father in all family matters. For those children without an appropriate replacement father, the mother appealed either to the memory of a father ("What would he think?") or the threat of his reaction upon returning home. Even Frau Iller, who had Uncle Kalle to act as a father figure, used this technique when reproaching Gustav for his bad behavior, asking him if he will not be ashamed of himself "when Father returns." Neither incarceration nor death could hold back the influence of these children's fathers.

Still, some children had no fathers, no one to invoke the memory of their fathers, and no viable father substitutes. These children would find or be found by an inappropriate father figure, or so assumed many adults. Orphaned Willi, in *Somewhere in Berlin*, falls victim to Herr Birke's black market schemes, learning to lie and steal. In *Germania anno zero*, without a healthy father to instruct him, Edmund falls in with a group of juvenile delinquents and then under his former teacher's spell, ultimately committing the ultimate sin of patricide. His sister Eva brings shame onto the family by spending her evenings flirting with Allied soldiers, father figures who want to make her their lover—a sin compounded by their status as the victorious enemy. When Inge turns to the city administration for help in *Die Kuckucks*, the bureaucrats threaten to break up the family, placing the children in orphanages and foster homes. In the eyes of the state, the fatherless Kuckert siblings, raised by the young, single Inge, do not even constitute a real family worth maintaining. Worse, we learn, the rowdy children disrupt their neighbors' peace. Only a father can turn this unruly bunch into a family safe from harm and respectful of social mores. By the end of the film, a "good" father, Inge's new boyfriend, has stepped in and saved the family from the "bad" father figure, a gigolo who pretended to be a wealthy man ready to help the young Kuckerts. The absence of a father was thus a double threat to children. Not only did they not have his guidance, but also were vulnerable to misguidance from unsavory male characters.

Some adults—at least males—needed father figures as well. Herr Iller in *Somewhere in Berlin*, for instance, was not emotionally equipped after the war to reassume his fatherly role. His family could help him regain his physical strength, but the male family friend Uncle Kalle was instrumental in helping Herr Iller become a father again to young Gustav. This mentoring relationship suited both men. Uncle Kalle had lost his son, and coped with his grief by throwing himself into his work and advising his friend Herr Iller and the young Iller boy. This arrangement reflected the film's antifascist lesson that physical reconstruction of the nation would heal postwar wounds. But there is also a subtext that fathers needed sons, or at least surrogate sons, as well. In this case, when Uncle Kalle was no longer a surrogate father to Gustav after Herr Iller's return from the war, the family friend was then duty-bound to act as a father figure to the despairing Herr Iller. Fathers would help reconstruct the nation, but only if they were competently fulfilling their responsibilities. Just as for their sons, society stepped in to provide surrogate fathers to these faltering men. It is the scene of dozens of neighborhood boys helping rebuild the Iller garage, with Gustav at the head of the gang, that ultimately pushes Herr Iller firmly back into fatherhood. In a twist on the typical familial division of labor, the children devote all their energy to keeping the father as head of the household. In the public mind, fathers were so important to the national reconstruction project that all of society's members—children and adults—had a responsibility to restore these men to their proper stations.

Homecomings

If popular images of family reunions presumed that only sons cared about their fathers, the reality was quite different. As young people's writings attest, both boys and girls dreamed of the day their fathers would return. Like Gustav in *Somewhere in Berlin*, sons and daughters believed that their fathers would rebuild their lives and the nation according to antifascist principles. And, like Gustav, children struggled to reconcile memories of an omnipotent father with the changed figure who appeared on their doorsteps. Still, they did their part to put their fathers back as head of the family.

In order to help the father return to his role, families maintained the home in his absence as best they could. The elementary pupil Giesela Mischok, for instance, recounted in a school essay how her father had been called to work in an armament factory during the war.[30] He taught his wife how to do small repairs so that their shop could stay open; he completed the major repairs on his free weekends. When Herr Mischok returned home permanently, the family began work immediately to rebuild the damaged family business. Mother and child were placeholders for the father. In such instances, children understood that they and their mothers were completing work that the fathers, under normal circumstances, would undertake. Similarly, pupils noted those moments in their lives when their fathers were not there for the family as they should have been. The fifth-grader Hannelore, for instance, painted a picture of a fatherless family that was defenseless against the Russians and then faced her own compatriots plundering her family's apartment after the war.[31] Her essay betrays a sense of desperation mixed with accusation: although the war was over, her father could not yet be counted on to keep them safe. Worse, when the family began the work of reconstruction, her father could "unfortunately" not help them, since he was "still out in Russia." There is no doubt that even a young girl in the Soviet zone would have understood that her father's presence in the Soviet Union was a result of his status as a soldier for the German, and thus wrong, side. Hannelore's father had failed the family multiple times, by leaving them vulnerable and by not being there to help them rebuild their lives. He could thus not make the successful transition from a soldier for the Nazis to a father/builder for the antifascists.

Families so identified themselves through the father that even a brief absence could destroy their sense of well-being. In an ironic twist to the scenario of an absent father, the family F briefly lost their father after he had returned home.[32] They had survived the war with occasional absences of the father for work and military service. Then, after the war, occupation soldiers caught him stealing coal for his family, and the military court sentenced him to five months' imprisonment. The mother petitioned for his early release, asking the court to recognize that he had been trying to help his freezing family. While the family waited for the court's decision, the mother and seven children sank into despair, ceasing to care about themselves or the household. When the father returned after only four weeks, the interviewer described them as "metamorphosed": "The children cheered, the mother was dressed nicely again, and the apartment was picked up." The presence of the father outweighed all other problems. Although he was in poor health and without work, the mother could find the energy to complete daily tasks again. "Now there is someone here again who will care for us," she stated, referring to him alternately as "my husband" and "our father." As head of the household, his actual ability to provide for his family was less important than their impression that "now everything would get better" with him home.

Children also believed that their fathers would bring relief to their difficult living situations. A 1946 play, "Wiederaufbau Prenzlauer Berg" ("The Reconstruction of Prenzlauer Berg") set in the working class district of Prenzlauer Berg in Berlin, and written by Christel and six of her classmates at a girls' middle-school, weaves together memories and fantasies of an ideal family, using the return of the father as a symbol of the return of normalcy. The play opens with the happy scene of Christel and her family celebrating the father's homecoming. We learn that the father returned the previous evening, a year after the end of the war. Exhausted and disoriented, upon arriving he had not noticed the luxuries of electricity, gas, and running water. After a night's sleep, however, he was able to fully appreciate the progress toward reconstruction of his city. As he explains, "I was just happy that it [our home] was still there!"[33] The doorbell then rings; Christel's friends Brigitte and Sonja arrive to work on their homework assignment about the reinstatement of public utilities during the previous year. Christel's brother Horst, clearly continuing a surrogate father role, instructs them to write a play comparing their experiences of 1945 with 1946, thus illustrating the slow but steady progress toward normalcy. But it is late, he admonishes, and their father is tired— they will need to work on their assignment the next day so that the father can sleep.

Although gas, water, and electricity are the ostensible main actors in the play, it is the figure of Christel and Horst's father who is actually at the center of family activity. Without doubt, the partial reinstatement of basic utilities connoted a significant step toward achieving normalcy for postwar society and signaling the defeat of Nazism. But the clearest sign of the family's improved life is the return of the father. The moment he arrives, his family begins to define themselves through him. Only by outlining their lives since the end of the war for him can they understand the progress they and their society have made. The family, in particular the children, bask in his praise, lovingly explaining every detail of their household routine to him. Still, they acknowledge, not everything is as it should be: utilities are still rationed, and the family must sometimes do without gas in order not to exceed their quota. The father, too, is not entirely recovered, retiring to bed early to recover from the stress of the war and his internment. Acknowledging this situation, the children declare their day over when his day is over: the son notices the father's fatigue, and the children's activity stops so that he can rest and recover his strength. They have waited his entire absence to see him reinstated as head of the family. They will continue to devote all their energies to this goal, even sacrificing their own needs.

The final war and postwar years were not just a hardship to be endured; for Germans in the Soviet zone, the many abnormal situations signified a complete loss of civilization that egregiously disrupted family life. Act Two of the play depicts a different family in May 1945, overwhelmed by the darkness brought about by the lack of electricity. The father is completely impotent in this situation. He must ask the son for matches, and has entirely abdicated responsibility for providing a light source. "Do you not have any other candle?" he asks his family helplessly. When a small piece of candle is finally located, it dies out quickly, and they all sit down "resignedly" in the living room:

> FATHER: It is really as if time had been turned back 100 years.
> SON: 100 years? People *had* petroleum and candles then, but we are even further back in culture now. I imagine that the lighting was like this in the Middle Ages.
> DAUGHTER: The Middle Ages were not *that* dark either, they had torches and pitch fires!
> SON: I think that it was only *this* dark for cave dwellers, which we have now become. (italics in the original)

The father then distances himself from the Nazis, sarcastically announcing that they can thank the *Führer* for having led them to such "glorious times." Powerless to resist Hitler's plans, and now powerless to improve his family's life, the father predicts that the occupation authorities will do their best to restore electricity to the city as quickly as possible. In the next second, as if invoked by the father, the radio and light turn on. Water and gas are not yet restored, but electricity heralds a return to civilization, and thereby the father's return to power.

Christel and her friends clearly valued the reunification of the traditional family, even though it is unlikely that all the girls lived in one. Several textual clues point to the young playwrights as having different types of family situations—one whose father did not leave during the war, one with a father returned from the war, one without a father. In the pictures that accompany the play, for instance, only mothers and children are shown—no fathers. It is likely that some of the girls had experienced the scene they described of a father's joyful return. Others in the group, however, might still have been waiting for their fathers a year after the end of the war, or lost them to war casualties or to illness. In the Western zone city of Hamburg, to provide an example of another heavily bombed city, the magazine *Benjamin* in 1947 claimed that three out of ten children did not have a father at home, due either to death or internment.[34]

To complicate father-child relationships even further, not all homecomings proceeded as smoothly as Christel described her father's. The presence of a father did not always mean that he could live up to the image of a strong head of household. Many fathers returned home wounded and unable to perform previous tasks, at least temporarily. Typically, families struggled with readapting to having their fathers home, as the fathers struggled to understand the changes that had taken place in their absence. Very young children lacked memories of their fathers, in some cases putting both parent and child in the position of having to get to know a total stranger. Fathers often found it difficult to accept the intimate relationships that had developed between their wives and children, and struggled with feelings of jealousy.[35] Many families noted the insatiable hunger of the father upon his returning home, which sometimes led to bitter fights over food in the family.[36] The father who returned home sometimes seemed to be almost a different person, at least initially. In a 1947 letter to her friend, the young Brigitte Reimann, later a leading author of the GDR, described such a scene. She, too, wrote with enthusiasm of the return of her father from a Soviet prisoner of war camp to their home in Burg, a small city of 20,000 residents in Saxony-Anhalt. However, she and her three younger siblings had not been prepared for the figure that greeted them at the train station. Their scrawny, louse-covered, and now bearded father elicited shyness and fright in the children, particularly in the younger ones. Much like *Somewhere in Berlin*'s Gustav, Herr Reimann's children did not initially recognize him. Even his language had been distorted and rendered unfamiliar by the war experience, explained Brigitte, "so anxious, so sick and—I don't know, just so terribly strange!"[37] His strange way of speaking so disturbed Brigitte that she could not bring herself to call him *Vati* (Daddy). Worse, he had become "rather meticulous," so that the household had to be run under tighter control. With little or no resources to deal with the problems of the psychological and physical strains of war— even shell-shock, or posttraumatic stress disorder—World War II veterans and their families such as the Reimanns were left alone to cope with the father's return home as best they could.

None of the surprises about their father's condition dampened Brigitte and her siblings' determination to appreciate the reunited family structure. Like DEFA's young film heroes and Christel and her friends, Brigitte worked hard at bringing her father back to his usual place in the family. The family prepared the best meal they could afford, including their carefully guarded canned meat reserve. They had coincidentally baked his favorite dessert, gingerbread, a talisman that in

Brigitte's retelling seemed to presage the father's arrival. Her sister Ursula helped Brigitte accept their father's unfamiliar accent and syntax. The father, too, worked to reintegrate himself into the family. Immediately upon arriving, he discarded the clothes that Brigitte claimed were nothing but "old rags, a rough Russian coat, and a tri-cornered Russian cap," and that caused the apartment to smell like "delousing treatment and *Russki*."[38] As soon as he shaved off his "disfiguring beard," the youngest daughter no longer tried to hide from him. Brigitte and her siblings saw past these many traumatic moments into a future that looked much like the prewar past. Despite the father's continuing problems—double pneumonia, malnutrition, calluses, a fractured kneecap, swollen feet, his new insistence on discipline and order, and his "terrible anxiety," Brigitte could report that she and her siblings had grown used to having their father around again. She referred to him now as *Vati*, and it seemed to all of them as if "he had always been here and was never gone." And yet he had been gone, and everything had changed. Only the family's desire to imagine a continuity of family life remained constant.

* * *

The traditional family unit during and after the war had shifted perceptibly from the prewar days. Loss or physical incapacity of family members, a change in the family's financial status, forced relocations of the family home due to either shifting borders or bomb damage, and different work routines based on job changes all contributed to a new home environment. Still, even more trying situations, such as the death of a father or his internment, did not bring about major changes in families as might be expected. In many cases, the father continued to be present in his absence, whether or not he was expected to return home. The majority of Soviet zone family members, parents and children, worked hard to maintain their families.[39] Popular culture portrayed fathers interacting only with their sons, underlining the importance of this masculine relationship in emerging ideas about rebuilding the nation. Despite losing a war, fathers were heroes. As such, they led the way for the rest of their compatriots to make the transition from Nazis to antifascists.

CHAPTER 7

Reestablishing Traditions

The antifascist democratic material that filled "new school" classrooms with educational thought resonated with echoes of multiple pasts. Educational administrators in the Soviet zone did not attempt to create a total rupture with German history. Rather, they sought to establish the Soviet zone as the rightful inheritor of Germany's past. History—reaching back to the Weimar Republic, the failed revolution of 1848, the French Revolution, and even the Renaissance—could be used to position the Soviet zone as the logical successor to the German nation.[1] The past, as represented by historical traditions, commemorations, and icons, was something to be applied judiciously in the antifascist democratic reconstruction of Germany.

This chapter discusses the use of an antifascist and German approach to history and traditions in the Soviet zone. Because of the school's unique position as an official authority on history, the language of "culture" and "heritage" became particularly important for defining the new society's past through history lessons, both in and outside of actual history instruction. The "new school" evolved into a central laboratory for the emerging debate over a German heritage in the emerging socialist state, whose constructors found themselves vacillating between two versions of the German nation. One included a harmonious, unified vision of Germany. The other sought to disinherit the Western zones from the family estate, or, as the writer Erich Kästner described Germany, the "ruined farm estate (*Bauerngut*)."[2] In each case, new celebrations in the Soviet zone drew on traditional and historical German commemorations and holidays. History thereby provided a means of positioning the Soviet zone as the true German nation.

* * *

The Antifascist Turn

The cultural heritage claimed by communists and social democrats in the postwar era had distinct bourgeois roots. Neither the SMAD nor the Germans were interested in eliminating the classicist tradition that had characterized the past century of German intellectual life. On the contrary, they embraced it. Leading political figures in the Soviet zone reestablished the bourgeois cultural standards with which they had grown up—although not without a twinge of guilt. Alfred Kurella, the leading SED official for cultural policy in the GDR, once confessed that he found Thomas Mann more interesting than Bertolt Brecht.[3] He was not alone. Cultural life in the Soviet zone underwent a "springtime" that reawakened many slumbering traditions, rather than inventing new ones.[4] "Twelve years of dictatorship are, when you look back on them, not a long time"; one historian later wrote, "and the regular audience of theaters and museums took up the epoch before '33 without further ado . . . This audience didn't see those [postwar] years as a radical new beginning, but rather as the repair of an industrial accident."[5] But the "taking up" (anknüpfen) of prewar traditions was not seamless.[6] Goethe and Schiller, rechristened as antifascist heroes, had to be forced into new roles. Such reappropriations of the national culture resulted largely from the regeneration of a prewar antifascist culture.

The Hungarian Marxist Georg Lukács laid the foundation for an antifascist culture with his writings in the late 1920s and 1930s,[7] though his later influence in the Soviet zone and GDR remains a matter of some dispute.[8] His discussion of German literature under imperialism was published in 1945; his other works appeared in rapid succession throughout the postwar years.[9] Lukács criticized the irrationalist tradition in Germany that rejected the Enlightenment, and he established Goethe and Schiller as the centerpiece of an antifascist aesthetic. This move determined the emerging canon of cultural expressions in the Soviet zone and GDR up through the 1960s.[10] It also positioned the antifascist democratic Soviet zone as the logical inheritor of the French Revolution.[11] "The fighters change, their experience has increased," wrote Heinrich Mann for the cultural journal Aufbau in 1945, "the goal is, as always, freedom. The French Revolution still moves in us as well."[12]

German classicism and the Enlightenment found a happy union in their role as an antifascist cultural heritage.

The antifascist canon was easy to define as long as classicism meant only Goethe. Other elements of the cultural heritage had to be considered more judiciously. In the northern German city of Neubrandenburg, for instance, city architects agreed in 1955 not to reconstruct the city center, with particular reference to the cultural house, in an "untrue Romanticism." The art and ideology of Romanticism focused on spirit and emotion, not intellect, positioning it at odds with the rational and intellectual heritage of the Renaissance values so prized by antifascist reformers. Architects drew instead on the "Nordic Renaissance," whose "reserved [*gebunden*] and cool posture," according to the lead architect, better represented the "character of the Northern German as well as the geographic milieu."[13] As is evident from the architect team's accounts, members of the cultural community were passionately committed to the idea of linking the old with the new, but not all traditions were honored equally. Literary traditions offered as many confusing decisions for the antifascist canon. Fairy tales, for instance, could not easily be included. The unquestionable imprint left by the Brothers Grimm upon the German language and cultural traditions could hardly be ignored, yet the brothers' connection to German Romanticism—and therefore irrationalism, in this interpretation—made them unacceptable for inclusion in the new antifascist tradition. The Central Institute for Pedagogy thus initially decided not to publish Grimms' fairy tales because they were too reactionary and bourgeois for socialists' tastes.[14]

This decision did not stand long. The struggle to define an appropriate cultural heritage for the first socialist state turned into a struggle to wrestle the rights to this cultural heritage away from the Western zones. As David Bathrick has observed, "The fight, in other words, was nothing less than the rightful claim to the national heritage, and it was clear with the increasing emphasis upon national identity in the GDR that the Grimms would have to be included in the accepted canon."[15] It is furthermore hard to imagine that parents were not already telling the well-known tales to their children, at bedtime for instance, or perhaps even still owned prewar copies of the fairy tales. In cases in which adults recounted stories remembered from their own childhood, they returned to the tradition of an oral storytelling culture from which the Brothers Grimm first gathered the tales.

Thus the canon of antifascist literature expanded to include German folklore, paralleling areas such as architecture that began to look back to German (and not Soviet) traditions for visions of the future. The Central Institute for Pedagogy overturned its decision against the Grimms and "rehabilitated" the fairy tales, publishing a two-volume edition of them in 1952, reprinted multiple times throughout the GDR. The editors of the now-acceptable tales included the educational reformer Hans Siebert (KPD/SED), who played a large part in the creation of the Soviet zone and GDR school and its curriculum. The resulting work illustrates an "antifascist turn" in its alignment of parts of German cultural traditions with socialist goals. The values portrayed in the fairy tales now emphasized labor and the moral superiority of peasants. The redrafting of the Grimms' fairy tales occurred within a larger "rescue operation for the democratic, humanist, nationalist tradition."[16] But traditional, bourgeois German values remained even in this sanitized version: The story of Red Riding Hood, for instance, ends with the law of maternal authority over all matters of children's safety: "Red Riding Hood then thought: You won't your whole life long ever again leave the path and run into the woods when your mother has forbidden it."[17] The antifascist turn that allowed fairy tales to become part of the socialist cultural heritage in the GDR had started with more traditional, even bourgeois, aspects of German history.

The Uses of History

Antifascist educators' uncertainty about the future of the entire German nation led to their sometimes contradictory presentation of the Soviet zone's relationship to German history. At times, their history narratives proceeded from an assumption of impending national unification, while at others they focused on division and Soviet zone demarcation (*Abgrenzung*).[18] Educational administrators deliberated over which elements of history to use or discard in creating a usable past for history instruction. They debated what these decisions suggested to the outside world about Soviet zone goals, and worried about which versions of the past would provide the best orientation for the future. This tension between using history to bring the two halves of Germany together or to establish the Soviet zone as the true successor to Germany marked educational discussions throughout the postwar period. By 1948, however, as a result of external political events, history in the Soviet zone more clearly served to identify the antifascist half of Germany as the legitimate German nation.

This use of history as a legitimating function was not unique to the Soviet zone. The construction of a public history played a fundamental role in the self-rationalization of both halves of postwar Germany.[19] Yet history lessons were not an exercise in indoctrination; such a view wrongly overemphasizes political motivations—that is, Soviet communist ones—that ostensibly manipulated the substance of historical events in political narratives.[20] The postwar debates in the Soviet zone about the future of the German nation included many considerations, some of them specifically targeted against the construction of a Soviet Marxist-Leninist model. Especially in the crucial postwar years, debates in the Soviet zone about Germany's future were of a multipartisan, ambivalent character. Scholars who have written on such uses of the past and of history as a school subject have assumed that the Soviet zone implemented a contorted version of historical events in order to achieve clearly defined common goals. With few exceptions, they focus on the Marxist-Leninist approach used in the teaching of history in the Soviet zone.[21] But other important approaches informed pedagogical thoughts. Educators were not only reading Marx, but also the Weimar reform pedagogue Georg Kerschensteiner. The appeal of such concepts as Kerschensteiner's "work school" was not pedagogical habit that lost ground to Soviet theories.[22] Rather, reform pedagogy, antifascist convictions, and Marxism-Leninism interacted throughout the Soviet zone, and later the GDR, to create a unique GDR pedagogy as well as an identifiable GDR approach to history.[23] True to the reform pedagogy tradition, history instruction in the Soviet zone was remarkably pupil-centered.

The drafting of the history curriculum for the "new school," which represented one important aspect of the antifascist turn in history construction, had two purposes. The first was to teach pupils a different, antifascist worldview, one that praised democratic historical events rather than militaristic ones. "Pupils must recognize," as the introduction to the eighth-grade history curriculum for the Soviet zone stated, "that the reactionary and military forces in Germany twice pushed our people into the catastrophe of a world war, and that only the development of democratic forces can ensure the future of our people."[24] Insisting that young people had to be given "ample opportunity to make opinions and judgments," educational administrators wanted pupils to be able to "make sense of the present from the past," so that they might better understand "Germany's present responsibilities."[25] A considerable portion of the history curriculum focused on Germany's role in the two world wars—the eighth grade curriculum, which covered the Industrial Revolution in England up to the present, devoted 8 of 120 instructional hours to World War II, including a section on "the deceptive non-aggression pact with the Soviet Union."[26] At no point, however, did educators hint to pupils that they should be ashamed of Germany for the nation's role in World War II. Pupils instead were to learn to see Germany's past critically, and then help the nation learn from these events.

The second objective of history instruction was to contextualize German history within world history, as the introduction to the history curriculum for the Soviet zone explained: "The object of history instruction is first and foremost the history of the German people, although in the context of world history."[27] This perspective, hoped educational reformers, would pave the way for a process of normalization of German history and German society. For example, the general instructional goals in the 1946 Greater Berlin curriculum started by emphasizing Germany's positive contributions to world history. "History instruction should educate youth to true democratic national consciousness that is based upon pride in *those* achievements of our people which have served the advancement of humanity."[28] The curriculum directives then stated that history instruction "should also educate them [youth] in the spirit of international community by teaching them to understand and respect other peoples and by demonstrating how the developments of our *Volk* have been affected by other nations' developments and achievements."[29] In other words, all nations, not just Germany, had neglected their common responsibility toward keeping their members firmly anchored in the "international community." This concept of community deflected attention away from Germany as an isolated actor and signaled its desire to be regarded as a normal country. Germany's wartime actions, in this interpretation, were partly the result of historical determinist processes, but also the logical result of negligent international actions.

At times, the work of antifascist historical construction represented a struggle between the German Educational Administration and the SMAD. One such example was the Educational Administration's decision to commemorate May 10, 1933, marking the day when Nazis had staged a large book burning of the works of so-called degenerate authors, many of them Jewish.[30] The first evidence of the day's commemoration in the Soviet zone was in 1947.[31] In the agenda for an educational meeting, Herbert Volkmann (KPD/SED), director of DVV's department for art and literature,[32] referred to the new commemoration as the "Day of German Literature." His explanation of the significance of the day mentioned the international literature burned by the Nazis, but his main theme emphasized helping youth understand the significance of the great German works that were burned. In a strong statement, Volkmann proposed that pupils gather in the school auditorium that day to hear a lecture about those German authors who chose death or exile rather than submitting to the Nazis.

A poster for the May 10 commemoration in 1948 revealed a change in the purpose of this anniversary. A pile of burning books in the middle of the poster declared May 10 to be the "Day of the Free Word," flanked on both sides by an overwhelmingly German list of authors, such as Kurt Tucholsky, Heinrich Mann, and Thomas Mann, whose works had been burned in 1933.[33] Out of the flames rose a book with the inscription, "Forbidden, Burned. Our intellect (*Geist*) did not burn up." Germany's wounded but surviving culture defined the day's significance. A year later, in 1949, a poster showed one book with the phrase, "Day of the Free Book," and a dove flying out of the book.[34] May 10 had begun as a day to educate pupils and the Soviet zone public about the cultural loss inflicted upon Germany with the banning of German authors. It also made role models out of German authors who had dared to defy the Nazis, apparently without specific reference to their political or confessional affiliations. Authors from other countries received mention, but they were not featured as key actors. Two years later, in 1949, May 10 had lost its specifically German reference. Instead, the commemoration promoted the right of authors to express themselves freely (ironically a right not to be enjoyed by most authors in the GDR). The 1949 event demonstrated a more general day of commemoration, one that did not attempt to elevate German literature above that of other countries.

The transformation of the "Day of German Literature" to the "Day of the Free Book" marked one of the cultural battles that Berlin lost to Moscow. Norman Naimark has argued that Moscow dictated the form and content of public ceremonies and rituals in the Soviet zone.[35] Even commemorations of German heroes such as Goethe or the communist martyr Ernst Thälmann were identifiably "Sovietized." But commemorations in the Soviet zone should also be read beyond the level of Soviet influence. German educational administrators often attempted to placate the Soviet Union by showing that the "new school" integrated Soviet culture in its curriculum, while actually concerning themselves more about the promotion of German culture. This dichotomy resulted at times in a schizophrenic educational program. It externally emphasized respect for the Soviet Union but internally focused on creating a "German consciousness." The Soviets appreciated and respected many German Classical figures, but they frowned on obvious homages to elements of German culture that could be interpreted as nationalist. Soviet zone educational administrators were nonetheless persistent in their desire to create an antifascist, German heritage without direct references to Soviet influence.

The original motivation for the commemoration of May 10 served a specific German purpose. Not only was May 10 a "Day of German Literature" to reclaim those parts of indigenous culture rejected by the

Nazis, but educators also hoped that it would contribute to the creation of antifascist role models for Soviet zone youth. These goals broke two unwritten rules about German interactions with Soviets. First, they elevated German culture above Russian culture, potentially offending Moscow. The significance of this threat is evident in educational administrators' constant attempts to demonstrate their respect for the Russian literature and language. At a 1947 conference of educational administrators, Paul Wandel, director of the DVV, warned in ominous tones that it was necessary for Germany's "national existence to create an especially close connection to the Russian people and especially to the Russian language."[36] He insisted that foreign languages and cultures be treated equally: "We do not acknowledge a division between East and West. Instead we know that great cultural treasures are slumbering in the East and the West that must be opened for our children, for people who tomorrow will have the possibility of communicating with French, English and Russian people and follow their literature."[37] Changing tone slightly, he went on to say that it was especially important to ensure better teacher training for Russian instructors. A colleague protested the desire to accord all occupational languages the same weight, complaining that, "It wouldn't occur to anyone in the English zone to give Russian as a first [foreign] language. So why should we have English?" Wandel retorted sharply, "Because we are fundamentally different from the others." The discussion continued in this manner, teetering between the philosophical and practical importance of teaching all foreign languages and concessions that Russian should be, after all, the principle foreign language. In the end, decisions to promote Russian were based upon arguments about appearances. As Max Kreuziger, DVV representative for the school division, explained, "We have to understand that this directive will remain on paper only. It's probably important politically, but practically there won't be any possibility of executing it."[38] The need to demonstrate a special respect for Russian culture outweighed all other considerations, thus evidencing the difficult situation that German administrators found themselves in when attempting to promote German culture.

The first 1947 concept for May 10 also more generally honored German disobedience to totalitarian authority. Such praise did not survive SMAD scrutiny. Though the explicit reference was the totalitarian Nazi regime, the implicit moral, which alluded to antifascist values gleaned from the Spanish civil war, could also easily be understood to include a totalitarian military administration, such as the one Soviet zone Germans lived under. This interpretation presented a lesson to young people that the Soviets could not have endorsed. Moreover, antifascist democratic commemorations were not allowed to be publicly pro-German. Nor could reclaiming the works of German Jewish authors have found resonance in the SMAD just when Moscow was beginning its anti-Semitic "anti-Cosmopolitan" campaign and was purging Jews from important positions. A suggestion, in whatever context, to prefer death to censorship could not have found approval with Soviet authorities. May 10 was a past that German educational reformers ultimately could not continue to use in their construction of an antifascist democratic history.

The past had other, more everyday uses, however. For example, it could be invoked as a warning against inappropriate behavior. Certain practices and even language usage caused educational administrators and intellectuals to worry that Nazi ideology was still present at the political and everyday levels. For instance, a 1946 letter published in the Berlin newspaper *Horizont* claimed that an applicant for a *Neulehrer* course had not been accepted because the individual in question did not belong to the SED. This charge was a common but serious one, seldom supported by concrete evidence.[39] When city councilor Otto Winzer asked the paper for further details, the journalist responded that confidentiality issues prevented disclosures. He explained that the incident had in fact taken place in Potsdam and was thus outside of Winzer's jurisdiction. Winzer sent copies of this correspondence to the Allied educational commission with the observation that unfounded stories constantly appeared in *Horizont*. He added, "It can justifiably be asked whether these sorts of methods already used by the Nazis serve to renew the schools or to educate pupils to become upstanding, sincere democrats."[40] Thus, after an allegation that the press was using Nazi intimidation tactics, he strengthened his complaint by implying that such activities harmed the reeducation of young people. On the same day, Winzer fired off a letter to *Horizont*, complaining of slander. If the event had occurred in Potsdam, he fumed, then there was no reason to lead readers to think that the incident had occurred in Berlin. "This underhanded method is not new," he fumed. "It was used extensively by Goebbels and Hugenberg in their slander of republican authorities before 1933, and it also belonged to the propaganda arsenal of the Hitler Youth."[41] Reproaching the letter writer for using Nazi tactics, Winzer demonstrated that the struggle for "democratic antifascism" entailed a vigilant and aggressive awareness of the past in order to overcome it.

Such conceptualizations of the past as something unpleasant rested on the implicit assumption that new ideas meant progress, and that progress was preferable to the status quo of any earlier epoch. Social projects with roots in previous eras could be used to rationalize contemporary policies by demonstrating historical determinism, which came to dominate GDR historiography. By using the past as a marker from which to move forward, educational reformers could label undesirable historical elements as outdated. Clinging to wreckages of the past was thus construed as counterproductive. This tendency created significant confusion for educational administrators as well as pupils when elements of the past would not lie quietly. One of the best examples of this phenomenon was that of religious instruction, a central question in the *Einheitsschule* project of creating a unified German nation. As the author of a contemporary newspaper article insinuated, certain elements of the past were best kept locked away. The unnamed author launched a colorful attack on proponents of religious instruction and accused them of historically impeding German unity: "Behind these false speeches about national unity is in fact hidden nothing but the desire to maintain the division of the school system in confessional schools, to preserve 'that old ghost' haunting the closed-off upper stories."[42] Ghosts are not generally desirable residents of emerging societies. They connote forgotten and moldy rooms, filled with trunks of another generation's souvenirs. Confessional schools and religious instruction, since they were identified with past eras, did not fit the development of an antifascist, socialist, secular Soviet zone. The active inclusion of the church in schools seemed to prevent society from moving into the future. As the self-proclaimed progressive half of the German nation, the Soviet zone could not afford traces of social elements that its leadership deemed reactionary, including the church. In practice, however, Soviet zone administrators cooperated with Catholic and Protestant churches and allowed the use of classroom space for religious instruction if it did not take place during regular school hours. Rhetorically, proponents of the *Einheitsschule* insisted that those in favor of religious instruction in the schools clung to a past that had never seen true German unity.

The past found a variety of expressions in the Soviet zone. It established pupils' role in constructing their own interpretation of history, and it helped negotiate a balance between Soviet and German conceptions of acceptable culture. The past could be viewed as negative, something to move beyond; but it also contained progressive ideas that could now be implemented. Antifascist educators did not always succeed in creating a public history according to their ideas, especially where Soviet influence necessitated an adjustment of commemorations that seemed too nationalist. The most important use for the history that evolved in the "new school," however, whether in curricula or in commemorations, was to maintain a focus on an antifascist democratic interpretation of German history.

A Century of Revolution: Remembering
1848 in the Antifascist Classroom

Using the past thus entailed locating antifascist role models and events in history that demonstrated Germany's historical destiny of moving toward socialist humanism. To this end, educators had already begun the careful work of creating various festivals to commemorate acceptable past icons even before the war was over.[43] Celebrations and pedagogical heroes helped anchor the Soviet zone's educational programs historically and defined a concrete task for the school in the work of reeducation. Alongside schools' celebrations of Heinrich Heine's 150th birthday[44] or the 120th anniversary of Beethoven's death,[45] two festivals dominated the discussions of commemorations in the Soviet zone. One was the historically significant year of 1848. The other was the 200th anniversary of Goethe's birth, centered in Weimar, his primary city of residence.

The year 1848 offered an almost endless array of choices from which the Soviet zone could select an appropriate cultural heritage. The "new school" was perhaps the institution that most clearly claimed the legacy of 1848. Other attempts in the Soviet zone to commemorate 1848 were less successful. Divided between the symbols of the "parliament and the barricade"—the Paulskirche revolution and the violent street demonstrations of the "March Days"—1848 got caught in a battle of the appropriation of symbols in East and West.[46] Because policymakers were unable to decide how to make sense of the entire year, most anniversary celebrations did not attract large crowds and could not be turned into any real cult of memory in either the Soviet zone or the Western zones. Even the socialist newspaper *Neues Deutschland* reported that the "people's demonstrations" to commemorate various dates throughout 1948 were poorly planned, with the trade union remembering to announce the November 30 event "literally at the last minute."[47] Ultimately, the anniversary year 1948 came to mark the beginning of separate paths for the GDR and the Federal Republic.[48]

Things looked different inside the school building. Educators in the Soviet zone pointed to the failed revolution of 1848 as an event they were now destined to realize. Their version of the "failed liberal-bourgeois revolution" in Frankfurt provided a different interpretation of the revolution, one that put the school at the center of 1848 plans. The key goal of Paulskirche, according to this narrative, had been a socially unified Germany, not only a political unity. The 1948 speeches commemorating the 1848 revolution spoke of the age-old desire to unite Germany through the school system and change Germany's historical downturn. A century later and in the schools of the Soviet zone, believed educational reformers, this dream could now be achieved under the banner of socialist humanism. The idea of 1848 deflected attention away from the recent Nazi past, placing it alongside the many missed opportunities that could now be realized. "That is the main goal. If we make this task of the '48 Revolution our own, then the Revolution of 1848 will not have been in vain. Then 1948 will and should bring about a period of historical ascension for Germany."[49] Educational discussions highlighted how essential educators had been in helping formulate concepts of unity in 1848. Soviet zone educational administrators emphasized the role of the proposed *Einheitsschule* in mid-nineteenth-century projects to bring about true unity for Germany: a unity of all citizens that resulted from eliminating all categories of division—politics, religion, and even gender—although the last was more controversial. References to educators in 1848 also alluded to a strong sense of cooperation among teachers in their work to bring about a revolution, thereby criticizing the rifts between different groups of educators, or between political parties. Fulfilling the past a century later placed the Soviet zone and its educational system historically at the culmination of a century of revolution.

Educational programs like that of the *Einheitsschule* in 1848 allowed Soviet zone educators to point to the idea that their school and social plans had German roots. In a preparatory meeting for the 1948 German-wide school minister conference in Stuttgart in the Western zones, a speaker discussed the need to publicize the objectives of earlier German educational reform programs: "The fact that the democratic unity school was an old demand of democratic school reformers—and therefore not in the least suggested by the SMA [the Soviet military administration]—can be proven beyond a doubt with historical documents."[50] Of course, not all documents could serve this purpose. Participants at the Stuttgart meeting learned that the 1848 plans most closely mirrored Soviet zone goals, whereas drafts of programs such as the 1819 Süvern curriculum plan, which had accepted confessional schools, contradicted their present secularist goals. Even documents from 1848 had to be selected with discrimination. Party programs from that period included bourgeois-democratic sentiments no longer in tune with Soviet zone political aspirations, and statements of support for the *Einheitsschule* from intellectuals needed to be found to counteract the assertion that this had only been a movement among elementary school teachers. Educators in the Soviet zone understood that they had to take control of their historical traditions as quickly and confidently as possible to ensure their utility in constructing an antifascist democratic narrative.

But they did not have the autonomy to act alone, as they well understood. This problem was all the more obvious because of the presence of foreign military governments that retained the right to veto any plans. The approval of curricula and teaching materials represented a sizable degree of compromise and negotiation among many participants. The proposed history curriculum for Berlin in 1947 illuminated this phenomenon.[51] The Germans present at the meeting considered the possibility that the proposed curriculum should not be too detailed. In this way, the Allied commission responsible for approving it would have fewer opportunities to object. They alternatively debated it needing to be longer, so that the Allies would have a better sense of the German educational planners' historical concept and plans for teaching it. One of those educators present, Herr Klesse, added another worry: "I could imagine that the Russians and French would be very skeptical of a draft about 1848 that you can turn any which way. If we let a general historical-political line shine through, we are going to have some difficulties."[52] Lack of approval by the Soviets or, for the case of Greater Berlin, the Western Allies, did more than stall the implementation of a new curriculum; it opened up the possibility that the occupying powers might doubt Germans' ability to take charge of their own social reforms.

A 1948 school board conference in Dresden succinctly announced the triple meaning for the memorial year 1848. It included the March Days of the failed liberal Frankfurt Paulskirche Revolution; the birth of the workers' movement with the publication of the Communist Manifesto; and the first meeting of the *Allgemeine Deutsche Lehrerversammlung* (All-German Teachers' Assembly), which had proposed a "progressive school program."[53] The conference proudly claimed Dresden as the site of the visionary teachers' meeting, in which participants a century earlier had called for a unified school system fully integrated into the state apparatus. As it turned out, Dresden's educational ministers asserted too much. The revolutionaries had actually convened in Eisenach and issued their demands a month later.[54] Dresden's honor was a slightly different one. Throughout 1848, several areas of Germany witnessed the founding of local teacher associations that called for educational and societal reforms. In August 1848, 900 Saxon teachers attended the second meeting of the teacher association in Dresden. This meeting appealed to all teacher associations to join together in the creation of a General Teacher Association.[55] When German educators began to work together as a unified group, it certainly counted among the key dates in German educational history. But it was not the one that most clearly emphasized the school's role in national unity.

Even in this slightly misguided case, antifascist educators understood the implications of linking the All-German Teachers' Assembly in Eisenach to the Frankfurt Revolution. The clear message was that the school had been part of German revolutionary struggles, and that it should continue to play a significant political role in restructuring society. Antifascist educators then took advantage of every moment during 1948 to explain the necessity of realizing 1848's programs. At the 1948 Dresden conference, education administrator Viehweg insisted that it should not just be a year for remembering, but also for reflection (*Besinnung*). He informed his audience that every teacher meeting, conference, or other related activity should have a brief and "(naturally psychologically cleverly presented) moment of reflection" about the meaning of the revolutionary year.[56] At a conference in December 1947, Soviet zone educational ministers unveiled twenty illustrated panels delineating the developments leading up to 1848 for display throughout the zone.[57] One panel depicted the goals from 1848 that 1948 would finally realize: "land reform, the defeat of capitalist monopoly (*Großkapital*), and democratic school reform." Other panels noted important historical facts: in 1847, Germany had more kilometers of railway than either England or France; in 1833/1834, Germany began its slow march through the new capitalist economic order with the founding of the customs' union (*Zollverein*). Thuringia's education minister Marie Torhorst suggested a separate panel for the All-German Teachers' Meeting in Eisenach, and another minister saw the need for three more panels to show political developments up to 1947. "[W]e have to say that now, 100 years later, we must accomplish everything that we failed to do during the entire last century."[58] A sense of determination and a desire to control the Soviet zone's destiny characterized the majority of these speeches and undertakings, but these statements also bespoke an uneasy feeling of desperation. To make 1848 unfold a century later, educators would need the full engagement of the Soviet zone population.

Accordingly, the month of March, when the "March Days" of Paulskirche had occurred, saw a large number of "revolution festivals" in most schools and classrooms. School choirs sang the 1848 revolutionary song "Die Gedanken sind frei"; elementary classes recited poems from the nineteenth-century poet of German unification Hoffman von Fallersleben; and graduating classes in 1948 listened to speeches about their responsibility to implement 1848's goals.[59] The actual lessons about 1848 that pupils heard in the classroom, however, often worried educational administrators. An educational committee, upon observing an "older gentleman" teach a lesson on 1848 at an upper secondary school, stated that his reactionary approach to history suggested that he had clearly not accepted democratic ideals. This failure, the administrators feared, indicated that he would be incapable of winning pupils over to concepts of democracy or freedom.[60] Pupils had ample opportunity to hear the "correct" antifascist interpretation of 1848, though. For the entire year of 1948, pupils and teachers were bombarded from all sides with the same antifascist revolutionary message. No one could have escaped the numerous exhibits, presentations, and posters commemorating 1848, all claiming that 1948 was the year to achieve the democratic goals of 1848.[61] Even during the last months of 1848, educational administrators reminded teachers that the commemorative year had not yet ended. The first history teachers' journal, which appeared shortly before the fall semester of 1948, was a double issue devoted to 1848, including a report on how student teachers at Berlin's teachers' college had celebrated the year.[62] By the end of 1948, educational administrators had not only incessantly presented their message of 1848 to the educational community; they had also suggested that seemingly everyone had participated in the commemorations.

But educators in the Soviet zone were unable to create a truly united German teaching front. In February of 1948, the first combined East–West zonal teachers' meeting took place in Stuttgart, reminiscent of the Eisenach conference a century earlier in its desperate hope for societal change. The meeting was overshadowed by polarizing differences of opinions before it even began, leaving it as unsuccessful as its predecessor had been a century earlier. Educators in the Soviet zone, while hoping for an opportunity to find common ground, also understood that they were fighting an uphill battle. The thorough preparations undertaken by the Soviet zone for the conference show that Soviet zone representatives expected a struggle. Even the title of a report written in January 1948 on the anticipated arguments from the Western zones against the Soviet zone school reform referred to "our school reform" instead of making general arguments against the *Einheitsschule*. The author clearly acknowledged the limited support that existed in the Western zones for what everyone seemed to agree were in fact primarily Soviet zone educational programs.[63] He also complained that the "new school" suffered from the same kinds of criticisms as had earlier school reforms: "All arguments being raised today against our 8-year elementary school were also raised against the Weimar elementary school."[64] He insisted, however, that Weimar school reforms had led to an improvement in educational achievement. Yet such statements were vulnerable to the retort that perhaps these reforms did in fact lead to catastrophe by creating a terrain that was open to Nazi propaganda.

Political events soon overshadowed western support or rejection of the "new school." In June 1948, the Western zones accepted currency reform as part of the United States-sponsored Marshall Plan (a major financial program for war-torn countries that ultimately excluded communist-ruled countries).[65] The Soviet zone intellectual Alexander Abusch described the currency reform as "the intentional consequence of a determined policy" of promoting a West German state.[66] West Berlin quickly began to disengage itself from East Berlin institutions, including from the joint school administration. In November 1948, Ernst Wildangel reported furiously that the carriages of eight typewriters in the Berlin central school board had been removed, rendering them useless, and that his official car was never returned after being driven to the workshop for repairs.[67] In a confused discussion about rumors of the establishment of a separate West Berlin educational administration, the departmental directors made arrangements for how to protect their files and remaining possessions, such as typewriters, if not the car. Wildangel well understood how much damage could be inflicted upon his organization's ability to administer schools if goods and personnel continued to seep out of East Berlin. He remarked, "Although they have practically paralyzed us, I would like to prevent it from happening technically."[68] Educators began 1948 with the hopes of finding more inroads toward national unity, and finished the year with plans for how best to protect typewriters from being stolen for use in the new West Berlin school administration. At the end of the century 1848–1948, German unity remained unresolved.

Goethejahr 1949

The final year of the Soviet zone era, 1949, was a pivotal year of increased demarcation between Western and Soviet zone cultures—even while claims of a single, larger national culture continued to be heard. The "Goethe Year" of 1949 continued the Soviet zone tendency of demanding cultural unity, while working toward a viable antifascist state if unity should prove impossible. In the first weeks and months following the war, finding acceptable cultural icons on which to base a cultural reconstruction campaign had been a delicate task. Only a small number of intellectual heroes could be accepted unquestionably. Johann Friedrich von Goethe stood at their head. In East and West, Germans turned to him as a shining example of native genius that survived the war unscathed. His 200th anniversary conveniently provided Weimar, where Goethe resided for the majority of his adult life, with the occasion to claim one of Germany's unquestioned cultural heroes for the Soviet zone. With Goethe societies throughout the world—the English Goethe Society was founded in 1886, one year after the first one at Weimar[69]—Germans could rest assured that the western world approved of their selected cultural icon. The first images of Goethe promulgated in the Soviet zone reflected his Germanness, and their primary function centered around presenting Goethe as proof that Germany's past was not barbaric. Later, Goethe appeared as a specific antifascist democratic icon, a forerunner of socialist thought. Goethe offered the Soviet zone with a German hero to celebrate, one who could be used to promote either German unity or Soviet zone ideology, depending upon the desired political message.

The various brochures and books published in commemoration of the Goethe Year bestowed a hollowed status on the poet. For a society officially separating itself from Christian iconography, Goethe provided a substitute. More specifically, postwar Germans needed a new secular figure to worship, one to replace the cult of Hitler.[70] A wealth of materials provided Germans with guidelines for the correct appreciation of Goethe as well as the opportunity to share stories of how Goethe had affected them. These books and pamphlets presented Goethe not only as an author to be read, but as a hero to be praised. The handbook "The Way to Goethe," published by the Schwerin House of Culture, gave instructions on how best to commemorate Goethe, including guidelines on how to sing songs he had written and find appropriate music to accompany his poetry—preferably the music of Bach, Franz Josef Hayden, or Beethoven.[71] Only German music would be appropriate to complete the image of a thoroughly German icon. The front cover of the brochure included the subtitle, "Germans tell how they found Goethe," a language reminiscent of religious testimonies. The reverent essays and notes in the brochure elaborated on "what Goethe can teach us," continuing in this spiritual vein. Other examples treated Goethe as an almost holy figure, as well: a photograph of a school class visiting Weimar in 1949 shows the pupils in a half-circle, gazing with adoration at Goethe's portrait, the light positioned to increase his other-worldly appearance. Goethe personified a secular savior of the German nation, and the festivities in 1949 praised him accordingly.

The references to Goethe had a clear political use, as well. He conveniently embodied the tension of a "natural" single nation that suffered under the imposed political division of the country, a familiar situation in Germany even in his time. As a pamphlet for Goethe Year explained, "Goethe's life achievement consists above all in the fact that amidst the inner division of Germany, he was able to portray national unity through linguistic and intellectual matters."[72] Goethe thus taught that discussions about cultural elements could be used as an auspex for discussions about politics, a practice that Germans in East and West often agreed with. The Western zone Bremen cultural minister Christian Paulmann, for instance, insisted in 1948 that the realization of cultural objectives was a precondition of achieving political objectives.[73] But Marie Torhorst put political unity before cultural unity. "The newly begun cooperation among Germany's educational ministers." she stated, "can only assure cultural unity, and I emphasize this, after political unity has been achieved."[74] In this variation, Torhorst still implied that a common culture was the most desirable aspect of German unity, but she put politics first. The DVV director Paul Wandel did not embrace either of these points of view, choosing to be franker about the real goals of talks about German unity and about the likely price to be paid for settling for cultural cooperation only. Speaking candidly, he exposed the rhetoric of focusing on culture as an unacceptable substitute for political and economic unity. In this view of a desirable German future, cultural unity could only be seen as an integral part of a larger political project, and not as an intermediate step. With this vision, he specifically rejected an interpretation of culture as being the "natural" bond that tied Germans together ethnically or spiritually.

By 1948, Wandel had ample grounds to suspect that the image of a culturally unified Germany primarily served the purpose of diverting public attention away from increasing political division between East and West. The noticeable pulling away transpired on both sides, although each zone accused the other of working against unity. Here, too, the lines of demarcation often coincided with the areas claimed as common culture. In the Soviet zone, the national figure of Goethe developed into a figure who would have been unrecognizable in the Western zones. A new volume of Goethe's works published in 1949 for Soviet zone school use, for instance, claimed to let a different Goethe speak than the one seen heretofore: "Not the Goethe that the bourgeois world has claimed as theirs until now. Nor the stiff privy counselor, the minister, but rather the man and poet Goethe as he stood in his time."[75] The editor explained how the selected writings described the poet's belief that the bourgeois pursuit of capital threatened true humanity, so that he had actually anticipated socialist humanist thought.[76] Similarly, a campaign poster for the cultural association *Kulturbund* asked voters to "work in Goethe's spirit" by casting their votes for the *Bund*.[77] These presentations of Goethe aimed at constructing a Soviet zone-Goethe, not one for general German use.

Politicians in the Soviet zone continued to proclaim the historical necessity of German unity even after the construction of two German states.[78] But the passionate speeches of politicians and intellectuals hinted more at their fear that this future was slipping away, at least temporarily. It was already too late by 1949 to use German heroes to construct the kind of German unity that Soviet zone politicians had desired. No one in a position of power could overlook the extent of German division in Goethe Year 1949. Very few areas in the political or cultural realm enjoyed coordination between the zones. In 1948, the London Conference had announced the western Allies' intention to create a West German state, and the Marshall Plan divided the country economically. Increasing division was particularly evident in school policies, in which zones refused even to recognize each other's school diplomas.[79] Given such daily aggravations in the struggle for unity, the themes of Goethe and the nation appear at best to have substituted for a more overt discussion of division, and at worse to be naive.

In spite of rhetoric about a national culture, by the time Goethe Year 1949 came around, the Soviet zone had begun to claim Goethe primarily for its own uses. Happily for educators, it was an easy task to construct multiple versions of the writer. In addition to poetry and novels, he had published political and philosophical essays that set the desired tone of public meetings in the Soviet zone. His essay "Über Produktivität und Genialität" opened the inaugural meeting of a *Volkshochschule* (community educational institution) in Berlin-Prenzlauer Berg.[80] A teacher applicant writing an essay opposing corporal punishment based her reflections on Goethe's views on civilization.[81] Schoolbooks even demonstrated how similar Goethe's progressive social views were to those of Marx and Engels.[82] Engels' skeptical view of Goethe was ignored,[83] as were Moscow's and some Germans' attempts to present the poet as a "German Pushkin" or "German Gorki."[84] In 1946, for instance, the educational journal *die neue schule* offered examples of how to use Goethe's works in classrooms to teach Germany's cultural past. The commentary surrounding the proposed lessons on the poem "Mailied" or regarding Goethe's correspondence with Schopenhauer did not offer overtly philosophical or political interpretations of Goethe's work. The focus was trained on the German Goethe. German educators in the Soviet zone did not tolerate having their heroes dictated to them from the Soviet Union or elsewhere. Nor were they willing to turn over Goethe to the West. A large literature honored Goethe's contributions to German culture, all of it marked by a Soviet zone consciousness.[85]

The association of Goethe with Weimar gave the Soviet zone an opportunity to locate him firmly within the physical territory of the Soviet zone, thus strengthening their claim on his heritage and a larger cultural tradition. Weimar had been home to dozens of artists and intellectuals at the center of German classicism—Goethe, Friedrich Schiller, and Franz Liszt. Later, under Walter Gropius, Weimar witnessed the birth of the Bauhaus movement.[86] Six kilometers outside the city had been the concentration camp Buchenwald, in which many heroes of the antifascist movement had died, including the socialist leader Ernst Thälmann, one of the first martyrs of the antifascist struggle. The celebrated "Goethe Oak," under which, as legend has it, the famous author composed many of his works, had been located near the Buchenwald prisoners' barracks. Until it was chopped down, the great oak reportedly provided inspiration for prisoners. Buchenwald had also been the home to an underground, multipartisan educational planning group that included Walter Wolf, Hermann Brill, Werner Hilpert, and Ernst Thape, who had formulated a detailed school program for use after the war.[87] That the Soviets had also used Buchenwald as a forced-labor camp for German political prisoners—many of whom were executed, as was discovered in 1990—was not a matter of public discussion. Weimar's image remained pure. Located in the heart of the Thuringian forests, whose beech and oak trees further stood as symbols of an age-old German attachment to nature, Weimar provided the perfect setting to celebrate the anniversary of one of Germany's most famous sons, who was now part of the antifascist democratic heritage.

A Communist Christmas

Quite a different kind of holiday provided the KPD/SED with an opportunity to promote socialist unity and national reconstruction: Christmas. One of the most traditional holidays on the German calendar, the event could not be ignored by socialist planners, in or out of the school. In fact, many groups in the Soviet zone participated in Christmas festivities, from socialist organizations to city councils. In all cases, references to the holiday's Christian origins were absent, while the antifascist message of reconstruction rang clear. Initial attempts to entirely rewrite the Christmas holiday as a socialist one were not successful, however. Instead, public depictions of Christmas identified it as a specifically German holiday, one centered on the nuclear family, consumerism, and the unique nature of the Soviet zone as the true Germany.

The school was an important site for discussions and decisions about Christmas. It was not only that schools taught large numbers of children about appropriate antifascist celebrations; it was also that many religious holidays traditionally dictated school calendars. The schoolboard was thus an obvious recipient of correspondence for supporters of Christmas holidays. Indeed, correspondence from a conference of bishops to the SMAD, forwarded to the German Educational Administrations, petitioned for the need to allow a zone-wide plan for Catholic pupils to have time off from school for additional religious holidays not mentioned in an earlier agreement (Christmas seemed to have already been decided as a school holiday).[88] The bishops conference based its request on a double political rationale: First, the schools in the Western zones had those religious holidays free, and the Catholics of the Soviet zone would resent not having equal and fair treatment for their children. Second, and most interesting, the letter noted that "even in the Hitler period" German children had the right to be freed from instruction on church holidays. The Educational Administration did not take the bait, refusing to justify itself vis-à-vis either the Western zones or the Nazi years. Instead, the director merely responded in a separate letter (dated after Christmas), that individual states *(Länder)* could set their own school calendars, and that children already enjoyed the right to be absent from school for any religious reason.[89]

But such logic did not lend itself to adults' and children's actual desires about celebrating Christmas. Perhaps the most explicit antifascist discussion of Christmas and children in the Soviet zone was one begun by the Organization of the Victims of Fascism.[90] Tellingly, the group's attempt to focus on a socialist connection to Christmas did not find support in the broader community. Organization members involved in the child-advocacy group "Save the Children" decided in the early fall of 1945 to organize a party to mark the first Christmas after the Nazi years. In order to garner support for it at a meeting, they shared the moving story of a wartime Christmas holiday in the women's concentration camp of Ravensbrück. In the narration, the committee members made clear the link between antifascism, love for children, and Christmas. "For weeks," the story went, women collected smuggled materials in order to "lovingly" create toys for the children, and then spent an entire day secretly preparing a festival, baking and cooking holiday treats. Since the preparations violated camp regulations and put the women in danger of punishment, Christmas itself became a socialist holiday, its celebration constituting a daring antifascist act. But the religious overtones of the holiday did not disappear, even if they were no longer about Christian salvation. Appropriating the Christian symbol of light for socialist awakening, a speaker described the effect of the scene on those children who had never seen a lighted Christmas tree: "Then they walked into the brightness of the light," the party offering the young attendees a kind of antifascist divine salvation. The group's attempt to recreate this atmosphere failed in their postwar Christmas party. According to a January report by the group, they had not begun their preparations until too late—in October, the cultural subcommittee had not given them enough support, and the unanticipated expense of the party had drained their treasury.[91] The real problem, though, was the failure of the group to understand that the general population would not accept a Christmas celebration with political overtones. Christmas was not a time to be reminded of political sacrifice or persecution.

Instead, Germans wanted a return to the prewar joys of Christmastime. Here, in case anyone was unsure of how best to celebrate Christmas, advertising made it clear: with gifts. The department store Konsum presented colorful scenes of shopping in the wonderland of communist consumerism. Girls skipped merrily along, bearing presents for the family, while children sat in awe among stacks of boxes Christmas morning, joyful parents looking on. As such images made clear, gift-giving accomplished several important postwar goals for Soviet zone society. First, most gifts involved a purchase, which helped strengthen the severely damaged postwar economy. Cities' advertising of Christmas markets demonstrates that the financial aspect of the holiday was an important one. Second, Christmas celebrations brought together parents and children, recreating the traditional nuclear family. Popular images of happy children opening up Christmas gifts connoted a bread-winning father who was no longer incapacitated by war injuries, and a mother who was able to shop once again for luxuries, instead of only waiting in long lines to exchange ration cards for necessities like flour or butter. Finally, a Christmas celebration offered a return to normalcy, with annual events practiced as they had been before the disruption of war.

In an example of this schema, the sociolinguist Victor Klemperer kept a detailed account of the 1945 Christmas activities organized in his hometown of Dölzschen, near Dresden. The difference between the first postwar Christmas and the previous year's holiday was impressive: a "pre-Christmas" party, organized by the KPD and SPD, offered a history of the celebration and a criticism of Jesus' life as a role model.[92] But the interpretation was more bourgeois than Marxist: speakers observed critically that Jesus had not begun a family. In planning entertainment, organizers avoided religious carols, choosing instead non-Christian favorites. At the actual Christmas party later that month, Klemperer noted with satisfaction that a choir nonetheless sang Christian Christmas songs despite the KPD and SPD's attempt to eliminate such religious overtones.[93] A decorated Christmas tree stood in the corner, albeit without lights, but Klemperer appreciated that at least "individual little candles burned at the tables." He was particularly delighted by the generous provisions, which included the traditional sweet bread "*Stollen*" and so many other treats that he was able to take many samples home. Klemperer embraced a traditional Christmas celebration in his private life as well, rejoicing that he had been able to give his wife a gift "for the first time in years."[94] The Klemperers also exchanged presents with their neighbors, who had invited them for Christmas Eve dinner. Such largesse made little financial sense, given both families' dire economic situations, showing the weight of these gestures to their sense of well-being. Throughout the holiday period, Klemperer painted a picture of a community that was almost desperate to find solace in its age-old rituals. Socialism may have been a renouncement of capitalism, but it did not reject consumer culture, particularly as it upheld German traditions.

Still, consumer culture in the Soviet zone was not without rules. Some toys were inappropriate for antifascist children, particularly war toys. In order to educate parents on better selections for their children, the socialist children's group Young Pioneers sponsored a Christmas market for acceptable Christmas gifts, advertised by the question, "What shall I give my child?" (see figure 7.1). The poster for the market showed the typical antifascist gender roles: a young boy in his father's arms, playing with a toy industrial crane, his sister kneeling at their feet with a book and a doll.[95] Both children wore their Young Pioneer kerchiefs, proof of their good behavior that merited their socialist-Christmas-German presents. Clearly, the right toys encouraged children to participate in the roles foreseen for them in the reconstruction of the nation: boys should learn the importance of clearing away rubble and rebuilding neighborhoods (regardless of the actual gender division of such activities), while girls tended to childcare, and the occasional literary pursuit, the antifascist father-hero looking on all the while. Christmas thus offered an important opportunity to remind adults and young people of their antifascist responsibilities.

182

Figure 9 "What Should I Give My Child?" Poster for an Exhibit of Appropriate Antifascist Christmas Gifts.

Sources: "Was schenke ich meinem Kinde," Berlin, 1949, *Plakate der SBZ/DDR* DHM CD-ROM, Inv. P 90/2376

City councils and people's solidarity associations made the connection between Christmas and the standard of living explicit. Christmas concerts throughout the Soviet zone benefited the cities' reconstruction projects. Local groups played an overwhelmingly German program of traditional favorites like Bach and Beethoven in order to raise funds to rebuild the venerable old structures of museums, city halls, and churches (see figure 10).[96] In this manner, a city's inhabitants could buy themselves the greatest postwar gift of all: a return to the city's traditional image. But Christmas was not an event to be worried about only in the winter; rather, its organization and planning took place the entire year. A February 1947 poster celebrating the reinstatement of Saxony's factory productions noted proudly the availability of Christmas trees, decorations, and gifts in that region again.[97] Wooden nutcrackers, toys, and advent scenes would help create the desired German Christmas atmosphere again—one that concentrated on goods available only in the Soviet zone that had once graced all of Germany's mantels. Like other commemorations, Christmas could be remarketed as a celebration of Soviet zone uniqueness.

Teachers also worried about being able to offer children a proper Christmas. Additional instructors came to classrooms to teach girls how to sew Christmas gifts, such as dolls, and "Save the Children" organized a host of Christmas activities. Adults fretted over how to make sure every child received at least one gift, and reached out to the community at large to help out in this endeavor. Colorful posters, such as the one for the 1945 Dresden Christmas market, proclaimed "A Christmas Present for Every Child," tempting adult consumers with images of Christmas past, including a toy train and a wooden angel with candles for the tree.[98] Here, too, images of familial harmony seemed to emanate from the objects themselves, with light bathing pleased recipients in serenity. But lofty goals of creating such scenes in every household faced very everyday obstacles, however, such as the school whose party had to be cancelled after thieves took off with most of the presents. Less idealistic adults worried more about practical gifts for children, such as blankets, shoes, and coal. While political programs designed to alleviate children's malnourishment and freezing living conditions elicited only limited sympathy among the general populace, the call to provide basic necessities received broad support when introduced as a Christmas project.

Christmas advertising and programs suggested that a traditional Christmas with candles on the Christmas tree, toys for all children, and beaming wooden nutcrackers was within the Soviet zone's reach if everyone would only work together. The focus on the heritage of the holiday

Figure 10 Christmas 1947.

Note: This poster advertises a traditional Christmas market full of nostalgia: Wooden Christmas ornaments and glowing candles atop a christmas tree.

Source: "Weihnachten 1947," *Plakate der SBZ/DDR*, DHM CD-ROM, Dresden, 1947, Inv. P 90/6222.

and the centering of it firmly within the borders of the Soviet zone lent Christmas an air of Germanness mixed with socialist humanism. The apolitical holiday turned suddenly, if implicitly, political. With this strategy, the KPD/SED successfully coopted one of the most significant Christian holidays, redesigning Christmas as the rebirth of the antifascist German nation.

*　*　*

The ghosts of Germany's pasts—from Goethe, Paulskirche, and Saint Nicholas to a host of previous educational ideas—had been summoned by the decision to distance the Soviet zone from its Nazi past and, later, from the Western zones' school programs. The Soviet zone interaction with these specters called constant attention to the German nature of educational and cultural reforms, but also to their specific antifascist democratic character. Antifascist educational administrators seemed genuinely to hope for reconciliation with the West, but they also prepared the legitimacy of their half of Germany by laying claim to historical events and characters. In spite of the many optimistic and emotional speeches about the school's contribution through culture to German unity, other, more pessimistic voices had been heard even before the decisive establishment of two German states in 1949. Discussing the upcoming 1948 teachers' conference in Stuttgart, one educator expressed doubts about how much schools could even accomplish: "[T]he school is the institution that suffers most under the sociological shifts and that most clearly reflects them," he argued. "It is therefore a microcosm, or better said, a micro-chaos."[99] During the conference itself, an educational minister from the Western zone region Baden-Württemberg expressed surprise at how quickly the four zones had grown apart in all areas, especially in educational matters.[100] Educational administrators in the Soviet zone were not surprised by such signs of division. If they had hoped to change the course of political tendencies, they nonetheless had been constructing an antifascist history that would position the Soviet zone as the legitimate German nation state.

The debates surrounding appropriate historical curricular material overlapped with decisions about suitable cultural traditions for the Soviet zone as the heir to German traditions. Educators in East and West who meanwhile continued to strive for German unity did so while observing the daily construction of new obstacles to this goal. Many of these individuals did not give up hope, but neither did they see the only possible outcome of their projects as German unity. As they selected

German traditions for schools to teach the youth of the Soviet zone, the purpose of this exercise developed a primarily legitimating function for an antifascist democratic Germany. Some attempts to appropriate German history, such as the commemoration of May 10, apparently failed in the face of Soviet opposition. In other instances, including 1848, Goethe Year, and Christmas, the icons of a usable German past found an antifascist turn in one half of the nation. Consequently, antifascist educators' reconstruction of a "common" German national culture ultimately reified the nation's division.

Conclusion: Redemption through Reconstruction and Beyond

On May 31, 1950, the young GDR citizen Brigitte Reimann wrote a letter to a girlfriend in the Federal Republic, expressing both jealousy and pride in the reconstruction of her own part of Germany: "Sometimes I could really envy you guys over there a little— but only sometimes, like whenever the work starts really pouring down on our heads. Good grief, if you knew how much we work over here! But we'll do it, you can count on that!"[1] The sixteen-year-old Brigitte and her skeptical friend Veralore in the West both knew what "we'll do it" referred to: the construction of a model antifascist, democratic, socialist-humanist German state. By the end of the year, however, the first cracks appeared in Brigitte's avowals of loyalty to and love for the GDR. One day during school, one of her classmates was arrested, and no one was even certain who had done the arresting.[2] Brigitte insisted that the boy could not possibly be guilty of disloyalty to the state. The boy's mother had no news of his fate, and the class attempted unsuccessfully to locate him. Brigitte claimed that she cried desperately for days at school and even had to be physically pulled away by her friends when she began to scream and curse at the commandant who visited the school. This became the first instance of many expressions of angry disappointment in the socialist state throughout her life. "So you see—and then something in me broke," she wrote to Vera. "A belief, if you want to call it that. Why would anyone stain a great thing with—perhaps with the blood of someone still half a child?"[3] This combination of pride in the antifascist project with a continual reminder of the brutality of the regime underlies any assessment of the GDR as it slowly emerged out of the postwar years to become a key figure in the Soviet bloc.

On October 7, 1949, the Soviet Occupation Zone ceased to exist and the German Democratic Republic was born. The period of "antifascist democratic transition" had officially ended for the school; the era of Marxist-Leninist socialist pedagogy had begun. Germany's internal borders, however temporarily, had been clearly drawn. Pupils like Brigitte competed openly with their peers in the West for the claim of living in the better Germany. They also expressed hurt and anger when their new state betrayed their trust. The antifascist classroom had closed its doors, but pupils carried its lessons into the GDR.

Redemption through Reconstruction

The "new school" was part of a massive ideological and physical reconstruction project that aimed to locate the Soviet zone securely in the stream of history. Indeed, the "new school" was not entirely new. A large part of the personnel and pupils remained in place, as well as most institutional connections between the school and society. As a constellation of historical pedagogical and ideological practices, antifascist education was a project for realizing educational reforms dating back to the French Revolution. In turn, the "new school" could claim a dignified and progressive heritage. Part of this legacy involved the historical project of German unification. The antifascist school sought to unify the German nation with an ambitious program of national education, thereby defining membership in the nation in rational, political terms with a learnable, cultural base. At the heart of these concepts was the uniform schooling of citizens in the *Einheitsschule*, so as to erase class, confessional, and gender divisions.

As is clear from the "new school," culture and tradition are powerful forces in schooling as historical agents in programs of social change. Despite popular misconceptions, the school is not a puppet of the state. At the heart of a school's mission is to teach young people to think and learn. Pupils' successful acquisition of critical thought, however, ultimately frustrates, at least to some degree, the attempts of societies to control narrowly pupils' understandings of values and norms. Schooling provides for an interaction between adults and children to construct jointly a vision of their nation, as is visible in the Soviet zone case. Young people have surfaced throughout this work as conscious actors in the reconstruction of their nation and everyday lives, aware of their potential contributions to the emerging new nation as well as their right to question its tenets. By requiring and enabling mandatory, universal school attendance, the emerging East German state provided its younger citizens with the possibility of helping define society. The internal organizational structures of schools also determined the limits on the Soviet zone political elite's ability to enlist the "new school" in SED service. Lag times for producing and delivering new curricula and textbooks are factors that affect all school systems, and that consistently disadvantaged rural areas over urban centers. Diverse pedagogical approaches by teachers also gave pupils differing classroom experiences; here the tensions between the Weimar-era *Altlehrer* and the newly trained *Neulehrer* must be noted. Finally, pupils' own individual experiences influenced how they understood and incorporated school lessons into their beliefs about their place in the new Germany.

Beyond the framework of referencing historical projects, what was the mission of the "new school"? First and foremost, Soviet zone educational reformers genuinely desired a unified, antifascist, democratic German nation, although a wide difference of opinion existed among them about the likelihood of this outcome. They agreed that the only viable path to this goal lay in a unified national educational system. Its classrooms would produce shared experiences and bodies of knowledge, enabling in turn the creation and maintenance of a unified political and economic system. Suprapolitical decisions and circumstances beyond the direct control of German policymakers in the Soviet zone combined with the emergence of a clear Soviet zone school program, causing the educational experiences of the Soviet zone school community to be distinguishable from those of the Western zones. The "unity school" of the Soviet zone succeeded in educating its pupils to become Germans who were proud of the German nation, but the nation with which they were learning to identify had a distinct antifascist perspective on daily life.

This reevaluation of the school in the Soviet zone changes significantly the traditional view of the role of antifascism in postwar German life. The antifascist education that was elaborated and practiced in schools was not a state ideology dictated by Moscow for controlling its satellites. Educational reformers in the Soviet zone consciously fostered the German nature of their antifascist educational reforms. They were unable, however, to forget the Soviet Union, which lurked as a shadowy presence and interfered irregularly. Nonetheless, rejecting both Soviet and non-German Western zone school models, antifascist educational reformers selected those aspects of German educational history that most clearly demonstrated a socialist-humanist nation as a logical consequence of the redefined past. At the same time, they fought passionately for a unified German nation. They dated their historical reforms to a time even before World War II. By defining German unity as a political, cultural, and class construction, they claimed to be fulfilling a more progressive version of the 1848 desire for unification, with allusions as well to the French Revolution. Determined moreover to rebuild the national culture that they did not adequately defend against Nazism, antifascist educators initially agreed to put aside political differences for the good of the nation. Their growing sense of desperation at the apparent hopelessness of their cause did not dampen their commitment to national unity. It did push them to realize their programs in the Soviet zone, where they at least could influence the system, all the more strongly. Their plans for German unity thus became the means of creating a specific Soviet zone experience that pulled the eastern half of the nation away from the western half.

Antifascism was central to an emerging German national consciousness that was specific to the Soviet zone. In the absence of clear temporal or spatial borders, antifascism provided a means to map that part of Germany ideologically. Pupils, teachers, and parents worked together to reframe their personal memories of wartime experiences and their collective memories of Germany's past, using the narrative of antifascism to position themselves and their half of Germany as the corrective to Nazism and fascist ideology. The possibilities for accessing the nation through local channels allowed communities to partake in the national educational experience with local variations. Pupils who lived in border areas that changed occupying hands were only one example of specific local experiences that were not shared by the entire zone. Postwar conditions also exacerbated the traditional disparity between rural and urban schools. A destroyed city such as Dresden provided young people with a different atmosphere for learning about the new nation than did smaller rural towns in Brandenburg or Mecklenburg. Those areas generally saw little physical damage, but the waves of refugees from the East that came through their towns and classrooms, as seen in the case of Gunter de Bruyn's school, restructured significantly their demographic landscape. The national consciousness that Soviet zone Germans were offered and that they helped to construct actually comprised multiple experiences in the new nation, and thus multiple definitions of an antifascist, democratic Germany.

The school was in a unique position to facilitate this process. As a central and recognizable building in most communities, it was the local space of a national educational system. Membership in the new nation was won in a building whose construction and maintenance was supported at the regional level. In cities and villages where the school building had survived total destruction during the war, which was the case in the majority of communities, its familiar presence represented the possibility to adults and children of returning to a normal life. The school building did not become part of monumental reconstruction projects, nor was it the object of innovative architectural designs, despite educational reformers' constant references to erecting an entirely new building. Teachers, parents, and pupils arranged for replacement windowpanes, swept out the dirt, and made curtains to hang in the classroom. A clean, refurbished building represented the link between past educational programs and new curricular directions, the physical symbol of educational reformers' determination to position the Soviet zone as the true inheritor of the German nation.

The school was an important stage for adults to discuss the reorganization of German society. Debates about appropriate girls' and boys' education were debates about new gender roles, and coeducation as a national strategy represented a major realigning of traditional social relationships. Antifascist education meant that women were no longer disenfranchised from the nation, although this subject has not been adequately investigated.[4] At the same time, images of youth pointed to adults' ambivalent feelings about Germany's future. Coeducation symbolized, in theory, the possibility of erasing gender boundaries. Yet gender stereotypes remained strong. Girls were seen as more vulnerable to dangerous political movements, while boys continued to be the object of disciplinary actions. Young people's perceived immaturity also worried adults. Although they asked young Germans to take on the responsibility for the new nation, adults also feared that putting Germany's future in children's hands could only lead to disaster. Adults portrayed children not only as innocent and energetic, but also as corrupted and vulnerable. The intertwined themes of venereal disease and national health did not surface in the same discussions about fraternization, but they were topics being talked about in similar circles at similar points in time. Their overlap points to contemporary veiled references about consensual and nonconsensual sexual relations with Soviet soldiers, but only further investigation into medical and police records can determine if this connection was being made by Germans at the time, rightly or wrongly.

"No institution is more important to a people's being than its schools and the ways it has devised to raise its children," wrote Sterling Fishman in 1995 in his criticism of persistent western ignorance about the former GDR school system.[5] Schools alone, however, cannot determine the personalities and attitudes of a nation's citizens. Instead, they must cooperate with other institutions or suppress them. In the Soviet zone, the school led the way in organizing social life, both locally and nationally.

Because it involved parents in the schooling of their children in a variety of ways, families did not feel threatened by the state's near-monopoly on education. The institution of the family had remained relatively intact after the war, not ceding additional educational responsibilities to the school. The cooperative stance on both sides thus proved crucial for the success of the "new school."

An "antifascist turn" taken by educational reformers from the Enlightenment to socialism provided key reference points for antifascist education. The socialist cultural heritage combined Enlightenment rationality and German classicism, creating literary and cultural traditions in line with socialist narratives and goals. Goethe was the most obvious figure in this constellation, but the canon in its broadest sense expanded to include a broader variety of works than the favorite sons, like Gottfried von Herder and Heinrich Heine.

Statements about the leading role of the SED in Soviet zone educational policy must be nuanced.[6] As evidenced by a variety of educational discussions and decisions, the composition of the antifascist teaching administration and staff cannot be reduced to questions of party affiliation. SED membership did not imply a single educational philosophy. Former socialists and communists traditionally disagreed on the desired social arrangement for Germany, as well as on specific pedagogical issues. But even here, the lines cannot be so clearly drawn. Many communist educational reformers had worked closely with their social democratic colleagues before the war, and there were numerous examples of educators whose initial membership had been the SPD before they joined the KPD. It is difficult to determine the shades of teachers' political ideologies. The pedagogical struggles between teachers, particularly between *Altlehrer* and *Neulehrer*, demonstrate that they, no more than educational administrators, were not a uniform group. Articles and class essays by teachers and teacher candidates point to a mixture of political and pedagogical beliefs combined with a strong commitment to the teaching profession.

The question then remains: were antifascist educational reformers entirely out of step with the SED elite's plans for a socialist Germany? Were they naively convinced of the power of schooling to overcome political division? The answer here necessitates a longer view. Four days before the 1948 currency reform in Berlin, Heinrich Deiters (SPD/SED), head of teacher training in the Soviet zone, wrote a colleague in Vienna that he hoped that the currency change would bring about "clarity," with Berlin and the Soviet zone experiencing a new economic upturn.[7] In June 1950, he wrote to a colleague in Stuttgart, insisting on the importance of continuing the dialogue between eastern and western educators in order to bring about a new, unified Germany.[8] Twenty years later, in an unpublished autobiographical essay, his idealism at the time is even more evident. He had assumed that the creation of two states was a preparation for a subsequent new single state.[9] Further, expressing frustration with the state of SED politics, he had "held tightly to the hope" that the future unification of Germany would bring about the harmony of the "idea and reality" of socialist politics.[10] Deiters firmly believed in the power of education to enact major social change. Antifascist educators did not see themselves as dependent upon Moscow's or Berlin's plans for Germany; they were convinced of the righteousness of their cause, and of the strength of the "new school" to change all of Germany. There is strong evidence that this educational culture persisted throughout the GDR; only months after the opening of German-German borders in 1989, a group of GDR teachers traveled to Hamburg to compare educational systems and discuss reform strategies.[11]

I am no apologist for the GDR and am glad of its disappearance. Nonetheless, arguments that its government was fundamentally illegitimate and similar claims that recent social and cultural historiographical trends do not recognize that the GDR was a dictatorship are too often political interpretations that should not play a role in a study of the formation and execution of educational policy in the first years of postwar eastern Germany.[12] I am unconvinced that the question of state legitimacy brings us any closer to understanding the complicated nature of political and social relations between the Federal Republic and the GDR. Certainly, many conservative West Germans and U.S. citizens refused to acknowledge the GDR as a separate state, and the drafters of the 1949 West German Basic Law—that state's constitution—extended (West) German citizenship to those Germans unable to participate in the constitution's ratification (i.e., those Germans residing under Soviet occupation). By 1972 at the very latest, with the West-East German Basic Treaty, the community of nations—including the United States—had accepted the GDR as a sufficiently legitimate state for political, economic, and social interaction on multiple levels. Its admittance in 1973 to the United Nations alongside the West German state should alert historians to the questionable political stance of interpreting GDR society exclusively vis-à-vis the complex question of state legitimacy. Post–1989 reflections on the degree to which the GDR could steer its own policy must furthermore take into account the fact that the GDR government was headed by Germans, and that it was East German police and military units that arrested and killed those East German citizens who attempted to flee the state. Historians must also not forget that a similar ongoing historiographical and political argument about legitimacy also existed until 1990 for the Federal Republic that, after all, remained in practice an occupied country and whose government answered to its western occupational authorities. Certainly, the United States, Great Britain, and France did not assert their authorities over the Federal Republic and West Berlin in the violent manner of the Soviet Union over its satellites, but we would be lacking in our historical responsibilities to deem one country more legitimate than the other based on political beliefs. International political law did not do so, and to now invalidate the international community's and West Germany's implicit recognition of the GDR as a separate state is an historical interpretation informed by ideology rather than evidence. The Federal Republic and the GDR had a complicated and ambiguous relationship. While it is true that East Germans who fled to West Germany throughout the cold war and immediately after the 1989 *Wende* (transition) could immediately receive West German citizenship and some support to help them begin their lives in their new state, the Federal Republic did not have a ready policy for the eventuality of all GDR residents claiming West German citizenship.[13] The massive confusion that followed the opening of GDR borders is testament to this citizenship policy having been largely an abstract ideal. There is no doubt that the GDR state violated human and citizen rights and that its government was ossified, corrupt, and unfailingly sadistic. But these facts should not be allowed to reconstruct the four and a half years of the Soviet zone that preceded the GDR as a period of "failure" or as a dress rehearsal for the GDR's later policies.

It would moreover be a mistake to make a direct connection between the long-term failure of the East German regime and the educational policies and practices of the school system. If one were to argue in that vein, however, it would be necessary to point out the failures of the West German school system to address social tensions that later erupted in schools and universities in 1968 and beyond, causing massive political upheaval and a social-wide questioning of that government's legitimacy. Moreover, as I have mentioned earlier, the GDR initially looked to the U.S. comprehensive school system as a role model. The U.S. educational reformers hoped in vain to convince the West German educators to do the same. Although U.S. politicians devoted their "educational" policies to limiting the encroachment of communism rather than reconceptualizing the role of education and civil enculturation, educational policymakers were more successful at introducing student governments into

West German schools.[14] There is no need to hold up the Soviet zone and later GDR educational systems as worthy of direct emulation; nonetheless, there was much to recommend these schools, such as a first step toward total coeducation (not an occurrence in the Federal Republic until the late 1960s and early 1970s), an early, postwar attempt to democratize rather than merely denazify pupils, and a true commitment—however unevenly implemented—to equalizing educational opportunity for all German citizens. These commendable goals were set between the years of 1945 and 1949.

Was, then, the school successful in creating antifascist, democratic, socialist, German citizens? The response here depends upon the scope of the question. If the desired information refers to whether young people became automatons for the SED elite, the answer is an unambiguous no. Pupils adapted themselves to the multiple contradictions of their situation, but continued to comment on confusing developments critically in their schoolwork. They found multiple, viable means of protest against programs they disliked, whether by not choosing Russian language instruction, or by attending dance activities instead of political-cultural evenings. In other words, in spite of difficult and painful daily struggles for survival, they continued to lead children's lives, which included an awareness of their own agency in determining their everyday lives. Historians or sociologists writing about young people living in traumatic times often describe them as having to grow up quickly, becoming adults too soon.[15] Children in the Soviet zone definitely had to learn physical and emotional survival skills that do not correspond to many people's childhoods, including assuming responsibilities traditionally associated with adult life. The hardships they endured forced them to become aware of the political and social environment in a way that otherwise would not have been necessary. To remark that their world had been turned upside down would be an understatement. Parts of the city in the Soviet zone posed hazards to children unthinkable in peaceful times, such as the young boy who watched his friend blown up by unexploded munitions left in an abandoned car.[16] Children in the postwar era, however, played in traffic and dangerous construction sites out of curiosity and a search for adventure. After experiencing disruptions in their childhood, they set out to reclaim it. Children who watched the Soviets Friedrichshain bunker in Berlin blow up in the summer of 1946 ran over to it immediately afterwards, partly to gather wood for fuel from the ruins, partly to play on the hill of rubble.[17] By connecting a child's act of play to the adult one of gathering fuel, these young people began the work of claiming and reappropriating their own experiences and lives.

If, however, the question is directed at finding out if young eastern Germans thought of themselves as young Germans who were different from their peers in the other half of Germany, the answer is yes. Pupils constructed their part of Germany with pride, and internalized significant elements of the antifascist interpretations of the past and the future, particularly its focus on the victim-turned-victor status of Soviet zone Germans. Required by the state to attend school, young people seized the opportunity to take an active part in extolling the virtues of their half of Germany. They separated their religious lives from their school lives, and rethought negative images of their Soviet liberators. Blaming capitalist economies for social injustice, they praised the worker as the hero of the system. They retained gender-specific behavior to a large degree; although even here, the coeducational school system provided for some long-term changes. Their teachers, too, participated in and reflected the construction of antifascist education. Intense pedagogical debates were obvious signs of their commitment to their profession, but they left numerous other traces of their private work in the public antifascist project. In their essays and in written comments on pupils' assignments, they consistently demonstrated a sense of the importance of their task in reeducating Soviet zone youth, and also their own struggles to ensure that they were teaching the right lessons.

The real answer to the question of how successful the school was can only be a qualified one. The antifascist classroom was a German, socialist-humanist learning environment, redefined hourly and daily by its inhabitants. This atmosphere was perhaps less rigidly socialist and too recognizably German than its observers in Moscow or even some offices in Berlin might have desired. But it was also considerably more socialist and not German enough for many observers in the West. While some members of the Soviet zone community worried about German division, others continued to expect unification. Most Germans in the Soviet zone school, though, young or old, were most concerned with the reconstruction of their daily lives and the reestablishment of a connection to their nation, regardless of its form. Considerations about the long-term ideological and geo-political effects of Allied Command and German policies took second place to policies that emphasized immediate change or improvement. In this process of constructing the everyday environment according to a perceived antifascist framework, whether in the classroom or in other familiar arenas, German antifascist national consciousness was born in the first months after the war. It would take decades for GDR citizens to elaborate it—and its fault lines—fully.

Beyond 1949

I have discussed the emergence of an antifascist collective consciousness among Soviet zone young people; the obvious follow-up question is, what happened after that? What was the legacy of the "new school"? How did those children and adolescents at the end of the war, who were in their mid-thirties for the violent suppression of the Prague Spring in 1968, and in their fifties when the state with which they had grown up collapsed in 1989, react to the turbulent political and social events that accompanied their lives? Similarly, what of their children? What did young people born into GDR socialism during the cold war inherit from the first antifascist generation? Some tentative answers to these questions help illustrate not only what happened to antifascism, but also allow us to view those first antifascist years as formative for a GDR national consciousness.

The end of the Soviet zone administrative period thus meant an end for the "new school," whose main purpose had been to replace National Socialist thinking with antifascist, democratic, socialist-humanist thinking. The next phase was to create a true socialist society, and educational policymakers needed a different type of school to achieve these goals. Indeed, the end of the optimistic "new school" was already being spelled out in the second half of 1948. As the currency reform laid the groundwork for a separate West German state, eastern German educators began planning for an educational system that no longer taught "unity" as a desirable goal. Although no concrete plans were initially in place for a new educational system, administrators noted the need for a new policy direction as early as the summer of 1948.[18] The next years would see an increasing embrace of Marxist-Leninist principles and Soviet pedagogy, with the SED at the head of all school policy, and a rejection of progressive Weimar-era reforms as bourgeois. Educational policy became a formal instrument within the socialist planned economy, which ended the school's function as a center of humanist learning. Educational administrators also implemented Soviet-style reforms, mirroring developments in other social institutions. By 1951, Russian lessons were mandatory. Marxist-Leninist ideology was the basis for all classes, in particularly German, history, and social studies. Beginning in the mid-1950s, the eight-year obligatory school program was extended to ten years, at which point students could continue in vocational training or further education. By 1956 these schools received the new designation of "polytechnic middle schools." Two years later, the construction of a truly polytechnic institution was undertaken more fully, making the GDR school system the pride of the Communist bloc for its commitment to erasing the separation between physical and mental labor and to serving as a key component in the education of socialist citizens. The final reforms were in place in 1965, including an additional optional two-year upper secondary school track that prepared students for university admission.

If at first glance little seemed to remain of the revolutionary "new school" by the last years of the GDR, the reality is more nuanced. Coeducation, free schooling, and secular instruction remained two hallmarks of East German education. The commitment to a "unified" education in these areas did not diminish during the half-century of the GDR's existence. In other areas, education was no longer as unified.[19] Although most pupils attended the ten-year general polytechnic school, a few did not. Mandatory school attendance actually ended after the eighth grade, and a few pupils could decide at that point to leave the school and receive vocational training. A small minority—less than 3 percent of pupils—did not enroll in the traditional polytechnic school program, instead attending "special classes" or "special schools" in certain areas, including classical and modern languages (in particular Russian), mathematics, sports, and sciences. Created to support advanced education for gifted students, the elitist nature of these special schools and classes left them open to criticism for the threat they posed to the unified education principle. The dream of an entirely unified educational system that would school all Germans identically, allowing them equal access to the nation, was never fully realized. As an ideal, however, the concept of a unified school continued to surface through the last decade of the GDR's existence.

Perhaps most enduring was the sense of antifascist "Germanness" that the "new school" helped foster among its pupils, but also a feeling of the normalcy of their lives. In essays and interviews from later years throughout the GDR, themes echo many of those ideas written by Soviet zone pupils at the end of World War II. Gone by the first decade after the war, of course, was the daily struggle with hunger and cold. Young people's focus on their family lives had not changed, though. Even in times of more economic prosperity, children looked to their parents as important actors in their lives. In 1964, for example, a group of pupils wrote essays about what they wished their parents would do better. The majority complained that their parents worked too hard and thus had little free time for the children—a complaint not unknown to other modern cultures, but indicative of the continued importance of the family unit even under socialism, despite that state's focus on the public and the collective.[20]

Almost forty years after the end of the war, the conscious reflection upon what it meant to reside in the eastern half of Germany was present in young people's writings. In 1984 two East German authors collected over 1,500 essays about GDR life from pupils aged fourteen–nineteen years of age, about a hundred of which appeared in book form. One recurring leitmotif included the state as a caring and moral one. "The GDR as socialist state, yes, it is a real social state," mused the vocational student Katrin in 1984 about the opportunities for the elderly to find support in society.[21] Likewise, in interviews conducted with East and West Berlin children from 1981 to 1982, the children's psychologist Thomas Davey discovered that young people in the socialist half of Germany believed that their society was a fairer one than that of the Federal Republic. But the state was not the only institution that socialist young people paid attention to; they were also very concerned with their families. The above-mentioned Katrin, after naming different senior citizen organizations that the state provided, realized that her own grandmother still needed the family more than anything else: despite the girl's obvious identification with her state, she recognized her family as a key source of strength for her and her loved ones. Claudia, a tenth-grader, made the presence of the family against the background of a confident state even clearer.[22] She told of her grandfather's and great-grandparents' arrest by the Nazis and internment in concentration camps. Her great-grandparents were never heard from again; it was only a "fortunate coincidence" that her grandparents survived the war. There is mention of a Jewish wife in her narrative, but it is unclear whether Claudia or her family practiced any religion. What is clear is her belief that her "ancestors and those who shared their fate fought for and dreamed of a better world," one made possible by the Soviet Union's "liberation of the German people from fascism." This wording was more than an empty phrase. The idea of earlier generations having sacrificed themselves for the good of their children and grandchildren was at the heart of her moving essay, leading her to recognize the opportunities she enjoyed that her grandparents did not: "I live in socialism, and I expect a happy and fulfilled life." Her faith was a secular, historical, antifascist one that lay in her state and its political system. The family remained separate from but linked with the development of the GDR.

Within the family, the pupils' essays from 1984 suggest a continuation of the gender roles that their postwar peers knew. Although the GDR boasted of equality in schools, it is clear that formal educational policy alone did not resolve persistent discrepancies in gendered divisions of labor. In ninth-grade Bärbel's description of the hierarchy of family discussions at the end of the workday, it was the father who spoke a "word of power" that allowed him to always talk first, followed by the mother and then the children.[23] Throughout these essays, mothers bowed to paternal authority while playing the role of nurturer. Kristin, a twelfth-grader, described her mother as responsible for "harmony and happiness" in the family.[24] Similarly, when it was time for a family picnic, Jens's mother packed the basket with coffee and juice and baked a cake, a role that only she—along with other mothers—could adequately fulfill in everyone's mind.[25] Other gendered traditions remained, as well. Mothers such as Steffen's baked Christmas cookies, filling the apartment with a wonderful smell "for days"; the "highlight" of the holiday was the father's search for the perfect Christmas tree—even in a country in which girls participated in military drills in schools, no one expected a mother to wield an axe in a winter forest.[26] In case there was any doubt about the masculine nature of this holiday excursion, Steffen described vividly how the father "arm[ed] himself with saw, file, and chisel" for this project. Scholarship in the GDR itself bore out findings that young people still learned that the traditional nuclear family was the ideal: a 1989 study by a female GDR citizen of the textbooks for the third through eighth grades demonstrated that working women were shown primarily in traditional female professions, such as the service industry and teaching. Further, according to the research findings, pupils read about young and older women who were helpless, of only average intelligence, and weak. Their male counterparts were assertive and held important career positions.[27] A half-century after the GDR's claims that it would solve the inequalities between men and women through coeducation, pupils still wrote uncritically of women being responsible for the domestic sphere, while men took care of discipline and physical labor. Neither the unity school nor its successors eliminated the "woman's question"; similarly, pupils at the end of the GDR did not reflect on these topics critically any more than had Soviet zone pupils.

One of the most articulate young voices on life in the Soviet zone and GDR was the author Brigitte Reimann, whose correspondence with her girlfriend in Hamburg I have cited earlier. Throughout her adolescence, Brigitte defended the GDR against Veralore's "misconceptions." Early in her adolescent life, Brigitte's patriotism shined through her letters. Career success in the Soviet zone did not result from membership in the SED, she indignantly insisted; the "smoking factory chimneys" and "waving cornfields" that she looked upon with pride had come about because of her contributions to the reconstruction of her country.[28] Brigitte had always been an excellent student, and she learned the lesson of differentiating eastern/socialist from western/capitalist culture quite well: Disparagingly, she once complained that her boyfriend, supposedly a good FDJ functionary, nonetheless chewed western chewing gum, wore western boogie-woogie shoes and thought that Truman and Stalin were both good men in their own ways.[29] In passages such as these, written to her friend in the newly established Federal Republic of Germany, Brigitte was not only stating her ideological stance to Veralore; she was trying to convince her friend of which system was better.

Yet no matter how clearly Brigitte Reimann identified herself as a proud citizen of the GDR, her feelings about the state were ultimately quite complicated. From her teenage years onward, Reimann had experienced the private and public horrors of living under the SED regime, and she had struggled her entire life with her desire to trust her society while feeling betrayed at its random and even brutal injustices. In 1950, when she wrote to Veralore about the classmate whom the Soviet secret police arrested during school hours under suspicion of conducting subversive activities, Reimann noted self-accusingly that when a similar incident had occurred at another school recently, she had assumed the boys in question were guilty. Here, though, she realized that this boy could not possibly have been an enemy of the state: he was, she wrote without irony, simply "too dumb" to pose a threat to anyone.[30] Almost two decades later, her diary entry from November 1968 told of another episode of her "desperation" and "fits of hate" regarding her feelings for her regime after she learned of her own government's role and subsequent concealment of the violent suppression of the Prague Uprising in Czechoslovakia: "I was a well-intentioned fool."[31] But Reimann, like many East Germans of her generation, clung to hope in a better socialist future that would heal the wounds inflicted by the government. In 1972, the final year of her life, the forty-year old Brigitte Reimann wrote a last letter to Veralore, and once again explained to her friend and to herself why she had not and could not give up her belief in a socialist state: "[B]ut I'm forgetting again that you haven't lived here in years and haven't experienced certain political events that have helped shape us, and that you probably can't imagine that one can't yet write about these events with the necessary harshness," she wrote, continuing a lifelong habit of alluding to violent state crackdowns and oppression only cryptically. "But in all these years it has been demonstrated again and again that stories that once were taboo some time later make it out into daylight, to the public. That just belongs to our difficulties in socialism, and one can find them tragic or not, accept them or not—by the way, I still have a strong belief in reason."[32] Still, on her deathbed, she continued to claim that in her country, reason still reigned over ideology, albeit all too often not immediately in the face of crises.

For the first generation of postwar eastern Germans like Brigitte Reimann, at least, the Soviet zone school had succeeded in making them conscious and proud of the community in which they had grown up and helped build. The educational system had failed, however, to convince East Germans like Reimann that her political beliefs and her government were one and the same. The state had consistently betrayed her and its people; pride in her nation and socialist ideals remained. And these difficulties, too, paraphrasing her final thoughts on the subject, we can view as tragic or not, and accept them or not, depending upon her (and our) view of the entire antifascit-socialist project. But as we continue the work of understanding the half-century that comprised the Soviet zone and GDR, it would be a grave mistake to ignore the formative experiences such as schooling in the lives of young Brigitte Reimann and her peers.

Notes

Introduction: Redemption through Reconstruction

Notes to Pages 1–3

1. The practice of using upper or lower case for the words "east," "west," "eastern," and "western" follows the logic of whether I am referring to a specific geographical location (e.g., East Berlin, the East). I have tried to follow scholarly conventions in all cases.

2. Siegfried Baske, "Allgemeinbildende Schulen," in *Handbuch der deutschen Bildungsgeschichte vol. VI/2 1945 bis zur Gegenwart: Deutsche Demokratische Republik und neue Bundesländer*, ed. Christoph Führ and Carl-Ludwig Furck (Munich: Beck, 1998), 159–160. Baske noted that western scholarship has typically used 1948 as the actual end of the first phase of educational policy in the Soviet zone. In theory this periodization is logical; in practice, schools did and even could not make sudden changes in curricula or teaching practices, as will be clear throughout the following chapters.

3. Even recent scholarship that has rejected traditional periodizations of the Soviet zone/GDR as informed by political ideology rather than historical evidence has not acknowledged the importance of the first postwar years in the formation of an East German consciousness. See Corey Ross, *The East German Dictatorship: Problems and Perspectives in the Interpretation of the GDR* (London: Arnold, 2002).

4. "Gemeinsamer Aufruf der KPD und SPD zur demokratischen Schulreform," October 18, 1945, in *Zwei Jahrzehnte Bildungspolitik in der Sowjetzone Deutschlands*, vol. 1, *1945–1958*, ed. Siegfried Baske and Martha Engelbert (Heidelberg: Quelle and Meyer, 1966), 5–7.

5. Ibid, 5.

6. I have maintained the use of quotation marks around "new school" to emphasize its contemporary use as a specific terminology.

7. Ursula Reimann, LAB/STA 134/13, 181/2, n.d. [May 1946], no. 433. Note that the LAB/STA collection is now housed entirely in the Landesarchiv Berlin. All translations, unless otherwise noted, are my own.

8. Eva Schmude, born May 26, 1932, "Wiederaufbau unserer Schule," LAB/STA 134/13, 181/2, n.d. [May 1946], no. 434.

9. Waltraut Motzigemba, born March 26, 1932, seventh grade, "Zustand im Mai 1945–April 1946," LAB/STA 134/13, 181/1, [May 1946], no. 41.

10. Brian Puaca, "Learning Democracy: Education Reform in Postwar West Germany, 1945–1965," (Ph.D. diss., University of North Carolina—Chapel Hill, 2005); " 'We learned what democracy really meant': The Berlin Student Parliament and School Reform in the 1950s," *History of Education Quarterly* 45.4 (Winter 2005): 615–624; "Missionaries of Goodwill: Deutsche Austauschlehrer und –schüler und die Lehren der amerikanischen Demokratie in den frühen fünzigen Jahren," in *Demokratiewunder: Transatlantische Mittler und die kulturelle Öffnung Westdeutschlands 1945–1970,* ed. Arnd Bauerkämper, Konrad H. Jarausch, and Marcus M. Payk (Göttingen: Vandenhoeck & Ruprecht, 2005), 305–331; "The Pen is Mightier than the Sword? Student Newspapers and Democracy in Postwar West Germany," unpublished ms. (Christopher Newport University), September 2005.

11. Wolfgang Mitter, "Continuity and Change: A Basic Question for German Education. Present-day Questions in Education in the Federal Republic of Germany Against a Historical Background," [originally written 1986] in *Education in Germany: Tradition and Reform in Historical Context,* ed. David Phillips (London: Routledge, 1995), 44–45.

12. Annie Lacroix-Riz, "Politique scolaire et universitaire en Allemagne occupée," in *Kulturpolitik im besetzten Deutschland 1945–1949,* ed. Gabriele Clemens (Stuttgart: Franz Steiner, 1994), 143–148. For a more traditional interpretation, see Roy F. Willis, *The French in Germany: 1945–1949* (Stanford: Stanford University Press, 1962).

13. Rainer Hudermann, "Kulturpolitik in der französischen Besatzungszone—Sicherheitspolitik oder Völkerverständigung? Notizen zu einer wissenschaftlichen Discussion," in *Kulturpolitik im besetzten Deutschland,* ed. Gabriele Clemens (Stuttgart: Franz Steiner, 1994), 183–199.

14. Scholars have generally acknowledged the Western zones' refusal to make structural reforms to the school system as a larger societal rejection of modernity, but more research remains to be done on the concept of the Soviet zone and GDR educational system as an embrace of modernity. See the entries by Hans J. Hahn, "Education System: FRG," 172 and "Education System: GDR," 175–176 in *Encyclopedia of Contemporary German Culture,* ed. John Sandford (London: Routledge, 1999); and the special *Zeitschrift für Pädagogik* issue on pedagogy and modernity: Jürgen Oelkers, ed., *Zeitschrift für Pädagogik,* 28. Beiheft, "Aufklärung, Bildung und Öffentlichkeit: Pädagogische Beiträge zur Moderne," (Weinheim and Basel: Beltz, 1992).

15. Annie Lacroix-Riz, "Politique scolaire et universitaire en Allemagne occupée," in *Kulturpolitik im besetzen Deutschland 1945–1949,* ed. Gabriele Clemens (Stuttgart: Franz Steiner, 1994), 131.

16. For examples of German schools' historical failures and successes, see Karl A. Schleunes, *Schooling and Society: The Politics of Education in Prussia and Bavaria, 1750–1900* (New York: Berg, 1989) and Steven R. Welch, *Subjects*

or Citizens? Elementary School Policy and Practice in Bavaria 1800–1918 (Melbourne: University of Melbourne, 1998).

17. Yet historical studies of youth in the postwar years too often rely solely on adults' sources to describe young people's experiences. This tendency is not limited to the postwar era. One explanation for this phenomenon might be that generational studies in general have only recently begun to attract interest among historians, who have thus not yet adequately explored potential source bases. See the collection by Mark Roseman, ed., *Generations in Conflict: Youth Revolt and Generation Formation in Germany 1770–1968* (Cambridge: Cambridge University Press, 1995), especially the introductory essay by Roseman, "Introduction: Generation Conflict and German History," 1–46.

18. See for instance Stephen Harp, *Learning to be Loyal: Primary Schooling as Nation Building in Alsace and Lorraine, 1850–1940* (DeKalb: Northern Illinois University Press, 1998); Mona L. Siegel, "Lasting Lessons: War, Peace, and Patriotism in French Primary Schools, 1914–1939," (Ph.D. diss., University of Wisconsin—Madison, 1996).

19. The techniques and theories that have been developed in the field of memory and oral history studies proved invaluable to my analysis of pupils' and adults' memories of World War II as they were recorded in the first years of the Soviet zone. See Petra Gruner's use of oral history to narrate individuals' and groups' changing self-perceptions. Petra Gruner, *Die Neulehrer: Ein Schlüsselkonzept der DDR-Gesellschaft?* (Ph.D. diss., Humboldt University—Berlin, 1998). See also the interweaving of oral history and archive work in Laird Boswell, *Rural Communism in France, 1920–1939* (Ithaca: Cornell University Press, 1998), especially his reflections on interviews as an historical source, 15–17.

20. The thorough methodological reflections on soldiers' letters from the front have guided my thoughts about evaluating pupil essays. See Bernd Ulrich, *Die Augenzeugen: Deutsche Feldpostbriefe in Kriegs- und Nachkriegszeit 1914–1933* (Essen: Klartext, 1997), 11–38.

21. Herta Hielscher, "Meine IIIb. Pädagogische Erfahrungen einer Junglehrerin," *die neue schule*, 2.1 (1947): 10–14, here 12. Note that the lower-case nouns in this and other educational periodicals of the Soviet zone reflected a conscious decision to democratize the language by simplifying stylistic rules, and are not typographical mistakes on those authors' or my part. See, for instance, "Correspondence from Marquardt of Schulabteilung" Re: Writing reform. BArch DR-2/73, 8.

22. Forty-one of the essays have been published in the edited volume by Annett Gröschner, *"Ich schlug meiner Mutter die brennenden funken ab": Berliner Schulaufsätze aus dem Jahr 1946* (Kontext: Berlin, 1996). Gröschner generally used pseudonyms for the young authors, so readers comparing the essays in her book to those same essays in the archive used may find different authors' names.

23. During presentations of this study, faculty members in many departments and universities expressed skepticism that pupils or university students ever bother reading teachers' comments. Such opinions do not take into account the potentially devastating or reassuring effects of a teacher's marginal comments in red ink.

24. Ernest Labrousse has defined the field of the everyday as the "third level," contrasting it to the "economy" and "society." These levels did not function independently of one another, but worked instead in a symbiotic relationship. See Ernest Labrousse, *Le mouvement ouvrier et les théories sociales en France de 1815 à 1848* (Paris: Centre de documentation universitaire, 1961) and Peter Schöttler, "Mentalitäten, Ideologien, Diskurse: Zur sozialgeschichtlichen Thematisierung der 'dritten Ebene'," in Alf Lüdtke, ed. *Alltagsgeschichte: Zur Rekonstruktion historischer Erfahrungen und Lebensweisen* (Frankfurt AM: Campus: 1989), 85–136.

25. Alf Lüdtke, "Einleitung: Was ist und wer treibt Alltagsgeschichte?" in *Alltagsgeschichte: Zur Rekonstruktion historischer Erfahrungen und Lebensweisen*, ed. Alf Lüdtke (Frankfurt, AM: Campus: 1989), 11–12.

26. I have kept the original term *Einheitsschule* and translated it as well throughout the study literally as "unity school" in order to capture its proponents' conscious emphasis on political unity. The structure of the school, however, is accurately referred to in educational literature as a comprehensive school.

27. Cf. Andrew Parker, Mary Russo, and Doris Sommer, eds. *Nationalisms and Sexualities* (New York: Routledge, 1992).

Chapter 1 Antifascism, Unity, and Division

1. Schulrat Viehweg, "Bericht über die Tagung der Schulaufsichtsbeamten des Bundeslandes Sachsen am 19. und 20. Oktober 1945," BArch DR 2/488, no. 34.

2. "Sofortprogramm der Kampfgemeinschaft gegen den Faschismus!" in *Der Aufbau. Organ der Kampfgemeinschaft gegen den Faschismus* [Bremen] 1 (May 6, 1945), reprinted in Peter Brandt, *Antifaschismus und Arbeiterbewegung: Aufbau-Ausprägung-Politik in Bremen 1945/46* (Hamburg: Hans Christian, 1976), 255–257.

3. Heinz Heitzer, *DDR: Geschichtlicher Überblick*. 4th ed. (Berlin [East]: Dietz, 1987).

4. Wolfgang Emmerich claimed specifically that the idea of a zero hour was a means of ignoring any real discussions of the Nazi past, embraced primarily by the Western zones. Wolfgang Emmerich, *Kleine Literaturgeschichte der DDR*. 2nd ed. (Leipzig: Gustav Kiepenhauer, 1996).

5. Otto Winzer, "Zur Kulturpolitik in Berlin," n.d. [1946], LAB/STA 120/3227, no. 1.

6. Helmut König, "Robert Alt: 1905 bis 1978—Erziehung und Gesellschaft— Grundthema seiner wissenschaftlichen Arbeit," in *Wegbereiter der neuen*

Schule, ed. Gerd Hohendorf, Helmut König and Eberhard Meumann (Berlin [East]: Volk und Wissen, 1989), 32; Tilo Köhler and Rainer Nitsche, eds. *Stunde Eins oder die Erfindung von Ost und West* (Berlin: Transit, 1995), 7–8.

7. For instance, Erika Broszy, 14. Volksschule, "Wie unsere Schule vor einem Jahr aussah und wie es jetzt ist," LAB/STA 134/13, 181/2, no. 525; Karl Sothmann, "Ein Jahr 'die neue schule': Kritische Bemerkungen zur Entwicklungen der Zeitschrift" *die neue schule*, 2.7 (1947): 1–3.

8. His wartime diaries have been translated into English: Victor Klemperer, *I Will Bear Witness: A Diary of the Nazi Years 1933–1941*, trans. Martin Chalmers (Random House: New York, 1988) and *I Will Bear Witness: A Diary of the Nazi Years 1942–1945*, trans. Martin Chalmers (Random House: New York, 1989). His postwar diary: Victor Klemperer, June 17, 1945, *Und so ist alles schwankend: Tagebücher Juni bis Dezember 1945* 3rd ed. (Berlin: Aufbau Taschenbuch Verlag, 1996), 7.

9. Klemperer, July 15, 1945, *Und so ist alles schwankend*, 56.

10. Stadtrat Otto Winzer, "Das erste Jahr des Berliner Schulaufbaus. Notwendige Bemerkungen zur materiellen Grundlage der Schulreform," LAB/STA 120/30, [ca. 1946], no. 303.

11. Johann Wolfgang von Goethe/Friedrich Schiller, *Xenien* Nr. 95–96, 1796. The *Xenien* were a series of epigrams written by the two great scholars on a range of critical subjects, from politics to contemporary literature.

12. "Protokoll, Gesamtkonferenz der Hauptschulräte," June 29, 1947, LAB/STA 120/124, no. 78.

13. See Benedict Anderson, *Imagined Communities: Reflections on the Origin and Spread of Nationalism*. 2nd ed. (London: Verso, 1991), ch. 10, 143–186 and Mark Monmonier, *How to Lie with Maps*. 2nd ed. (Chicago: University of Chicago Press, 1996), ch. 9, 123–138.

14. Karl Sothmann, "Ziel und Weg. Methodische Hauptstelle der neuen Schule bei der Deutschen Verwaltung für Volksbildung in der sowjetischen Besatzungszone II," May 17, 1946, DIPF/BBF/Archiv NL Karl Sothmann, 0.4.01., no. 1.

15. J.G. [Hans] Siebert, "Die Grundlagen und der Charakter der freien deutschen Kultur," speech for Kulturtagung, January 1944, DIPF/BBF/ Archive, NL Hans Siebert 0.4.07, fo. 123, no. 13.

16. Prof. Dr. Bousquet, report on French educational system at Magistrat, Berlin, "Protokoll über die Pädagogische Tagung am 16. November 1945," LAB/STA 120/6, no. 21.

17. Ibid.

18. Ernst Wildangel, "Protokoll über die Pädagogische Tagung am 16. November 1945," LAB/STA 120/6, no. 25.

19. Dr. Bousquet, "Protokoll über die Pädagogische Tagung am 16. November 1945," no. 25–26.

20. Franz Albrecht, "Ernst Wildangel—1891 bis 195–'Du bist Lehrer, du kannst unsere Schulen in Gang bringen!'" in *Wegbereiter der neuen Schule*,

ed. Gerd Hohendorf, Helmut König and Eberhard Meumann (Berlin [East]: Volk und Wissen, 1969), 267–273.

21. Victor Klemperer, June 20, 1945, *Und so ist alles schwankend*, 15.

22. The problem with Wildangel's position is immediately apparent: his background was unique. As such, it could not contradict one of the main criticisms made then as now against the "ideological lie" of antifascism: that the majority of Soviet zone/GDR citizens who believed themselves to be "antifascists" had not in fact ever resisted fascism. Dan Diner, "On the Ideology of Antifascism," *New German Critique* 67 (Winter 1997): 123.

23. The series was entitled, "Vom Schulwesen in anderen Ländern." See, for example, Ludwig Peters, "Schulreform in Finnland," *die neue schule*, 3.4 (1948): 8–9. *die neue schule* also regularly reported on educational developments in other countries in its editorials and short notices.

24. Examples of Siebert's and Koebbel's wartime and postwar activities in England can be found in DIPF/BBF/Archiv NL Hans Siebert, 0.4.07, fos. 121 and 137; and AdJb NL Eberhard Koebbel, N Koebel 106 and 107, correspondence 1946–1948.

25. See, for instance, Eric Hobsbawm, *The Age of Extremes: A History of the World, 1914–1991* (New York: Pantheon, 1994). See Anson Rabinbach's excellent discussion of Hobsbawm's interpretation, Anson Rabinbach, "Introduction: Legacies of Antifascism," *New German Critique* 67 (Winter 1997): 6–7.

26. Dan Diner, "On the Ideology of Antifascism," 123 and 131. See also Alan L. Nothnagle, *Building the East German Myth: Historical Mythology and Youth Propaganda in the German Democratic Republic, 1945–1989* (Ann Arbor: University of Michigan Press, 1999) and Antonia Grunenberg, *Antifaschismus—ein deutscher Mythos* (Reinbek bei Hamburg: Rowohlt, 1993).

27. Rabinbach, "Introduction: Legacies of Antifascism," 3.

28. The SMAD permitted the establishment of two other parties in 1948, the National Democratic Party of Germany (NDPD) and the Democratic Farmers' Party of Germany (DBD). Neither appeared to be active in educational activities. See Dietrich Staritz, "National-Demokratische Partei Deutschlands (NDPD)," 574–594 and Berndhard Wernet-Tietz, "Demokratische Bauernpartei Deutschlands," 584–594, in *SBZ Handbuch: Staatliche Verwaltung, Parteien, gesellschaftliche Organisationen und ihre Führungskräfte in der Sowjetischen Besatzungszone Deutschlands 1945–1949*, ed. Martin Broszat and Hermann Weber (Munich: Oldenbourg, 1993).

29. Dietrich Staritz and Siegfried Suckut, "Einleitung," 435–439, in *SBZ Handbuch*, ed. Martin Broszat and Hermann Weber.

30. Dietrich Staritz, "National-Demokratische Partei Deutschlands," in *SBZ Handbuch*, ed. Martin Broszat and Hermann Weber, 574.

31. Siegfried Bahne, "Sozialfaschismus in Deutschland. Zur Geschichte eines politischen Begriffs," *International Review of Social History*, 10 (1965), 211–245.

32. Klaus-Michael Mallmann, *Kommunisten in der Weimarer Republik: Sozialgeschichte einer revolutionären Bewegung* (Darmstadt: Wissenschaftliche Buchgesellschaft, 1996), 68.

33. Jürgen Zarusky, *Die deutschen Sozialdemokraten und das sowjetische Modell: Ideologische Auseinandersetzung und außenpolitische Konzeptionen 1917–1933* (Munich: Oldenbourg, 1992), 15.

34. Grunenberg, *Antifaschismus—ein deutscher Mythos*, 22–25.

35. Zarusky, *Die deutschen Sozialdemokraten*, 282; Brandt, *Antifaschismus und Arbeiterbewegung*, 21.

36. Grunenberg, *Antifaschismus—ein deutscher Mythos*, 26.

37. Werner Müller, "Sozialdemokratische Deutschlands (SPD)," in *SBZ Handbuch*, ed. Martin Broszat and Hermann Weber, 461–462.

38. Mallmann, *Kommunisten in der Weimarer Republik*, 75.

39. Werner Müller, "Kommunistische Partei Deutschlands (KPD)," in *SBZ Handbuch*, ed. Martin Broszat and Hermann Weber, 444–445.

40. Anton Ackermann, "Demokratische Schulreform," November 4, 1945, cited in Siegfried Baske and Martha Engelbert, eds. *Zwei Jahrzehnte Bildungspolitik*, vol. 1, 7. For the *Kulturbund* and Johannes Becher, see Norman Naimark, *The Russians in Germany: A History of the Soviet Zone of Occupation 1945–1949* (Cambridge, MA: Belknap Press of Harvard University Press, 1995), 334 and 400–408.

41. Anton Ackermann, "Demokratische Schulreform," November 4, 1945, cited in Siegfried Baske and Martha Engelbert, eds. *Zwei Jahrzehnte Bildungspolitik*, vol. 1, 12.

42. Ibid., 9.

43. See the tensions between ethnicity and imagined ethnicity in Anthony D. Smith, *The Ethnic Origins of Nations* (Oxford: Basil Blackwell, 1986), 133, 141, 212, and especially, *National Identity* (London: Penguin Books, 1991; reprint, Reno: University of Nevada Press, 1991), 71; Anderson, *Imagined Communities*. Intriguing arguments have been made for seeing citizens of the GDR as a distinct ethnic group, including theories that national groups must create an imagined ethnic consciousness in which one does not exist. Such universal claims would be overstated. For the GDR as an ethnic community, see Marc Alan Howard, "Die Ostdeutschen als ethnische Gruppe? Zum Verständnis der neuen Teilung des geeinten Deutschland," *Berliner Debatte* 4/5 (1995): 119–131.

44. For a related discussion of *ius sangue* vs. *ius terre* in citizenship definitions, see Rogers Brubaker, *Citizenship and Nationhood in France and Germany* (Cambridge, MA: Harvard University Press, 1992).

45. Michael Lemke's otherwise thoughtful reflections on Sovietization conclude with the remarks that culture and everyday practices were not Sovietized in the GDR, and thus no "change of identity of Germans occurred." Michael Lemke, "Die Sowjetisierung der SBZ/GDR im ost-westlichen Spannungsfeld," *Aus Politik und Zeitgeschichte* B 6/97 (January 31, 1997): 53.

46. See Monika Kaiser, "Sowjetischer Einfluß auf die ostdeutsche Politik und Verwaltung 1945–1970," in *Amerikanisierung und Sowjetisierung in Deutschland 1945–1970*, ed. Konrad Jarausch and Hannes Siegrist (Frankfurt, AM: Campus, 1997), 111–114.

47. See the discussions on power relationships between colonizer and colonized in Thomas S. Popkewitz, ed., *Educational Knowledge: Changing Relationships between the State, Civil Society, and the Educational Community* (Albany: State University Press of New York, 2000), especially the chapter, "Globalization/Regionalization, Knowledge, and the Educational Practices. Some Notes on Comparatives Strategies for Educational Research," 5–7.

48. Werner Müller, "Kommunistische Partei Deutschlands (KPD)," in *SBZ Handbuch*, ed. Martin Broszat and Hermann Weber, 444–446. Siegfried Baske and Martha Engelbert, "Einleitung: Die Haupttendenzen der sowjetzonalen Bildungspolitik von 1945 bis 1965," in *Zwei Jahrzehnte Bildungspolitik*, vol. 1, xvii.

49. "Aufruf der Kommunistischen Partei Deutschlands," June 11, 1945, cited in Siegfried Baske and Martha Engelbert, *Zwei Jahrzehnte Bildungspolitik*, vol. 1, 1.

50. Correspondence from Karl Sothmann to Herr Hadermann, December 2, 1947, DIPF/BBF/Archiv NL Karl Sothmann, Nr. 0.4.01, fo. 5, no. 1.

51. [Helmut] Holtzhauer, "Politische Probleme der Gegenwart," Kreisschulratkonferenz, November 3–5, 1948, Dresden-Wachwitz, BArch DR 2/489, no. 337.

52. Heinrich Deiters, "Rückblick auf mein Leben," DIPF/BBF/Archiv NL Heinrich Deiters, 0.4.05, fo. 54, no. 9.

53. Staritz, "National-Demokratische Partei Deutschlands (NDPD)," 574–594 and Wernet-Tietz, "Demokratische Bauernpartei Deutschlands," 584–594 in *SBZ Handbuch*, ed. Martin Broszat and Hermann Weber.

54. Anton Ackermann, "Demokratische Schulreform," Nov 1945, cited in Siegfried Baske and Martha Engelbert, *Zwei Jahrzehnte Bildungspolitik*, vol. 1, 8, and 9.

55. "Gemeinsamer Aufruf der KPD und SPD zur demokratischen Schulreform," October 18, 1945, in Siegfried Baske and Martha Engelbert, eds. *Zwei Jahrzehnte Bildungspolitik*, vol. 1, 7.

56. "Gesetz zur demokratisierung der deutschen schule," in *pädagogik* 1, no. 1 (August 1946): 2.

57. Manfred Koch, "Parlamentarische Gremien und Verfassungsgebung. Beratende Versammlungen," 321 and 325; and Manfred Koch, "Landtage," 332–348, in *SBZ Handbuch*, ed. Martin Broszat and Hermann Weber.

58. Anton Ackermann, "Demokratische Schulreform," November 4, 1945, cited in Siegfried Baske and Martha Engelbert, eds. *Zwei Jahrzehnte Bildungspolitik in der Sowjetzone Deutschlands*, vol. 1, 9.

59. Paul Wandel, "Zur Demokratisierung der Schule," Speech for the First Pedagogical Congress, August 15, 1946, in Siegfried Baske and Martha Engelbert, *Zwei Jahrzehnte Bildungspolitik*, vol. 1, 32.

60. Christa Uhlig, "Zur Erarbeitung der bildungspolitischen Programmatik für Nachkriegsdeutschland," in *Kindheit, Jugend und Bildungsarbeit im Wandel,* ed. Heinz-Elmar Tenorth, *Zeitschrift für Pädagogik,* 37th Beiheft (Weinheim/Basel: Beltz, 1997), 411–412.

61. Correspondence from Präsidialabteilung to Landesregierung Brandenburg, Ministerium für Volksbildung, Herr Minister Rücker, October 21, 1947, BArch DR 2/4008, no. 216–217, here 217; and "Entschließung des 3. Pädagogischen Landeskongresses des Landes Mecklenburg am 25. und 26. Mai 1948," BArch DR 2/1394, no. 110.

62. Emphasis in original. "Die neue deutsche Schule," n.d. BArch DR 2/1080, no. 9.

63. Rolf Steininger, *Deutsche Geschichte seit 1945,* vol. 2, *1948–1955* (Frankfurt, AM: Fischer, 1996), 23.

64. Walter Dirks, "Einheit und Freiheit," *Frankfurter Hefte* (1948), reprinted in *Stunde Eins oder die Erfindung von Ost und West,* ed. Tilo Köhler and Rainer Nitsche, 105.

65. From Präsidinalabteilung to Landesregierung Brandenburg, Ministerium für Volksbildung, Herr Minister Rücker, October 21, 1947, BArch DR 2/4008, no. 216–217.

66. Vice president Erwin Marquardt, "Konferenz der Minister für Volksbildung der Länder und Provinzen," December 2–4, 1947, BArch DR 2/58, no. 3.

67. Max Kreuziger, "Entwurf für die Volksbildungsminister am 27./28. Jan. 1948, DVV, Schulabteilung," Berlin, January 21, 1948, BArch DR 2/72, no. 27.

68. Minister Holtzhauser, "Politische Probleme der Gegenwart," "Kreisschulratskonferenz vom 3.–5. November 1948," Dresden-Wachwitz, BArch DR /489, no. 337.

69. Paul Oestereich, "Protokoll, Pädagogische Tagung 1947," DIPF/BBF/ Archiv NL Karl Sothmann, 0.4.0.1., fo. 2, no. 124.

70. "Bericht über die soziale und sozialpädagogische Arbeit des Hauptschulamtes," Magistrat der Stadt Berling Abt. Volksbildungs, Hauptschulamt, Berlin, November 3, 1945, LAB/STA 120/30, no. 17.

71. Hans Siebert, "Sozialistische Kulturarbeit: 1. Jahresbericht über die Tätigkeit der Abteilung Kultur und Erziehung des Zentralsekretariats der SED, für die Zeit vom Mai 1946 bis August 1947," DIPF/BBF/Archiv, NL Hans Siebert, 0.4.0.7, fo. 24, no. 3.

72. Cf. "German Educational Reconstruction," DIPF/BBF/Archiv, NL Hans Siebert, October 1944, 0.4.0.7, fo. 14, no. 1–2.

73. Answers here have been made difficult by the overwhelming "dictatorship literature" that has ignored cultural realms of ideologies, assuming only a violent top-down application of political will. Dividing a society into those who "have" power and thus "produce" ideologies, and those who do not and thereby are "recipients" of the proffered viewpoints, misrepresents most societal structures and ignores what are now well-established studies of power relationships. By avoiding the "ogre" of reception, and equating

production with a small political elite, much of the historiography of antifascism has ignored those who actually developed its practice. Alon Confino, "Collective Memory and Cultural History: Problems of Method," *American Historican Review*, 102.5 (1997): 1395. See also a critique of the dictatorship model in Alan McDougall, *Youth Politics in East Germany: The Free German Youth Movement 1946–1968* (Oxford: Oxford University Press, 2004).

74. Mary Fulbrook, *The People's State: East German Society from Hitler to Honecker* (New Haven: Yale University Press, 2005), 29.

75. See, for instance, the recent work by Alan L. Nothnagle, *Building the East Germany Myth*. In chapter three in particular, Nothnagle dismisses antifascism as a myth controlled entirely by the state's propaganda machine.

76. Ibid., 114.

77. Confino, "Collective Memory and Cultural History," 1396.

78. It should be noted that the educators portrayed in this work did, in fact, have many similar experiences that could easily be interpreted as characteristic biographies. The differences, though, were still too great for GDR scholars. Gerd Hohendorf, Helmut König, and Eberhard Meumann, "Vorwort," in *Wegbereiter der neuen Schule*, ed. Gerd Hohendorf, Helmut König, and Eberhard Meumann, (Berlin [East]: Volk und Wissen, 1989), 12.

79. Eva Landler, "Käte Agerth: 1888 bis 1974—Im Herzen immer jung geblieben," and Heinz Warnecke, "Wilhelm Heise: 1897 bis 1949—Der neue Lehrer soll ein politischer Mensch sein," in *Wegbereiter der neuen Schule*, ed. Gerd Hohendorf, Helmut König, and Eberhard Meumann, 13 and 130, respectively.

80. *Lexikon der Pädagogik*, 3rd ed. (Freiburg: Herder, 1970–1971), s.v. "Gemeinschaftsschule," "Versuchsschule, Schulversuch;" Gustav Wyneken, *Schule und Jugendkultur*. 2nd ed. (Jena: Eugen Diederichs Verlag, 1914 [1913]).

81. See John Dewey, *Democracy and Education* (New York: Macmillan, 1916); Ellen Key, *Das Jahrhundert des Kindes*, (Berlin: S. Fischer, 1904); Maria Montessori, *Formazione dell'uomo* (Milan: Garzanti, 1949), and *La scoperta del bambino* (Milan: Garzanti, 1950, first published as *Il metodo della pedagogia scientifica*, 1909); Rudolf Steiner, *Die Erziehung des Kindes vom Gesichtspunkte der Geisteswissenschaft* (Berlin: Besant-Zweig of the Theosophical Society: 1907).

82. Detlef Oppermann's introduction to the autobiography of educational reformer Heinrich Deiters makes clear how 1945 constituted the possibility of an unexpected career move for Deiters. Detlef Oppermann, "Schule und Bildung zwischen Tradition und Umbruch. Die Lebenserinnerungen von Heinrich Deiters: Ein Einführung," in *Bildung und Leben: Erinnerungen eines deutschen Pädagogen*, ed. Heinrich Deiters, (Cologne: Böhlau, 1989), xxi.

83. Helmut König, "Robert Alt—1905 bis 1978," 31–38.

84. Ursula Basikow and Karen Hoffman, "Kurzbiographie von Marie Torhorst," in DIPF/BBF/Archiv NL Marie Torhorst, 0.4.13.2, Findbuch. According to Basikow and Hoffman, Torhorst left the Karl-Marx-Schule because of philosophical differences with Karsen, who later went on to work on education reforms in the Western zones until emigrating to the United States. See Gerd Radde, *Fritz Karsen: ein Berliner Schulreformer der Weimarer Zeit* (Berlin: Colloquium-Verlag, 1973).

85. Ibid. On a side note, Torhorst was consistently referred to using the male form of her office: "der erste weibliche Minister," a practice that became common in the GDR.

86. For a description of this incident, see Rudi Goguel, *Cap Arcona: Report über den Untergang der Häftlingsflotte in der Lübecker Bucht am 3. Mai 1945* (Frankfurt AM: Röderberg Verlag GmbH, 1972).

87. Of Alt's many essays, none dealt specifically with the Soviet Union. He focused more on historical educational figures such as the Swiss Johann Heinrich Pestalozzi and the Czech Johann Amos Comenius. Rudi Schulz, "Bibliographie der Arbeiten von Robert Alt," *Jahrbuch für Erziehungs- und Schulgeschichte*, vol. 5/6 (Berlin [East]: Akademie-Verlag, 1966), 427–432. Alt's one specific treatment of the Soviet Union was a lecture a few years before his death, "50 Jahre Sowjetunion—50 Jahre marxistisch-leninistische Pädagogik in Aktion," Festvortrag 1972, Akademie der Pädagogischen Wissenschaften der DDR, cited in König, "Robert Alt—1905 bis 1978," 36.

88. Marie Torhorst, "Sowjetschule und deutsche Schulreformer," *Schöpferische Gegenwart—Kulturpolitische Monatszeitschrift Thüringens*, 5 (November 1948): 296, DIPF/BBF/Archiv, NL Marie Torhorst, no. 1.

89. In contrast to Alt, whose wife and son were never included in biographical accounts, Torhorst and her sister's decision not to marry received frequent mention. Biographers praised her political accomplishments, but used her single status to explain her success. A career-oriented spinster could be more easily inserted into official narratives of antifascists, who did not include many women in their upper administrative ranks. Ursula Basikow and Karen Hoffman, "Kurzbiographie von Marie Torhorst," in DIPF/BBF/ Archiv NL Marie Torhorst, 0.4.13.2, Findbuch.

90. Geoff Eley and Ronald Grigor Suny, "Introduction: From the Moment of Social History to the Work of Cultural Representation," in *Becoming National*, ed. Geoff Eley and Ronald Grigor Suny (New York: Oxford Univrsity Press, 1996), 8.

91. Anthony Smith described the phenomenon of "socialist nationalism" as a mixture of "anti-fascist nationalism" and Marxism. Anthony D. Smith, *Nationalism in the Twentieth Century* (New York: New York University Press, 1979), 115–116.

92. Grunenberg, *Antifaschismus: ein deutscher Mythos*, 120–130.

93. Eley and Suny, "Introduction: From the Moment of Social History," 7.

94. Diner, "On the Ideology of Antifascism," 128.

Chapter 2 Setting up the School

1. See the statistical information on young people in Edeltraud Schulze, *DDR-Jugend: Ein statisches Handbuch* (Berlin: Akademie, 1995), 95; Günter Braun, "Daten zur demographische und sozialen Struktur der Bevölkerung," in *SBZ Handbuch: Staatliche Verwaltung, Parteien, Gesellschaftliche Organisationen und ihre Führungskräfte in der Sowjetischen Besatzungszone Deutschlands 1945–1949*, ed. Martin Brozat and Hermann Weber (Munich: Oldenbourg, 1993), 1070; Schulabteilung der Deutschen Verwaltung für Volksbildung in der Sowjetischen Besatzungszone, ed., *Die deutsche demokratische Schule im Aufbau* (Leipzig: Volk und Wissen, 1949), statistical appendix, 77.

2. Jürgen Diederich and Heinz-Elmar Tenorth, *Theorie der Schule: Ein Studienbuch zu Geschichte, Funktionen und Gestaltung* (Berlin: Cornelsen, 1997), 200.

   A vertically differentiated educational system refers to different "tracks" that educate pupils in different areas. These parallel tracks are often associated with ability, for example, academic versus technical tracks. A horizontally structured system is one in which all pupils attend the same courses according to age group, with the different tracks proceeding from one year to the next.

3. Dieter Noll, *Aufbau: Berlin 1945–58. Bibliographie einer Zeitschrift* (Berlin [East]: Aufbau, 1978), 14.

4. The "Gruppe Ulbricht" is the best-known of this Soviet-trained group of German communists, but was not the only group. See Rolf Steininger, *Deutsche Geschichte seit 1945*, vol. 1, *1945–1947*, (Frankfurt AM: Fischer, 1996), 156–167; Wolfgang Leonhard, "Die 'Gruppe Ulbricht'—Strategie und Taktik der Machteroberung 1945/46," in Friedrich-Ebert-Stiftung, ed. *Einheit oder Freiheit? Zum 40. Jahrestag der Gründung der SED* (Bonn: Friedrich-Ebert-Stiftung, 1985).

5. See, for example, the memoir and documents by Walter Wolf, *Wie gelang es dem illegalen deutschen Volksfrontkommittees im Konzentrationslager Buchenwald neue Vorstellungen zu erarbeiten, was in Deutschland nach der Zerschlagung der faschistischen Diktatur geschehen muß?* 1974, DIPF/BBF/Archiv, NL Walter Wolf, 0.4.0.3., fo. 42.

6. See for example the speeches held in London by Hans Siebert, DIPF/BBF/Archiv, NL Hans Siebert, especially fos. 11, 14, and 121.

7. This was the same day that schools opened in the British and American zones, and two weeks after the French zone. Bundesministerium für Gesamtdeutsche Fragen, ed., *Die Sowjetische Besatzungszone Deutschlands in den Jahren 1945–1955* (Bonn: Deutscher Bundesverlag, 1956), 21; F. Roy Willis, *The French in Germany: 1945–1949* (Stanford: Stanford University Press, 1962), 168; and Rolf Winkeler, "Das Scheitern einer Schulreform in der Besatzungszeit: Analyse der Ursachen am Beispiel der französisch besetzten Zone Württembergs und Hohenzollerns von 1945 bis 1949," in Manfred Heinemann, ed., *Umerziehung und Wiederaufbau: Die Bildungspolitik der*

Besatzungsmächte in Deutschland und Österreich (Stuttgart: Klett-Cotta, 1981), 213.

8. "Chronologischer Bericht des hausveraltenden Rektors 1.V. über die Wiedereinrichtung und den Neuaufbau der Schulen der Schulgruppe Eberswalder Str.," August 27, 1945, LAB/STA 134/13, 183, no. 312.

9. For instance, coeducation debates in the early twentieth century pointed to the linguistic origin of "coeducation" as from the United States, thereby indicating the concept's foreign and thus inappropriate character for Germany. Heinrich von Hähling Weihbischof von Paderborn, *Die Koedukation oder Die gemeinschaftliche Erziehung der Knaben und Mädchen, besonders in der Volksschule. Die richtigen Grundsätze in einer brennenden Frage.* 2nd ed. (Paderborn: Bonifacius-Druckerei, 1924), 4.

10. "Propaganda Broadcasts to German Youth and its Educators," September 29, 1942, DIPF/BBF/Archiv, L. 10–23, NL Hans Siebert, fo. 11, no. 2.

11. Franco Cambi, *Storia della pedagogia*, 6th ed. (Milan: Laterza, 1999 [1995]), 462–470.

12. Paul Wandel, "Zur Demokratisierung der Schule," Rede auf dem 1. Pädagogischen Kongress am 15. August 1946, in *Zwei Jahrzehnte Bildungspolitik*, vol. 1, ed. Siegfried Baske and Martha Engelbert, 36.

13. Karl Ellrich, "Die Entwicklung des Grundschulwesens in der sowjetischen Besatzungszone seit 1945," in *Die Deutsche Demokratische Schule im Aufbau*, ed. Schulabteilung der Deutschen Verwaltung für Volksbildung in der Sowjetischen Besatzungszone (Leipzig: Volk und Wissen, 1949), 9–17.

14. For information on further evolution of the *Einheitsschule* and the "unity" principle in the GDR, see D. Waterkamp, *Das Einheitsprinzip im Bildungswesen der DDR* (Cologne: 1985) and Hans J. Hahn, "Education System: GDR," *Encyclopedia of Contemporary German Culture*, ed. (London: Routledge, 1999), 176.

15. "Tätigkeitsbericht der Landesregierung Mecklenburg-Vorpommern in Schwerin, Ministerium für Volksbildung für das Jahr 1946," BArch DR 2/280, January 10, 1947, 31–33.

16. Karl Ellrich, "Die Entwicklung des Grundschulwesens," in *Die Deutsche Demokratische Schule im Aufbau*, ed. Schulabteilung der Deutschen Verwaltung für Volksbildung in der Sowjetischen Besatzungszone, 13–14; "Protokoll über die Gesamtkonferenz der Hauptschulräte," March 12, 1947, LAB/STA 120/124, no. 72. See also Jürgen Diederich and Heinz-Elmar Tenorth, *Theorie der Schule*, 200–204; Anke Huschner, "Vereinheitlichung und Differenzierung in der Schulentwicklung der SBZ und DDR. Zweige und Klassen mit verstärktem alt- bzw. neusprachlichem Unterricht im Schulsystem der SBZ/DDR (1945 bis Anfang der siebziger Jahre)," in *Zeitschrift für Pädagogik* 2 (March/April, 1997): 279–280.

17. "Protokoll über die Gesamtkonferenz der Hauptschulräte," March 5, 1947, LAB/STA 120/124, no. 47.

18. In 1959, the GDR parliament passed the "Law for the Socialist Development of the School System" that provided for the ten-year

polytechnical secondary school. The transition to this system had begun in the mid-1950s, and was completed only slowly. See Siegfried Baske and Martha Engelbert, "Einleitung," in *Zwei Jahrzehnte Bildungspolitik*, vol. 1, ed. Siegfried Baske and Martha Engelbert, xxx–xxxiv.

19. See Petra Gruner, *Die Neulehrer*.

20. Berlin maintained a special status as "Greater Berlin," administered by all four occupying powers. The *Hauptschulamt*, the main school administration in Berlin, was initially responsible for all of Berlin until after the currency reform in 1948 and the western Berlin decision to set up its own school administration. The KPD did not do well in elections in Berlin, and the "Law for the Democratization of the German School," applicable for all five regions of the Soviet zone, was not passed in Berlin, which had its own school law. See Lutz R. Reuter, "Administrative Grundlagen und Rahmenbedingungen," in *Handbuch der deutschen Bildungsgeschcihte*, vol. 6, bk. 2 *Deutsche Demokratische Republik*, ed. Christoph Führ and Carl-Ludwig Furck, (Munich: Beck, 1998), 37–40; Helga A. Welsh, "Deutsche Zentralverwaltung für Volksbildung," in *SBZ Handbuch: Staatliche Verwaltung, Parteien, gesellschaftliche Organisationen und ihre Führungskräfte in der Sowjetischen Besatzungszone Deutschlands 1945–1949*, ed. Martin Broszat and Hermann Weber (Munich: Oldenbourg, 1993), 229.

21. Andreas Herbst, Winfried Ranke, Jürgen Winkler, eds. *So funktionierte die DDR*, vol. 3, *Lexikon der Funktionäre* (Reinbek bei Hamburg: Rowohllt, 1994), 358. Wandel later fell out of favor with the SED, receiving a "strict Party reprimand" in October 1957. Broszat and Hermann Weber, eds. *SBZ Handbuch*, 1051.

22. Norman Naimark has noted the commonly accepted view that the Soviets apparently based their departmental divisions on the model proposed by the Allied Command, but Jan Foitzik has pointed out that the Soviets had extensive experience setting up military occupation governments in Rumania, Bulgaria, Hungary, and Poland. Norman M. Naimark, *The Russians in Germany: A History of the Soviet Zone of Occupation, 1945–1949* (Cambridge, MA: Belknap Press of Harvard University Press, 1995), 21–22 and Jan Foitzik, "Sowjetische Militäradministration in Deutschland (SMAD)," in *SBZ Handbuch*, ed. Martin Broszat and Hermann Weber, 13.

The SMAD Education Department was headed by Petr Vassilevi until January 1949, when Soviet accusations that he had not fulfilled his socialist duties brought about his replacement by his former deputy, I. D. Artjuchin. Also important was the charismatic Colonel Sergej Tjul'panov, who led the SMAD Information Administration and functioned as the main intermediary between Soviets and Germans. The extent of Tjul'panov's influence has been disputed, but his determined efforts to rekindle German cultural life provided an important stimulus in educational policy. Foitzik, "Sowjetische Militäradministration in Deutschland (SMAD)," 23; Naimark, *The Russians in Germany*, 372 and 457.

23. Correspondence, from P. Zolotuchin, Leiter der Abt. für Volksbildung der Sowjetischen Militärverwaltung in Deutschland, to Wandel. D. Zentralverwaltung für Volksbildung in der SBZ. BArch R2/636, Berlin, August 8, 1946, 129.

24. Gert Geißler, "Schulreform von oben: Bemerkungen zum schulpolitischen Herrschaftssytem in der SBZ/DDR," in *Erinnerung für die Zukunft II*, ed. Petra Gruner, 49–60.

25. Welsh, "Deutsche Zentralverwaltung für Volksbildung," *SBZ Handbuch*, 231–232.

26. Herr Ministerialdirektor [Wilhelm] Schneller, "Tagung der Bezirksschulräte," Meißen, March 12–13, 1946, BArch DR 2/488, no. 110.

27. "Tätigkeitsbericht der Landesregierung Mecklenburg-Vorpommern in Schwerin, Ministerium für Volksbildung für das Jahr 1946," BArch DR 2/280, January 10, 1947, 30.

28. Dietrich Staritz, *Die Gründung der DDR: Von der sowjetischen Besatzungsherrschaft zum sozialistischen Staat*. 3rd ed. (Munich: Deutscher Taschenbuch Verlag, 1995[1984]), 89.

29. Hermann Weber, "Freie Deutsche Jugend," in *SBZ Handbuch*, 674. See also Ulrich Mählert, *Die Freie Deutsche Jugend 1945–1949: Von den "Antifaschistischen Jugendausschüssen" zur SED-Massenorganisation: Die Erfassung der Jugend in der Sowjetischen Besatzungszone* (Paderborn: F. Schöningh, 1995) and Alan McDougall, *Youth Politics in East Germany: The Free German Youth Movement, 1946–1968* (Oxford: Oxford University Press, 2004).

30. Gert Geißler, "Schulreform zwischen Diktaturen? Pädagogik und Politik in der frühen Sowjetischen Besatzungszone Deutschlands," in *Bildung und Erziehung in Europa*, ed. Dietrich Benner, *Zeitschrift für Pädagogik*. 32nd Supplement (Weinheim: Beltz, 1994), 49–63.

31. See Freya Klier, *Lüg Vaterland: Erziehung in der DDR* (Munich: Kinder, 1990).

32. "Konstulation bei Frau Prof. Dr. Bordag[-Wettengel]," January 13, 1966, DIPF/BBF/Archiv, Sammlungsgut, Döbelner Konferenz, no. 144.

33. Staritz, *Die Gründung der DDR*, 93.

34. Ibid., 96.

35. Wilhelm Schneller, "Bericht Nr. 2," Dresden, n.d. [shortly after September 12, 1945], DIPF/BBF/Archiv, Sammlungsgut, Döbelner Konferenz, no. 15.

36. Wilhelm Schneller, "Bericht Nr. 5," Dresden, n.d. [ca. Fall 1945], DIPF/BBF/Archiv, Sammlungsgut, Döbelner Konferenz, no. 20.

37. The Swede Ellen Key is considered the founder of reform pedagogy, also known as progressive education, child-centered education, new pedagogy, etc., but she elaborated its concepts in Germany. For a synopsis of the movement in Germany, see Marjorie Lamberti, "Radical Schoolteachers and the Origins of the Progressive Education Movement in Germany, 1900–1914," *History of Education Quarterly*, 40.1 (Spring 2000): 22–48;

Sterling Fishman, *The Struggle for German Youth: The Search for Educational Reform in Imperial Germany* (New York: Revisionist Press, 1976).

38. As an American, Dewey was losing favor in the Soviet zone as well, although the Humboldt University professor of pedagogy Heinrich Deiters (SPD/SED) claimed in 1946 that examining Dewey's contributions to pedagogical practice belonged to the "most pressing and fruitful" tasks facing Soviet zone educators. Heinrich Deiters, "Deweys Bedeutung für die deutsche Pädagogik," speech held to teachers on December 24, 1946, DIPF/BBF/Archiv NL Heinrich Deiters, fo. 46. In 1949, Deiters again referred to Dewey as the "most important of American pedagogues in the last half century." Heinrich Deiters, "Rundfunkvortrag," January 4, 1949, DIPF/BBF/Archiv, NL Heinrich Deiters, fo. 47.

39. Kreisschulrat Richter from Plauen-Land, "Plan zur politischen Bildung der Lehrer," Kreisschulratkonferenz, Dresden-Wachwitz, September 20–24, 1949, BArch DR 2/489, no. 456.

40. See Robert Alt, "Über unsere Stellung zur Reformpädagogik," in *Pädagogik* 11, nr. 5/6 (1956): 345–367; and Baske, "Pädagogische Wissenschaft," in *Handbuch der deutschen Bildungsgeschichte*, 143–147.

41. "Konferenz der Kreisschulräte," Dresden-Wachwitz, November 29–30, 1947, BArch DR 2/489, no. 12.

42. "Berichte 1947," BArch DR 2/4743, no. 4.

43. Ibid.

44. Werner Müller, "Freier Deutscher Gewerkschaftsbund," in *SBZ Handbuch*, 626.

45. More research remains to be undertaken on the role of the FDGB and schools. See Gert Geißler, "Schulreform zwischen Diktaturen?" 49–63; Ulrich Gill, *Der Freie Deutsche Gewerkschaftsbund: Geschichte-Organisation-Funktionen-Kritik* (Opladen: Leske and Budrich, 1989), 49–168; Freunde der neuen Schule, "Arbeitsprogramm für das erste Jahr des Zweijahresplans," [1949], DIPF/BBF/Archiv NL Karl Sothmann, fo. 58.

46. "Tätigkeitsbericht der Landesregierung Mecklenburg-Vorpommern in Schwerin, Ministerium für Volksbildung für das Jahr 1946," BArch, DR 2/280, no. 31–41.

47. "Bericht über das Schulwesen in Sachsen," March 1949, BArch DR 2/280, no. 284.

48. "Berichte 1947," BArch R2/4743, no. 4.

49. Ibid.

50. Wilhelm Schneller, "Dis Ausbildung der Neulehrer," *Sächsische Volkszeitung* 29 December 1945, DIPF/BBF/Archiv, Sammlungsgut, Döbelner Konferenz, no. 117.

51. n.a., "Mit den Augen des Schülers," in *Schöpferische Gegenwart—Kulturpolitische Monatszeitschrift Thüringens* 5 (November 1948): 304, in DIPF/BBF/Archiv, fo. Marie Torhorst, Mappe 1.

52. Charlotte Diesel-Behnke, "Lebenslauf 1951," DIPF/BBF/Archiv, fo. Charlotte Diesel-Behnke.

53. Charlotte Diesel-Behnke, letters to Ministerium für Volksbildung in Weimar, July 27, 1949 and October 2, 1949, DIPF/BBF/Archiv, fo. Charlotte Diesel-Behnke.

54. "Fluktuation der Lehrer!" August 9, 1949, BArch DR 2/911, no. 77.

55. Dr. Rosenow, to Volksbildungsminterium, Schwerin, February 17, 1947, BArch DR 2/4702, no. 28.

56. "Bericht über den Stand der Volksbilding in Berlin für das Schuljahr 1947/48, Bezirksamt Köpenick von Gross-Berlin," July 17, 1948, LAB/STA 120/30, no. 51.

57. Ulrike Pilarczyk, "Veränderung des schulischen Raum-, Zeit- und Rollengefüges im Prozeß der Politisierung der DDR-Schule," in Heinz-Elmar Tenorth, ed. *Kindheit, Jugend und Bildungsarbeit im Wandel Zeitschrift für Pädagogik*, 117–118; 139–141.

58. See, for example, Brigitte Reimann, December 28, 1950, *Aber wir schaffen es, verlaß Dich drauf! Briefe an eine Freundin im Westen.* 2nd ed. (Berlin: Aufbau Taschenbuch Verlag, 1999), 91–92; see also Ulrike Pilarczyk, "Veränderung des schulischen Raum-, Zeit- und Rollengefüge," 139–141.

59. Naimark, *The Russians in Germany*, 467.

60. Schulabteilung der Deutschen Verwaltung für Volksbildung in der Sowjetischen Besatzungszone, ed., "Baulicher Zustand der Schulgebäude," *Deutsche Demokratische Schule im Aufbau*, 30–31.

61. "Rückblick und Ausblick auf das Schulwesen des Landes Brandenburg," Rede auf dem Pädagogischem Landeskongreß in Cottbus, May 25–26, 1948, BArch DR 2/1394, no. 34–37.

62. "Tagung der Bezirksschulräte am 4. und 25. April 1946 in Meißen," BArch DR 2/488, 211–212.

63. For instance, "Jahresbericht der Schulabteilung 1947, Land Thüringen Ministerium für Volksbildung, Weimar," BArch DR 2/1394, June 1, 1948, 55.

64. Paul-Ernst Hübner, "Hausaufsatz. Ein Jahr Wiederaufbau in Bezirk Prenzlauer Berg. Kampf mit dem Winter," LAB/STA 134/13, 182/2, May 1, 1946, no. 336.

65. Ursula Reimann, "Der Aufbau unserer Schule," LAB/STA 134/13, 181/2, no. 433.

66. Margot Müller, "Winterfestmachung der Schule," seventh grade, [ca. 1946], LAB/STA 134/13, 182/2.

67. Sonja Schönrock, "Ein Jahr Wiederaufbau im Bezirk Prenzlauer Berg. Unsere Schule." fifth grade, [1946] LAB/STA 134/13, 181/2, no. 527.

68. Gerda Faust, "Der Kampf gegen den Winter," fourth grade, [ca. 1946], LAB/STA 134/13; 182/2, no. 288.

69. Gottfried Uhlig, *Der Beginn der Antifaschistisch-Demokratischen Schulreform, 1945–1946. Monumenta Paedagogica*, vol. 2 (Berlin [East]: Akademie Verlag, 1965), 74.

70. Helga Hampke, "Kampf mit dem Winter," seventh grade, [ca. 1946] LAB/STA 134/13, 182/2, no. 290; Heini Dümmel, fourteen years old,

"Ein jahr Wiederaufbau. Kampf mit dem Winter," [1946], LAB/STA 134/13; 182/2, no. 313.

71. For instance, Christel Novak, Brigitte Klinge berger, Sonja Krakau, Bärbel Lisurek, Ruth Jenner, Ingrid Sorgenfrei, Siegrid Neumann, "Wiederaufbau Prenzlauer Berg," [May 1946], LAB/STA 134/13, 182/1, no. 105.

72. Paul-Ernst Hübner, sixth grade, "Hausaufsatz. Ein Jahr Wiederaufbau in Bezirk Prenzlauer Berg. Kampf mit dem Winter." 1 May 1946, LAB/STA 134/13, 182/2, no. 336.

73. Herta Hielscher, "Meine IIIb. Pädagogische Erfahrungen einer Junglehrerin" *die neue schule*, 2.1 (1947): 10.

74. Christel Novak et al., "Wiederaufbau Prenzlauer Berg," [May 1946], LAB/STA 134/13, 182/1, no. 107–108.

75. Annett Gröschner, "*Ich schlug meiner Mutter die brennenden funken ab*": *Berliner Schulaufsätze aus dem Jahr 1946* (Berlin: Kontext, 1996), 28–32.

76. "Schule mit neuem Geist," DIPF/BBF/Archiv, Sammlungsgut, Döbelner Konferenz, [ca. October 1945], no. 121.

77. "Bericht über die soziale und sozialpädagogische Arbeit des Hauptschulamtes," Berlin, November 3, 1945, Magistrat der Stadt Berlin Abt. Volksbildung, Hauptschulamt, LAB/STA 120/30, no. 1.

78. See Allied Command Order BK/O 45/173, October 23, 1945, in Berliner Senat, ed., *Berlin: Kampf um Freiheit und Selbstverwaltung 1945–1946* (Berlin [West]: Heinz Spitzing, 1961), 234.

79. New National Socialist curricula were not approved until 1937 for the elementary school in Nazi Germany and 1938 for upper secondary schools. Bernd Zymek, "Schulen," in *Handbuch der deutschen Bildungsgeschichte*, vol. 5, *1918–1945, Die Weimarer Republik und die nationalsozialistische Diktatur*, ed. Dieter Langewiesche and Heinz-Elmar Tenorth (Munich: Beck, 1989), 191.

80. Uhlig, *Der Beginn der Antifaschistisch-Demokratischen Schulreform*, 74. According to the British journalist Gordon Schaffer, by May 1947, 7,100,000 German textbooks had been printed, 170,000 Russian books, 300, 000 English books, and 3,760,000 arithmetic books, with history books presenting the biggest problem because of their ideological content. Gordon Schaffer, *Russian Zone* (London: Co-operative Press, 1947), 119.

81. Uhlig, *Der Beginn der Antifaschistisch-Demokratischen Schulreform*, 74.

82. Correspondence from Elternausschuß of the Tenth School in Berlin-Adlershof, to Hauptschulamt, October 6, 1948, LAB/STA 120/267, no. 41.

83. Uhlig, *Der Beginn der Antifaschistisch-Demokratischen Schulreform*, 74.

84. Naimark, *The Russians in Germany*, 454.

85. Prov. Ausschuss antifaschistischer Lehrer [PAAL], "Niederschrift über die Sitzung am 1.6.45, nach. 15 Uhr bei Kollegen Fritz, Dresden-Leuben, Dürrst. 18," DIPF/BBF/Archiv, Sammlungsgut, Döbelner Konferenz, 1945," no. 6.

86. Kathleen Southwell Davis, "Das Schulbuchwesen als Spiegel der Bildungspolitik von 1945 bis 1950," in *Umerziehung und Wiederaufbau: Die*

Bildungspolitik der Besatzungsmächte in Deutschland und Österreich, ed. Manfred Heinemann (Stuttgart: Klett-Cotta, 1981),154.

87. Ibid., 158–164.

88. Baske, "Grund und Rahmenbedingugnen," in *Handbuch der deutschen Bildungsgeschichte*, ed. Christoph Führ and Carl-Ludwig Furck, vol. 6, bk. 2, *Deutsche Demokratische Republik*, 13.

89. "Protokoll über die Gesamtkonferenz der Hauptschulräte," February 19, 1947, LAB/STA 120/124, no. 1–18, here no. 14.

90. See the lengthy process required by the Allied Command in the memo from Col. Ryan, Chief of Staff, Allied Kommandatura, "Berliner Schulen," September 4, 1945, LAB/STA 120/7, no. 3.

91. Davis, "Das Schulbuchwesen als Spiegel der Bildungspolitik," 159–160.

92. "Protokoll über die Besprechung der Obleute der Schulhelfer am 20.9.48, 1 Uhr im Hauptschulamt, Zimmer 201," DIPF/BBF/Archiv, NL Hans Löffler 0.4.34, pt. 1.

93. Falk Pingel, "National Socialism and the Holocaust in West German school books," in *Internationale Schulbuchforschung* 22 (2000): 12–13.

94. Horst Diere, "Die französische Revolution von 1789 bis 1795 und die Zeit Napoleons im Geschichtslehrbuch der DDR," in *Die Französische Revolution in den Geschichtsschulbüchern der Welt*, ed. Rainer Riemenschneider (Frankfurt, AM: Diesterweg, 1994), 329.

95. "Befehl des Obersten Chefs des SMA und Oberkommandierenden der sowjetischen Besatzungszonen in Deutschland Nr. 150," May 18, 1946, BArch DR 2/636, no. 40.

96. Karl-Heinz Günther and Gottfried Uhlig, eds. *Zur Entwicklung des Volksbildungswesens auf dem Gebiet der Deutschen Demokratischen Republik 1946–1949*, in *Monumenta Paedagogica*, vol. 3 (Berlin [East] Volk und Wissen, 1968), 33.

97. "Ministerbesprechung," Berlin, March 18–19, 1947, BArch DR 2/53, no. 37–38.

98. Johanna Zeitschel, "Lehr- und Lernmittel, Anschauungsmittel für den Geschichtsunterricht" *die neue schule*, 2.3 (1947): 3.

99. Dr. Werner Büngel, *Die Entwicklung des Brandenburgisch-Preußischen Staates bis 1786*–Lehrhefte für den Geschichtsunterricht in der Oberschule, nr. 3 (Berlin[East]/Leipzig: Volk und Wissen, 1946), 28–29.

100. Ruth Krenn, *Die industrielle Revolution in England*, Arbeitshefte für den Geschichtsunterricht in der Grundschule. Material für die Hand des Lehrers (Berlin[East]/Leipzig: Volk und Wissen, 1949).

101. Editor, "Einladung zur Mitarbeit," 1–2 and Richard Shallock, "Zum Beginn," 2–3, both in *Geschichte in der Schule* 2:1 (January/February 1949).

102. See, respectively, Paul Joecks, "Der Verfassungsentwurf im Geschichtsunterricht der Grundschule," *Geschichte in der Schule* 1, nr. 1/2 (July/October 1948): 20–21 and Ernst Niekisch, "Die deutsche Geschichtsschreibung, vorwiegend des 19. Jahrhunderts, und die

Gegenwart," *Geschichte in der Schule* 1, nr. 3 (November/December 1948): 21–24.

103. Ingeburg Schenk, "Eine junge Kollegin berichtet," *Geschichte in der Schule* 2, nr.1 (Jan./Feb. 1949): 46.

104. See the introductory essay for the educational journal *die neue schule*, Wandel, "Der neuen Schule entgegen" *die neue schule*, 1.1 (1946): 3.

105. For instance, Diederich and Tenorth, *Theorie der Schule*, 89.

106. "Konferenz Verband Länder und Provinzen, Kommission Fragen der Landschulreform," Berlin, August 30, 1947, no. 42–43.

107. "Rückblick und Ausblick auf das Schulwesen des Landes Brandenburg," "Rede auf dem Pädagogischem Landeskongreß in Cottbus," May 25–26, 1948, BArch DR 2/1394, no. 37.

108. Zentralsekretariat der SED, Abteilung Kultur und Erziehung, "Die neue deutsche Schule," n.d. BArch DR 2/1080, no. 6.

Chapter 3   Rebuilding the School

1. Norris Brock Johnson, "School Spaces and Architecture: The Social and Cultural Landscape of Educational Environments," *Journal of American Culture* 5 (Winter 1982): 79.

2. Herr Schulrat Viehweg comments, "Bericht über die Tagung der Schulaufsichtsbeamten des Bundeslandes Sachsen am 19. und 20. Oktober 1945," BArch DR 2/488, no.24.

3. "Bericht über die Tagung der Schulaufsichtsbeamten des Bundeslandes Sachsen am 19. und 20. Oktober 1945," BArch DR 2/488Ibid., no. 28–29.

4. "Referat des Ministerialrat Viehweg in der Lehrerkonferenz am 21. Juni 1947 in der Oberschule," "Döbelner Konferenz 1945, Pädagogisches Institut K.F.W. Wander Dresden Lehrstuhl Marxismus-Leninismus Abt. Geschichte der deutschen Arbeiterbewegung Dr. U. Mantzke," Saxony, DIPF/BBF/Archiv, Sammlungsgut, no. 112.

5. See the reflections on schools and architecture in the excellent volume by Jörn-Peter Schmidt-Thomsen, Helga Schmidt-Thomsen, Manfred Scholz, and Peter Güttler, *Berlin und seine Bauten*, vol. 5, bk. c, *Schulen* (Berlin: Ernst and Sohn, 1991).

6. Cf. Malcolm Vivian, John Seaborne, and R. Lowe, *The English School: Its Architecture and Organization*, vol. 2 (London: Routledge and Kegan, 1977); William Cutler, "Cathedral of Culture: The Schoolhouse in American Educational Thought and Practice since 1820" *History of Education Quarterly*, 29.1 (Spring 1989): 1–40; Henry Barnard, *School Architecture, or Contributions to the Improvement of School Houses in the United States*. 2nd ed. (Ann Arbor: University of Michigan Press; reprint 2001 [1848]).

7. Klaus Rasche, "1945–1990: Architettura e Urbanistica nella Germania Orientale. L'esempio di Dresda," *Storia Urbana*, 73 (September–October 1995): 132–134.

8. Dieter Langewiesche and Heinz-Elmar Tenorth, eds. *Handbuch der deutschen Bildungsgeschichte*, vol. 5. (Weinheim: Deutscher Studien Verlag, 1997).

9. Rasche, "1945–1990: Architettura e Urbanistica nella Germania Orientale," 133. Interestingly, Rasche claims that many architects found refuge in working on industrial and highway construction during the Nazi period as a way of opposing Nazism while maintaining contact with their careers.

10. Giuseppe Longhi, "Introduzione," *Storia urbana*, 73 (October–December, 1995): 5.

11. Steinniger [?] "Bericht über den Stand der Investitionsarbeiten, der in der Kulturversordnung der DWK vom 31.3.49 benannten Investionsvorhaben" July 15, 1949, BArchiv DR2/689, no. 48.

12. "Bericht über den Stand der Investitionsarbeiten," July 15, 1949, BArchiv DR2/689, no. 46–49.

13. Martin Kudlek, "Architecture," in *Encyclopedia of Contemporary German Culture*, ed. John Sandford (London: Routledge, 1999), 18.

14. Rasche, "1945–1990: Architettura e Urbanistica nella Germania Orientale," 131–152.

15. Friedhelm Fischer, "German Reconstruction as an International Activity," in *Rebuilding Euorpe's Bombed Cities*, ed. Jeffry M. Diefendorf (London: Macmillan, 1990), 134–137.

16. Karl Mang described the interior designs that did develop as an eclectic kitsch resulting from copying pieces of styles from around the world. Mang furthermore stated that this held true for bourgeois as well as working—class families (*borghese* and *classe operaia*). Karl Mang, *Storia del mobile moderno* (Rome: Laterza, 1991[1978]), 153.

17. Hilde Thurnwald, *Gegenwartsprobleme Berliner Familien: Eine soziologische Untersuchung an 498 Familien* (Berlin [West]: Weidmannsche Verlasbuchhandlung, 1948), 39–49.

18. Ibid., 38–43.

19. See the following posters from *Das politische Plakat der DDR 1945–1970* [CD-ROM]: "Leipzigs Männer und Frauen Wählt Liste 1 SED," after May 8, 1945, Inv. P 94/1986; "Vom Turm des neuen Rathauses herab grüßt der Stern von Leipzig," 1947, P 94/35; "Leipzigs Jugend hilft beim Aufbau!" after October 7, 1949, Inv. P 94/277; "Das neue Dresden 1946," 1946, Inv. P 94/41; "Bautzen im Aufbau 1945/49," 1949, Inv. P 94/280.

20. Discussions of living and working spaces, especially for workers, would soon dominate GDR architectural and public discourse. One of the best accounts of these debates is Brigitte Reimann's semiautobiographical novel *Franziska Linkerhand* (Berlin: Aufbau, 1998). Earlier publications of the novel represent the GDR censored form; this edition includes the previously removed passages and editorial essay.

21. "So oder So?" Oederan, 1949, in *Das politische Plakat der DDR, 1945–1970* [CD-ROM], Inv. P 94/1257.

22. "Jugend baut auf!" Leipzig, 1946, *Das politische Plakat der DDR, 1945–1970* [CD-ROM], Inv. P 94/21.

23. "Wir bauen Schulen," Erfurt, [ca. 1945–1949], in *Plakate der SBZ/DDR* [CD-ROM], ed. Deutsches Historisches Museum (Munich: Sauer, 1999), Inv. P 90/4507.

24. Rudy Koshar, *Germany's Transient Pasts: Preservation and National Memory in the Twentieth Century* (Chapel Hill: University of North Carolina Press), 208.

25. Jürgen Paul, "Reconstruction of the City Centre of Dresden: Planning and Building during the 1950s," in *Rebuilding Europe's Bombed Cities*. Ed. Jeffry M. Diefendorf (London: Macmillan, 1990), 173.

26. Klaus von Beyme, "Reconstruction in the German Democratic Republic," in *Rebuilding Euorpe's Bombed Cities*, ed. Jeffry M. Diefendorf (London: Macmillan, 1990), 190.

27. Schulabteilung der Deutschen Verwaltung für Volksbildung in der Sowjetischen Besatzungszone, ed., *Die Deutsche Demokratische Schule im Aufbau*. The statistical section was published as an appendix: Erich Oberhaus, "Statistik des Schulwesens der Sowjetischen Besatzungszone," and was renumbered 1–79.

28. W. J. T. Mitchell, "Representation," in *Critical Terms for Literary Study*, ed. Frank Lentricchia and Thomas McLaughlin (Chicago: University of Chicago Press, 1990), 13–15, 89.

29. "Tafel zu Tabelle 2, Schulen—Schüler—Lehrer," in Schulabteilung der Deutschen Verwaltung für Volksbildung in der Sowjetischen Besatzungszone, ed., *Die Deutsche Demokratische Schule im Aufbau* (Leipzig: Volk und Wissen, 1949), statistical appendix, 11.

30. Ibid., 5; "Der bauliche Zustand der Schulgebäude, Tabelle 15," and "Tafel zu Tabelle 15, Baulicher Zustand der Schulgebäude," 30–31.

31. Rudy Koshar cited the uneven damage suffered in the Soviet zone, with some regions almost destroyed and others having almost no physical signs of bombing. School buildings reflected this uneven destruction, but there was lighter percentual damage in rural areas. Koshar, *Germany's Transient Pasts*, 214.

32. "Tätigkeitsbericht 1945 des Volksbildungsamtes der Stadt Leipzig," BArch DR-2/1000, no. 18.

33. "Bericht des Volksbildungsministerums Sachsen über das Schuljahr 1947/48," BArch DR 2/1394, no. 27.

34. Manfred Scholz, "Schulen nach 1945," in *Berlin und seine Bauten*, ed. Helga Schmidt-Thomsen, Manfred Scholz and Peter Güttler (Berlin: Ernst & Sohn, 1991), 198.

35. Emphasis in original. Brian Ladd, *The Ghosts of Berlin: Confronting German History in the Urban Landscape* (Chicago: University of Chicago Press, 1997), 177.

36. For an excellent study on the origins and role of architecture in Nazi society, see Barbara Miller Lane, *Architecture and Politics in Germany, 1918–1945*

(Cambridge, MA: Harvard University Press, 1968). See also Rudy Koshar's discussion of the permanent change to the built environment effected by Nazi architecture programs. Koshar, *Germany's Transient Pasts*, 206–207.

37. For discussions of the uses of metaphors in social policy, see Donald A. Schön, "Generative Metaphor and Social Policy," in *Metaphor and Thought*. 2nd ed., ed. Andrew Ortony (Cambridge: Cambridge University Press, 1993), 137–163, 143–145; also, see Robert J. Sternberg, Roger Tourangeau, and Georgia Nigro, "Metaphor, Induction, and Social Policy: The Convergence of Macroscopic and Microscopic Views," 277–303 of the same volume.

38. Ernst Wildangel, "Ein halbes Jahr Schulaufbau," [1946], LAB/STA 120/30, no. 304.

39. Metonymy and metaphor are highly-contested fields of linguistic and literary theory. Metonymy has occasionally been interpreted as a subfield of metaphor. More often, metonymy is considered to be diametrically opposed to it. Here I define both concepts as having separate but overlapping functions. I am most interested in focusing on the common definition of part-whole relationships in metonymy as expressing "simple contiguous relationships between objects." See Raymond W. Gibbs, Jr., "Process and Products in Making Sense of Tropes," in *Metaphor and Thought*, ed. Andrew Ortony, 258; and the following essays in *Critical Terms for Literary Study*, ed. Frank Lentricchia and Thomas McLaughlin (Chicago: Chicago University Press, 1990): Thomas McLaughlin, "Figurative Language," 80–90; Louis A. Renza, "Influence," 186–202; Françoise Meltzer, "Unconscious," 147–162.

40. "Berichtsbuch der 4. Volksschule Berlin-Köpenick 1935–1947," October 29, 1945, SM/Do 92/141.

41. Karl Sothmann, "Die Rolle der demokratischen Organisationen und der Betriebe bei der Demokratisierung des Schulwesens," in *Die Deutsche Demokratische Schule im Aufbau*, ed. Schulabteilung der Deutschen Verwaltung für Volksbildung in der Sowjetischen Besatzungszone (Berlin [East]/Leipzig: Volk und Wissen, 1949), 57.

42. Memorandum, from P. Zolotuchin, to Wandel, Berlin, January 8, 1946, BArch R 2/636, no. 129–130.

43. For a discussion of the Soviets' understanding of the need to cultivate popular German support, see Norman Naimark, "Der Nationalismus und die osteuropäische Revolution 1944–1947" *Transit: Europäische Revue*, (Vienna) 15 (Fall 1998): 42–43.

44. Naimark, *The Russians in Germany*, 33–34.

45. Otto Winzer, "Das erste Jahr des Berliner Schulaufbaus. Notwendige Bemerkzungen zur materiellen Grundlage der Schulreform," [1946], LAB/STA 120/30, no. 303.

46. A more detailed comparison of Berlin's pre- and postwar East–West divisions remains to be undertaken. In the field of education, for instance, many of East Berlin's education reformers worked in the (politically) western region of Neukölln. Additionally, the figure of Paul Oesterreich is

interesting for his strange status between the Soviet zone/GDR and the Western zones/FRG. Oesterreich worked closely in Berlin with educational reformers in the Soviet zone, remained in the West throughout his life, but enjoyed a special status with financial and celebratory benefits in the GDR.

47. See Tony Judt's discussion of postwar continuations of prewar conflict, in Tony Judt, "Europas Nachkriegsgeschichte neu Denken," *Transit: Europäische Revue*, 15 (Vienna, Fall 1998): 10.

48. Herr Minister Direktor Schneller, "Konferenz der Kreisschulräte," Dresden—Wachwitz, November 29–30, 1947, BArch DR-2/489, no. 8.

49. Ibid., no. 9–10.

50. "Bericht des Kreisschulrates des kreises Randow in Boock), March 27, 1947, BArch DR-2/4702, no. 23.

51. "Chronologischer Bericht des hausverwaltenden Rektors 1.V. über die Wiedereinrichtung und den Neuaufbau der Schulden der Schulgruppe Eberwalder Str." August 27, 1945, LAB/STA 134/13, 183, no. 316. Debates about school districting have accompanied the evolution of mass schooling, and have always reflected local and national political decisions. For instance, Bavaria in 1873 established school districts to coincide with communal instead of parish boundaries, so that pupils from both Protestant and Catholic confessions attended the same school. The beginning of the "*Kulturkampf*'s most bitter round," this decision was revoked a decade later, a serious defeat for proponents of secularization. Karl A. Schleunes, *Schooling and Society: The Politics of Education in Prussia and Bavaria 1750–1900* (Oxford: Berg, 1989), 182–186.

52. See the discussion between local and general memories in Benoît Lecoq, "Le café," in *Les lieux de mémoire*, vol. 3, *Les France*, bk. 2, "Traditions," ed. Pierre Nora (Paris: Gallimard, 1992), 854–883.

53. Herr Min. Dir. Prof. Hoffmann, "Kritische Stellung zu den Gegenwärtigen Unterrichtsmethoden," "Protokoll über den Pädagogischen Landeskongreß im Maxim-Gorki-Haus in Schwerin am 25./26. Mai 1948," no. 98. Professor Ulrich Hoffmann was a CDU member from 1945 to 1946, after which he became an SED member. He held the position of director of the department for *Allgemeine Verwaltung im Ministerium für Volksbildung Mecklenburg* from 1947 until December 1948 and served in other political capacities. See Barbara Fait, "Dokumentation, Mecklenburg," 121, and "Führungskräfte in Staat, Politik und Gesellschaft: Biographische Daten," 934.

54. Herr Min. Dir. Prof. Hoffmann, "Kritische Stellung zu den Gegenwärtigen Unterrichtsmethoden," "Protokoll über den Pädagogischen Landeskongreß im Maxim-Gorki-Haus in Schwerin am 25./26. Mai 1948," no. 98.

55. Herr Stadtrat Riesner of Chemnitz, "Bericht über die Tagung der Schulaufsichtsbeamten des Bundeslandes Sachsen am 19. und 20. Oktober 1945," BArch DR 2/488, no. 46.

56. Günter de Bruyn, *Zwischenbilanz: Eine Jugend in Berlin* (Frankfurt, AM: Fischer, 1994 [1992]).

57. Ibid., 329 and 332.

58. Ibid., 353–355.

59. Thurnwald, *Gegenwartsprobleme Berliner Familien*, 111–112.

60. Klemperer, November 11, 1945, *Und so ist alles schwankend*, 211.

61. "Schulratskonferenz 3–5. Februar 1948 in Dresden-Wachwitz," BArch DR-2/489, no. 98.

62. Werner Kupke, born January 1, 1934, sixth grade, "Ein Jahr Wiederaufbau," May 9, 1946, LAB/STA 134/13, 181/1, no. 136.

63. Isolated cases of school meals existed in some schools in German history, especially as part of a social assistance program in times of need.

64. Käte Agerth, DIPF/BBF/Archiv, Nachlaß Käte Agerth 0.4.22.

65. "Schulchronik Bergfelde," October 11, 1945–September 1, 1949, Museum Kindheit und Jugend— Schulmuseum, 92/31 b.

66. A photograph of an elementary school class in 1949 of forty-five pupils shows children wearing sweaters, so that the weather was at least cool, but some of them are not wearing socks and two can be seen in the front row without shoes. Museum Kindheit Jugend—Schulmuseum, 12/01/01/09 SM/FO 88/150.

67. "Berichtsbuch der 4. Volksschule Berlin-Köpenick 1935–1947," October 16, 1945, Museum für Kindheit und Jugend—Schulmuseum, SM/Do 92/141.

68. "Sammlung von Apfel- und Birnenkernen zu Saatzwecken," signed Ellrich, Deutsche Zentralverwaltung für Volksbildung in der Sowjetsichen Besatzungszone, Berlin, October 25, 1946, BArch DR 2/3963, no. 90–91.

69. Ibid., no. 91.

70. See, for example, Doreen Massey, *Space, Place and Gender* (Cambridge: Polity Press, 1994); Philippe Boutry, "Le clocher," in *Les lieux de mémoire*, vol. 3, *Les France*, bk. 2, "Traditions," 60.

71. Peter Lundgreen addressed the modern European school as a site for learning as well as community life in his chapter, "La Scuola," in *Luoghi quotidiani nella storia d'Europa*, ed. Heinz-Gerhard Haupt (Rome: Laterza, 1993), 285–295.

72. "Protokollbuch der Lehrerbesprechungen der 2. Volksschule Berlin Prenzlauer Berg 1942–1949," September 4, 1946, Do/SM 92/55.

73. Klemperer, November 11, 1945, *Und so ist alles schwankend*, 179.

74. Helga Hampke, seventh grade, "Kampf mit dem Winter,"[1946], LAB/STA 134/13, 182/2, no. 290.

75. Margot Rosenthal, born July 9, 1931, "Ein Jahr Wiederaufbau im Bezirk Prenzlauer Berg," 17 May 1946, LAB/STA 134/13, 181/2, no. 351.

76. Wolfgang Banach, born March 3, 1933, sixth grade, "1 Jahr Wiederaufbau," LAB/STA 134/13, 181/1, no. 94.

77. Eva Schmude, born May 25, 1932, "Wiederaufbau unserer Schule," LAB/STA 134/13, 181/2, no. 434.

78. Vera Rietz, eighth grade, "Reinlichkeit und Gemütlichkeit," [1946], LAB/STA 134/13, 181/2, no. 445.

79. Ch. Tadday, seventh grade, "Dacharbeiten," [1946], LAB/STA 134/13, 181/2, no. 446.

80. Werner Weimar, seventh grade, "Unsere Straße," May 17, 1946, LAB/STA 134/13, 181/2, no. 524.

81. Brigitte Hassler, Oberschule class V, "Frauen räumen auf," April 8, 1946, LAB/STA 134, 182/1, no. 4.

82. Vera Benk, seventh grade, "Glasarbeit," [1946], LAB/STA 134/13, 181/2, no. 441.

83. Ruth Frah, sixth grade, Oberschule, LAB/STA 134/13, 181/2, no. 486.

84. See, for example, "Protokoll über die Gesamtkonferenz der Haupt- und Bezirksschulräte von Berlin am 13. November 1947," LAB/STA 120/125 no. 172.

85. "Bericht über die Tagung der Schulaufsichtsbeamten des Bundeslandes Sachsen am 19. und 20. Oktober 1945, BArch DR 2/488, no. 30.

86. Margot Müller, seventh grade, "Winterfestmachung der Schule," [1946], LAB/STA 134/13; 182/2.

87. Vera Benk, seventh grade, "Glasarbeit," LAB/STA 134/13, 181/2, no. 441.

88. Nora, "Entre mémoire et histoire: La problématique des lieux," xvii–xxiii, in Les lieux de mémoire, 7 vols., ed. Pierre Nora, vol. 1, La République (Paris: Gallimard, 1984). "Entre mémoire" is best known in its English translation by Marc Roudebush, "Between Memory and History: Les Lieux de Mémoire," Representations, 26 (Spring 1989): 7–25; Pierre Nora, "Présentation," in Les lieux de mémoire, vol. 3, Les France, bk. 2, "Traditions" (Paris: Gallimard, 1992), 13.

89. Herr Kreisschulrat Erler, Bad Schandau, "Kreisschulratskonferenz vom 3.–5. November 1948 in Dresden-Wachwitz," BArch DR-2/489, no. 324.

90. Lieselotte Walter, "Wie für uns in der Sonnenburger Schule der Kampf um Berlin begann," LAB/STA 134/13, 178, no. 3–9.

91. Ibid., no. 3.

92. Ibid., no. 4–5.

93. Ibid., no. 7–8.

94. "Chronologischer Bericht des hausverwaltenden Rektors 1.V. über die Wiedereinrichtung und den Neuaufbau der Schulen," August 27, 1945, LAB/STA 134/13, 183, no. 311.

95. Ibid., no. 312–313.

96. "Chronologischer Bericht des hausverwaltenden Rektors 1.V. über die Wiedereinrichtung und den Neuaufbau der Schulen," August 27, 1945, LAB/STA 134/13, 183, Ibid., no. 315.

97. "Leipziger Telegraf," Christmas 1947, written predominantly by Neulehrer, SM/DO 95/290.

98. Ibid.

99. Hans Joachim Linstedt, 15. Volksschule, "Wir sind ausgebombt!" January 22, 1946, LAB/STA 134/13, 180/2, no. 501.

100. Horst Prawitz, seventh grade, Schinkelschule Oberschule für Jungen, "Wie unser Haus, Stargarder Str. 1, Schönhauser Allee 70e, durch Artillerieeinwirkung verlorenging,"14 January 1946, LAB/STA 134/13, 183, no. 10.

101. Fritz Falbe, "Weshalb ich flüchten mußte" [1946] LAB/STA 134/13, 183, no. 324.

102. Hannelore Vandrei, 6. Klasse, 32. Volksschule, "Wie litt unser Schulhaus durch den Krieg," January 1, 1946, LAB/STA 134/13, 183, no. 308.

103. Fritz Falbe, "Weshalb ich flüchten mußte," [1946], LAB/STA 134/13, 183, no. 324.

104. Otto Winzer, "Das erste Jahr des Berliner Schulaufbaus. Notwendige Bemerkzungen zur materiellen Grundlage der Schulreform" [1946], LAB/STA 120/30, no. 303.

105. Thurnwald, *Gegenwartsprobleme Berliner Familien*, 112.

106. "Protokoll über die Gesamtkonferenz der Haupt- und Bezirksschulrät von Berlin am 13. November 1947," LAB/STA 120/125, no. 172.

107. Ruth Frahn, sixth grade, Mädchen-Oberschule, [1946], LAB/STA 134/13, 181/2, no. 486.

108. Hannelore Knuth, "Sprengungen," [1946], LAB/STA 134/13, 182/1, no. 400.

109. Hannelore Thal, 20. Volksschule, January 23, 1946, LAB/STA 134/13, 183, no. 307.

Chapter 4 Rubble Children and the Construction of Gender Roles

1. For instance, Andrew Parker, Mary Russo, Doris Sommer, and Patricia Yaeger, "Introduction," in *Nationalisms and Sexualities*, ed. Andrew Parker et al. (New York: Routledge, 1992), 1–18; Dieter Langewiesche, "Nation, Nationalismus, Nationalstaat: Forschungsstand und Forschungsperpsektiven." *Neue Politische Literatur* 40 (1995): 213; Jean-Claude Caron, "Young People in School: Middle and High School Students in France and Europe," and Luisa Passerini, "Youth as a Metaphor for Social Change: Fascist Italy and America in the 1950s," in *A Story of Young People in the West: Stormy Evolution to Modern Times*, ed. Levi Giovanni and Schmitt Jean-Claude, vol. 2. (Cambridge, MA: Belknap Press of Harvard University Press, 1997), 118 and 281–340, respectively; Cornelia Niekus-Moore, *The Maiden's Mirror: Reading Material for German Girls in the Sixteenth and Seventeenth Centuries* (Wiesbaden: Otto Harrassowitz, 1987).

2. The fourteen- to twenty-six year-old definition eventually evolved into a sociological and political category in the GDR that had cultural and legal bases, including participation in the secular rite of passage of the *Jugendweihe* and rights to certain legal benefits offered to young people. Edeltraud Schulze, *DDR-Jugend: Ein statistisches Handbuch* (Berlin: Akademie, 1995), 17–24.

3. Ibid., 95.

4. My own calculations, based on data in Edeltraud Schulze, *DDR-Jugend*, 26 and Günter Braun, "Daten zur demographische und sozialen Struktur der

Bevölkerung," appendix, in *SBZ Handbuch SBZ Handbuch: Staatliche Verwaltung, Parteien, Gesellschaftliche Organisationen und ihre Führungskräfte in der Sowjetischen Besatzungszone Deutschlands 1945–1949,* ed. Martin Brozat and Hermann Weber (Munich: Oldenbourg, 1993), 1070.

5. Sources vary in their statistics for the Soviet zone and GDR population. Differences can be contributed to irregular inclusion of East Berlin as well as extreme population fluctuations due to entering refugees, returning soldiers, and movement from the Soviet zone to the Western zones. The *SBZ Handbuch* and the *Deutsche Demokratische Schule im Aufbau* list the population of the Soviet zone in 1946 at 17,313,734 inhabitants, of whom 7,379,546 (42.6%) were male and 9,934,188 (57.4%) female. Although it is unclear from the tables if this number includes East Berlin, comparison with similar data suggests it does not. Braun, "Demographische und soziale Struktur der Bevölkerung," 1070, and Schulabteilung der Deutschen Verwaltung für Volksbildung in der Sowjetischen Besatzungszone, ed., *Die deutsche demokratische Schule im Aufbau* (Leipzig: Volk und Wissen, 1949), statistical appendix, 77. The figures here have been used both by the Federal Republic publication *DDR-Handbuch* [CD-ROM], s.v. "Bevölkerung," 1475 and Table 1, appendix, 1491; and the recent publication on GDR youth statistics by Edeltraud Schulze, *DDR-Jugend,* 26.

6. Report on elementary schools for school year 1946–1947, "Das erste Jahr der demokratischen Schulreformen in den Grundschulen der sowjetischen Besatzungszone," Berlin, July 17, 1947, BArch DR-2/1068, no. 2.

7. Schulabteilung der Deutsche Verwaltung für Volksbildung, *Die Deutsche Demokratische Schule im Aufbau,* statistical appendix, 19.

8. The figures for Greater Berlin were collected after the first academic school year; that is, before any structural reforms, which explains the different divisions of schools. The number of pupils does not always fully reflect the increase in population from refugees that occurred in the next years. "Die Berliner Schulen in Zahlen," LAB/STA 120/3227, no. 208–232.

9. Schulabteilung der Deutschen Verwaltung für Volksbildung, "Die Deutsche Demokratische Schule im Aufbau," statistical appendix, 77. The percentage of school-aged young people remained around 15% of the population throughout the GDR. The category fourteen- to twenty-six years of age made up ca. 18% of the population. Edeltraud Schulze, *DDR-Jugend,* 17–24.

10. For 1946, the figures are, respectively, 16.4% and 14.8%. Ibid., 27.

11. DVV, *Die Deutsche Demokratische Schule im Aufbau,* stastistical appendix, 42.

12. From Magistrat der Stadt Berlin, Abteilung Volksbildung, Hauptschulamt, Wildangel, March 11, 1946, LAB/STA 120/3292.

13. "Frauen, die unsere Kinder erziehen," *Tägliche Rundschau,* November 15, 1945.

14. Wilhelm Schneller, "Die Ausbildung der Neulehrer," *Sächsische Volkszeitung,* December 29, 1945, 117.

15. "Provinzialregierung Mark Brandenburg, Der Minister für Volksbildung, Wissenschaft und Kunst," Potsdam, March 14, 1947, BArch, DR-2/1394, no. 2.

16. William Schneller, "Die Ausbildung der Neulehrer," *Sächsiche Volkszeitung*, December 29, 1945, 117, DIPF/BBF/Archiv, 02.9.17, Sammlungsgut, Döbelener Konferenz.

17. "Frauen, die unsere Kinder erziehen," *Tägliche Rundschau*, November 15, 1945.

18. "Neue Lehrer für die neue Schule," Leipzig, 1945, *Das politische Plakat der DDR, 1945–1970* [CD-ROM], Inv. P 94/2465; "30,000 Neulehrer werden gebraucht," *Deutsche Volkszeitung*, December 26, 1945, 169.

19. "Die Lehrerin in unserer Zeit," *Neue Zeit*, December 12, 1945, 113.

20. See *DDR-Handbuch* [CD-ROM], s.v. "Frauenpolitik," 2470–2493. The feminization of the teaching force was perceived as a problem in Germany and the United States. See James Albisetti, *Schooling German Girls and Women: Secondary and Higher Education in the Nineteenth Century* (Princeton: Princeton University Press, 1988) and Ellen Condliffe Lagemann, *An Elusive Science: The Troubling History of Education Research* (Chicago: University of Chicago Press, 2000), 1–7.

21. "Werde Maurerhelferin . . ." Dresden, 1946, *Das politische Plakat der DDR* [CD-ROM], P 94/526.

22. "Ein Beruf für ganze Kerle. Werde Bergmann . . ." Berlin, 1946, *Das politische Plakat der DDR* [CD-ROM], Inv. P 94/390.

23. "Helfen. Jugendeinsatz beim Neuaufbau . . ." n.p., 1947, *Das politische Plakat der DDR* [CD-ROM], Inv. P 94/114.

24. Emil Knief, "Kommt die Koedukation?" in *Die Lehrergewerkschaft* 6, nr.1 (5 September 1947), STA/LAB 120/2520, no. 94.

25. "Der Kommunisten Ziel: Glückliches Deutschland", Leipzig, after May 8, 1945, *Das politische Plakat der DDR* [CD-ROM], Inv. P 94/1842.

26. "Weihnachten 1946 Arbeit und Aufstieg 1947 Konsum," Dresden, [1946–1947], *Das politische Plakat der DDR* [CD-ROM], Inv. P 94/619.

27. "Gedenkblatt zur Schuljahresfeier," Leipzig, [1948], SMDo 88/955.

28. "Rededisposition zur Hilfsaktion 'Rettet die Kinder!'," Berlin, [ca. Winter 1945] STA/LAB 120/2425, no. 44.

29. "Protokollbuch der Lehrerbesprechungen der 2. Volksschule Berlin Prenzlauer Berg 1942–1949," 28 January 1946, Museum Kindheit und Jugend—Schulmuseum, Do/SM 92/55.

30. For a discussion of "children as agents capable of rescuing adults from their own corrupting experiences," see Seth Koven's "From Rough Lads to Hooligans: Boy Life, National Culture and Social Reform," in *Nationalisms and Sexualities*, 365–388.

31. "Mütter, die Zukunft Eurer Kinder . . ." Dresden, 1945, *Das politische Plakat der DDR* [CD-ROM], Inv. P 94/1860.

32. Oswald A. Erich and Richard Beitl, 3rd. ed, *Wörterbuch der deutschen Volkskunde*, s.v. "Mohn" (Stuttgart: Alfred Kröner Verlag, 1974), 567–568.

33. Report from Volksschule director Häntzsche about teacher's comments, Dresden, May 29, 1945, DIPF/BBF/Archiv, Sammlungsgut, in folder 'Döbelner Konferenz.'

34. Report, "Pädagogische Tagung in Potsdam, vom 29–31. August 1945," BArch DR 2/488, no. 3.

35. "Konferenz der Kreisschulräte," Dresden-Wachwitz, November 29–30, 1947, BArch DR2/489, no. 55.

36. "Tagung der Bezirksschulräte," BArch DR2/488, April 24–25, 1946, no. 236.

37. "Konferenz der Kreisschulräte," Dresden-Wachwitz, November 29–30, 1947, BArch DR2/489, no. 51–52.

38. Norman Naimark noted that women in the French zone suffered similar atrocities by Moroccan soldiers, especially in the initial postwar period. Norman Naimark, *The Russians in Germany: A History of the Soviet Zone of Occupation, 1945–1949* (Cambridge, MA: Belknap Press of Harvard University Press, 1995), 106. See also Atina Grossmann, "Eine Frage des Schweigens? Die Vergewaltigung deutscher Frauen durch Besatzungssoldaten," *Sozialwissenschaftliche Informationen*, 24.2 (1995): 109–119.

39. "Der Tommy und das Fräulein," n.a., *Benjamin: Zeitschrift für junge Menschen*, (Hamburg) 1.5 (April 11, 1947): 20.

40. Martin Sinclair, "Schokoladenmädchen—Geschlechtskrankheiten," *Benjamin: Zeitschrift für junge Menschen*, (Hamburg) 1.7 (May 20, 1947): 14

41. Grossmann, "Eine Frage des Schweigens?" 110.

42. Ibid., 112–113.

43. All of the examination questions used "Russian" and not "Soviet," a practice not typical for official documents but not unusual in everyday language. "Prüfungsaufgaben zur schriftlichen Prüfung für den Acht-Monate-Kursus vom 4. Januar 1946," cited in Sylvia Mebus, "Zur Entwicklung der Lehrerausbildung in der SBZ/DDR 1945 bis 1959 am Beispiel Dresdens: Pädagogik zwischen Selbst- und Fremdbestimmung," vol. 2 (Habilitation, Griefswald, Germany: University of Greifswald, 1997), 415.

44. Naimark, *The Russians in Germany*, 93.

45. Report by school director Häntzsche and teacher Röhner from 64th Volksschule, Dresden, May 29, 1945, DIPF/BBF/Archiv, in Sammlungsgut, Döbelner Konferenz, 02.9./17, no. 4–5.

46. Ibid., 5.

47. Director of the Abteilung für Volksbildung der Sowjetischen Militärverwaltung in Deutschland, P. Zolotuchin, to Paul Wandel, DVV, Berlin, January 8, 1946, no. 129; and "Protokoll über die Gesamtkonferenz der Hauptschulräte," Berlin, June 29, 1947, no. 80–81.

48. Hildegard Borris, thirteen years old, 14. Volksschule, Klasse 8 b—Mädchen, "Unsere Gedanken zur Schule," DIPF/BBF/Archiv, NL Löffler, fo. 10, no. 2.

49. "Protokollbuch der Lehrerbesprechungen der 2. Volksschule Berlin Prenzlauer Berg," July 5, 1946, Do/SM 92/55.

50. "Tagung der Bezirksschulräte," Meißen, April 24–25, 1946, BArch DR2/488, no. 265.

51. "Berichtsbuch der 4. Volksschule Berlin-Köpenick 1935–1947," SM/Do 92/141, September 3, 1945.

52. "Protokoll, Gesamtkonferenz, Schulräte sämtlicher Bezirke," July 30, 1947, LAB/STA 120/125, no. 103.

53. "Bericht über den Stand der Volksbildung in Berlin für das Schuljahr 1947/48," Köpenick, July 17, 1948, LAB/STA 120/30, no. 55.

54. "Protokoll über die Gesamtkonferenz der Hauptschulräte," Berlin, June 29, 1947, LAB/STA 120/124, no. 69–72.

55. "Protokoll über die Gesamtkonferenz der Hauptschulräte," Berlin, April 23, 1947, LAB/STA 120/124, no. 130–132.

56. Manfred Ansurge, twelve years old, 13. Volksschule, Prenzlauer Berg, Klasse 6a, "1. Jahr Wiederaufbau," [May 1946], LAB/STA 134/13, 181/1, no. 91.

57. Thurnwald, *Gegenwartsprobleme Berliner Familien*, 319–322.

58. Ibid., 253.

59. Karl Ketteniß (Schulhelfer), "Meine Schüler in ihrer schulfreien Zeit," 13. Volksschule, Prenzlauer Berg, November 30, 1946, Museum Kindheit und Jugend—Schulmuseum), no signature.

60. Ibid.

61. Maja Riepl-Schmidt, "Die 'wertkonservativen' Erziehungskonzepte zur Vorbereitung einer 'mit Vernunft getragenen Nation' der Therese Huber (1764–1829)," in *Frauen und Nation*, Frauen und Geschichte, ed., (Tübingen: Silberburg, 1996), 93.

62. Jean-Jacques Rousseau, *Emile, or On Education*, trans. Allan Bloom (New York: Basic Books, 1979 [1762]), particularly book five.

63. For example, Olympe de Gouges, *Ecrits politiques*, with an introduction by Olivier Blanc (Paris: Côté-femmes, 1993).

64. Joachim Heinrich Campe, *Über einige verkannte wenigstens ungenützte Mittel zur Beförderung der Indüstrie, der Bevölkerung und des öffentlichen Wohlstandes*, cited in Christine Mayer, "Die Anfänge einer institutionalisierten Mädchenerziehung an der Wende vom 18. zum 19. Jahrhundert," in *Geschichte der Mädchen- und Frauenbildung*, vol. 1, ed. Elke Kleinau and Claudia Opitz (Frankfurt, AM: Campus, 1996), 373–374.

65. See, for instance, Bonnie Smith, *Ladies of the Leisure Class: The Bourgeoises of Northern France in the 19th Century* (Princeton: Princeton University Press, 1981).

66. Natali Stegmann has demonstrated how the Polish province of Posen promoted education in the late nineteenth century to increase national consciousness in women. This consciousness remained uniquely female and Catholic, and permitted women to become "politicized" and "nationalized" while rejecting "modernization tendencies." Natali Stegmann, " 'Je mehr Bildung, desto polnischer.' Die Nationalisierung polnischer Frauen in der Provinz Posen (1870–1914)," in *Frauen und Nation*, 165–177.

67. See, for instance, how the debates about women's rights and modernity have determined European perceptions of Turkey. Stefan Batzli, Fridolin Kissling, Rudolf Ziwmann, ed. *Menschenbilder-Menschenrechte. Islam und Okzident: Kulturen im Konflikt,* (Zürich: Unionsverlag, 1994). This discourse has been used by western and non-western societal reformers alike. See Frances Vavrus, "Governmentality in an Era of 'Empowerment': The Case of Tanzania," in *Educational Knowledge: Changing Relationships between the State, Civil Society, and the Educational Community,* Thomas Popkewitz, ed. (Albany: State University of New York Press, 2000), 221–242; Fanny Davis, *The Ottoman Lady: A Social History from 1718 to 1918* (New York: Greenwood, 1986), 50.

68. For an analysis of women's nation-building projects, which often ran parallel to the official men's version, see Linda Colley, *Britons: Forging the Nation 1707–1837* (New Haven: Yale University Press, 1992).

69. See Sabina Loriga, "The Military Experience," 11–36 and Jean-Claude Caron, "Young People in School: Middle and High School Students in France and Europe," 117–173 in *Story of Young People in the West.*

70. Siegrid Westphal, "Reformatorische Bildungskonzepte für Mädchen und Frauen—Theorie und Praxis," 135–151 and Andrea Kammeier-Nebel, "Frauenbildung im Kaufmannsmilieu spätmittelalterlicher Städte," 78–90, in *Geschichte der Mädchen- und Frauenbildung,* vol. 1, ed. Elke Kleinau and Claudia Opitz; Marianne Horstkemper, "Die Koedukationsdebatte um die Jahrhundertwende," in *Geschichte der Mädchen- und Frauenbildung,* vol. 2, ed. Elke Kleinau and Claudia Opitz, 206–207.

71. Marianne Horstkemper, "Die Koedukationsdebatte," in *Geschichte der Mädchen- und Frauenbildung,* vol. 2, ed. Elke Kleinau and Claudia Opitz, 205.

72. Figures for coeducation are difficult to interpret definitively. In Prussia, for instance, approximately 70% of pupils were coeducated in 1886, but this number does not differentiate between country and city. At the end of World War I, 97% of villages had coeducation, but only 30% of cities did. And of course none of this data takes into account gender-specific coursework, which was an important part of Nazi education. Ibid., 204–205.

73. R[udolf] Friemel, Rektor in Rixdorf-Berlin, "Trennung der Geschlechter oder gemeinschaftliche Erziehung?" *Pädagogisches Magazin* 327 (1908): 8–12.

74. Loriga, "The Military Experience," 21.

75. Bundesministerium für innerdeutsche Beziehungen, ed., *DDR-Handbuch,* 3rd rev. ed., s.v. "Nationale Volksarmee," in *Enzyklopädie der DDR* [CD-ROM], Digitale Bibliothek, vol. 32 (Berlin: Directmedia, 2000 [Berlin: Verlag Wissenschaft und Politik, 1985), 4627. Hereafter referred to as *DDR-Handbuch* [CD-ROM]. Note that page numbers refer to the CD-ROM pagination system.

76. See David Tyack and Elisabeth Hansot, *Learning Together: A History of Coeducation in American Schools* (New Haven: Yale University Press, 1990).

77. The later development of women's role in military service in the GDR presented a new aspect to this issue: in 1978, the GDR passed a law allowing

women to be conscripted in the defense of the GDR if necessary. *DDR-Handbuch* [CD-ROM], s.v. "Wehrdienst," 6900.

78. August Bebel, *Die Frau und der Sozialismus* (Hannover: Fakelträger, 1975 [1895]).
79. Paul Wandel, "Ministerbesprechung," March 18–19, 1947, Berlin, BArch DR-2/53, no. 75.
80. "Bericht über die Tagung der Schulaufsichtsbeamten des Bundeslandes Sachsen," October 19–20, 1945, BArch DR2/488, no. 30.
81. "Protokoll über die Sitzung des erweiterten Ausschusses für Frauenfragen," June 10, 1947, Berlin, BArch DR-2/576, no. 29–32.
82. Ibid., no. 32.
83. Eva Landler, "Käte Agerth—1888 bis 1974: Im Herzen immer jung geblieben," in *Wegbereiter der neuen Schule*, ed. Gerd Hohendorf, Helmut König, and Eberhard Meumann (Berlin [East]: Volk und Wissen, 1989), 13–19. Gert Geißler noted however that Agerth and others of similar political background, in contrast to their early roles in the communist resistance during the war, did not achieve any real positions of political influence in the GDR. Gert Geißler, "Schulämter und Schulreformer in Berlin nach Kriegsende 1945," in *Reformpädagogik in Berlin—Tradition und Wiederentdeckung*, ed. Wolfgang Keim (Frankfurt, AM: Peter Lang, 1998), 137–168.
84. "Sitzungen der Abteilungsleiter des Hauptschulamtes," December 20, 1948, LAB/STA 120/201, no. 213.
85. "Zur Ministerkonferenz in Januar 1947, DVV, Berlin, Betr: Arbeitstagung. Stellungnahme zu den Aufgaben der Mädchenerziehung in der Gegenwart," BArch, DR 2/53, no. 248.
86. Prov. Ausschuss antifaschistischer Lehrer [PAAL], "Niederschrift über die Sitzung am 1.6.45, nach. 15 Uhr bei Kollegen Fritz, Dresden-Leuben, Dürrst. 18," DIPF/BBF/Archiv, Sammlungsgut, Döbelner Konferenz, no. 6.
87. See Monika Gibas, "Vater Staat und seine Töchter: Offizielle propagierte Frauenleitbilder der DDR und ihre Sozialisationswirkungen," in *Parteiauftrag: Ein neues Deutschland: Bilder, Rituale und Symbole der frühen DDR*, ed. Dieter Vorsteher (Berlin: Deutsches Historisches Museum, 1996), 311–312.
88. DVV, Dr. Peters, "Referat Schulverwaltung und Koordinierung," Berlin, August 21, 1947, BArch DR2/4008, no. 226–227.
89. Internal correspondence in "Schulabteilung," from Möller-Krumbholtz to Wandel, "Referat: 'Lehrpläne, Naturwissenschaften, Mädchenbildung: Stand, Probleme, Einwände,' " Berlin, January 19, 1948, BArch DR-2/73, no. 43.
90. Ibid.
91. "Abteilungsleitersitzung," Berlin, June 23, 1947, LABSTA 120/139, pt. 2, no. 97–98.
92. Emil Knief, "Kommt die Koedukation?" in *Die Lehrergewerkschaft* 6, nr.1 (5 September 1947), STA/LAB 120/2520, no. 94.

93. Frau Dr. R. Kampf, Frankfurt, AM, "Die soziale und wirtschaftliche Lage in ihrer Bedeutung für das Problem der gemeinsamen Erziehung," Arbeiten des Bundes für Schulreform, *Dritter Deutscher Kongreß für Jugendbildung und Jugendkunde zu Breslau am 4.5. und 6. Oktober 1913* (Leipzig/Berlin, 1914): 79.

94. The German word for coeducation, *Koedukation*, came from English and quickly replaced terms such as *gemeinschaftliche Erziehung* or *gemeinsame Erziehung*. Heinrich von Hähling and Weihbischof von Paderborn, *Die Koedukation oder, Die gemeinschaftliche Erziehung der Knaben und Mädchen, Besonders in der Volksschule. Die richtigen Grundsätze in einer brennenden Frage.* 2nd ed. (Paderborn: Bonifacius-Druckerei, 1924), 4–8.

95. Dr. Friedrich Geisler, "Das amerikanische und das englische Erziehungswesen," unidentifiable source, n.d. [ca. 1946], DIPF/BBF/ Archiv, NL Löffler, fo. 10, no. 7–8.

96. "Eine zweite Rüge," in *Telegraf,* January 11, 1948; Heinrich Deiters, "Deutsche Schulreformen," in *Die Gegenwart,* August 24, 1946; BArch DR2/73, Internal correspondence in "Schulabteilung," from Möller-Krumbholtz to Wandel, "Referat: 'Lehrpläne, Naturwissenschaften, Mädchenbildung: Stand, Probleme, Einwände,'" Berlin, January 19, 1948, BArch DR-2/73, no. 43.

97. "Schulgesetz für Groß-Berlin," November 13, 1947, approved by Allied Command June 22, 1948, in *Dokumente zur Geschichte des Schulwesens in der Deutschen Demokratischen Republik,* pt. 1, *1945–1955* ed. Gottfried Uhlig, *Monumenta Paedagogica* (Berlin [East]: Volk und Wissen, 1970), 263–269; "Gesetz zur demokratisierung der deutschen schule," in *pädagogik,* 1.1 (August, 1946): 2–4.

98. A girls' upper secondary school remained in existence throughout the GDR in East Berlin.

99. "Protokoll über die Gesamtkonferenz der Haupt- und Bezirksschulräte von Berlin," April 1, 1948, LAB/STA 120/126, no. 42–46.

100. "Berichtsbuch der 4. Volksschule Berlin-Köpenick 1935–1947," February 18, 1947, SM/Do 92/141.

101. "Bericht zur Durchführung des Schulgesetzes," Berlin, June 25, 1949, LAB/STA 120/113, no. 44.

102. "Bericht über die Tagung der Schulaufsichtsbeamten des Bundeslandes Sachsen," October 19–20, 1945, BArch DR2/488, no. 30.

103. Heidemarie Kühn [Kemnitz], "Mädchenbildung im Schulsystem der DDR," in *Geschichte der Mädchen- und Frauenbildung,* vol. 2, ed. Elke Kleinau and Claudia Opitz, vol. 2 (Frankfurt AM: Campus, 1996), 434–435.

104. Walter Wolf, *Probleme des Kursunterrichtes (Untersuchungen, durchgeführt an Thüringer Schulen 1948 und 1949),* 1951, DIPF/BBF/Archiv NL Walter Wolf, fo. 39, no. 33.

105. Ibid.

106. "Konferenz der Minister für Volksbildung der Länder," Berlin, January 1948, BArch DR-2/73, no. 43.

107. Internal correspondence in "Schulabteilung," from Möller-Krumbholtz to Wandel, "Referat: Lehrpläne, Naturwissenschaften, Mädchenbildung: Stand, Probleme, Einwände," Berlin, January 19, 1948, BArch DR-2/73, no. 43.

108. Correspondence from Deiters to Bund demokratischer Lehrer und Erzieher Österreichs, Vienna, August 6, 1948, DIPF/BBF/Archiv, NL Heinrich Deiters, no. 2.

109. Dörte Gernert, "Mädchenerziehung im allgemeinen Volksschulwesen," 85 and Horstkemper, "Die Koedukationsdebatte um die Jahrhundertwende," 212 in *Geschichte der Mädchen- und Frauenbildung*. The encyclical's full translated title is *divini illius magistri*, Encyclical of Pope Pius XI on Christian Education to the Patriarchs, Primates, Archbishops, Bishos, and Other Ordinaries in Peace and Communion with the Apostolic See and to all the Faithful of the Catholic world." Cf. <http: www.vatican. va/holy_father/pius_xi/encyclicals/documents/hf_p_xi_enc_31121929_divini_illius_magistri_en.html>.

110. "Eine Erklärung der Falken, 'Christentum—Koedukation,' " *Benjamin: Zeitschrift für junge Menschen.* (Hamburg) 1.19 (November 1947): 6.

111. "Sitzung der Abteilungsleiter des Hauptschulamtes," LAB/STA 120/201, December 20, 1948, no. 213.

112. W. Pom., "Braucht der Lehrer einen Standpunkt?" *Tägliche Rundschau*, 1 (December 1945): 3.

113. Report by Herr Naumann of FDJ Dresden, "FDJ und Schule, Kreisschulratskonferenz," Dresden-Wachwitz, November 3–5, 1948, BArch DR 2/489, no. 351.

114. "Sitzung der Abteilungsleiter des Hauptschulamtes," June 23, 1947, LAB/STA 120/139, no. 97–98.

115. "Protokoll der Abteilungsleitersitzung," April 8, 1947, LAB/STA 120/2520, no. 80.

116. Ibid., no. 83.

117. Manfred Surge, born March 9, 1933, grade six, "1. Jahr Wiederaufbau," [1946], LAB/STA 134/13, 181/1, no. 91.

118. Christel Novak, Brigitte Klingeberger, Sonja Krakau, Bärbel Lisurek, Ruth Jenner, Ingrid Sorgentreig, Siegrid Neumann, "Wiederaufbau Prenzlauer Berg," girls' middle school, [1946], LAB/STA 134/13, 181/1, no. 103–117, and Sigrid Iggena, grade six, "Ein Jahr Wiederaufbau im Bezirk Prenzlauer Berg und seiner näheren Umgebung," LAB/STA 134/13, 181/1, no. 14; Sigrid Walther, grade seven, "Der Aufbau," [1946] LAB/STA 134/13, 181/1, no. 6.

119. H. Leipner, grade six, January 22, 1946, LAB/STA 134/13, 183, no. 103; Inge Weege, grade seven, "Die Ernährung Berlins," [1946], LAB/STA 134/13, 182/2, no. 432.

120. Christa Kynart, grade four, Berlin-Bernau, 1946, in "Arbeit und Alltag in der DDR," exhibit in the Staatsbibliothek Berlin, May 1998.
121. From, respectively, Marga Hellmuth to her schoolmate Ingeboark Gerigk in Wilsleben, March 3, 1947, SM/Do 93/152 and to Elfriede Schlegel from her female cousin in Müncheberg, 1946, SM Do 88/553.
122. Kiaz Kiewicz, [1946], LAB/STA 134/13, 180/1, no. 539.
123. Kurt Ortmann, grade five, thirteen years old, [1946] LAB/STA 134/13, 182/2, no. 413.
124. See, for instance, Ingrid Höll, grade four, [1946] LAB/STA 134/13, 182/1, no. 71.
125. W. Maaßen, grade four, April 30, 1946, LAB/STA 134/13, 182/1, no. 100.
126. Helga Katz, (elementary school pupil), [1946], LAB/STA 134/13, 180/2, no. 263.
127. Ingrid Welzel, Klasse 4, [1946], LAB/STA 134/13, 180/2, no. 506.
128. Giesela Mischok, born March 12, 1935, grade five, 32. Volksschule, "Wiederaufbau in Bezirk Prenzlauer Berg," May 8, 1946, LAB/STA 134/13, 182/2, no. 381.

Chapter 5   The Antifascist Narrative

1. Children begin to remember around the age of three. At five years of age, they can provide accurate information about past events. See Ulrich Neisser, "Self-Narratives: True and False," in *The Remembering Self: Construction and Accuracy in the Self-Narrative*, ed. Ulrich Neisser and Robyn Fivush (Cambridge: Cambridge Univ. Press, 1994), 16.
2. I would argue against the need to distinguish absolutely between terms such as collective memory and collective consciousness, since they are used interchangeably in most instances; nevertheless, Luhmann's attempt to define his terms rigorously if narrowly is intriguing. Niklas Luhmann, "Zeit und Gedächtnis," *Soziale Systeme* 2, no. 2 (1996): 307–330.
3. Neisser, "Self-Narratives: True and False," 11.
4. John Kotre, *White Gloves: How We Create Ourselves through Memory* (New York: The Free Press, 1995), 136.
5. Emphasis and double emphasis in original. Dorst, Muller-Krumbholz, Sothmann, "1. Entwurf: Richtlinien für die Didaktik und Methodik der deutschen demokratischen Schule," Berlin, May 18, 1949, DIPF/BBF/Archiv NL Sothmann, fo. 20, no. 1.
6. Jürgen Diederich and Heinz-Elmar Tenorth, *Theorie der Schule: Ein Studienbuch zu Geschichte, Funktionen und Gestaltung* (Berlin: Cornelsen, 1997), 235–240.
7. "Lehrplan für die Grund- und Oberschulen in der sowjetischen Besatzungszone Deutschlands, Deutsch, Entwurf der am 1. Juli 1946 veröffentlichten 2. Fassung," in Joachim S. Hohmann, *Deutschunterricht in SBZ und DDR 1945–1962* (Frankfurt AM: Lang, 1997), 114.

8. Ibid.

9. "Die Forderung der Richtlinien der Allierten Erziehungskommission," in *Lehrplan für den Geschichtsunterricht an den Berliner Schulen* (Berlin [East]/Leipzig: Volk und Wissen, 1948), 3.

10. "Niederschrift über die Sitzung der Geschichtskommission am 1. Oktober 1947," Abteilung Volksbildung, Hauptschulamt, Berlin, October 25, 1947, LAB/STA, 120/177, no. 8.

11. L.B. Helmut Otzen, Untermassfeld, Meinigen, July 26, 1949, "Welche Anforderungen werden an die Persönlichkeit des Lehrers gestellt, und wie suche ich sie zu erfüllen? Arbeit zur 1. Lehrerprüfung," SM/Do 87/131.

12. "Protokoll über den Pädagogischen Landeskongreß im Maxim-Gorki-Haus," Schweren, May 25–26, 25/26 1948, BArch DR 2/1394, no. 80.

13. Wilhelm Schneller, "Die Reinigung der Schule vom Nazismus," *Sächsische Volkszeitung*, December 4, 1945, in Sammlungsgut, Döbelner Konferenz, 1945, DIPF/BBF/Archiv, 02.9./17, no. 120.

14. Freunde der neuen Schule, Arbeitsgemeinschaft demokratischer Organisationen, "Arbeitsprogramm für das erste Jahr des Zweijahresplans," [1949], DIPF/BBF/Archiv, NL Karl Sothmann, fo. 58.

15. Erik Braune, "Pädagogik," February 20, 1946, DIPF/BBF/Archiv, Nachlaß Hans Löffler 0.4.34., pt. 1.

16. Theodor Litt, letter to Heinrich Deiters, Leipzig, July 15, 1946, DIPF/BBF/Archiv, NL Heinrich Deiters fo. 2.

17. K. Dräger, "Zum heutigen Stand der Landschule," Berlin, March 26, 1947, DIPF/BBF/Archiv, NL Karl Sothmann, fo. 35, 3.

18. P. Zolotuchin, Leiter der Abteilung für Volksbildung der SMA in Deutschland, to Paul Wandel, May 15, 1946, memorandum allowing history instruction in schools again, BArchiv DR 2/6269, no. 115.

19. "Gutachten über das Programm der englischen Sprache," from Inspektor der Abteilung für Volksbildung der Sowjetischen Militaerverwaltung in Deutschland, Uljanow, to Wildangel, n.d. [1945 or 1946], BArch DR 2/636, no. 138–139.

20. Schaffer, *Russian Zone*, 123.

21. Robyn Fivush, "Constructing Narrative, Emotion, and Self in Parent-Child Conversations about the Past," in *The Remembering Self: Construction and Accuracy in the Self-Narrative*, ed.. Ulrich Neisser and Robyn Fivush (Cambridge: Cambridge University Press, 1994), 136–157.

22. Hilde Thurnwald, *Gegenwartsprobleme Berliner Familien: Eine Soziologische Untersuchung an 498 Familien* (Berlin [West]: Weidmannsche Verlagsbuchhandlung, 1948), 324–328.

23. Ibid., 333 and 321, respectively.

24. "Tätigkeitsbericht der Landesregierung Mecklenburg-Vorpommern in Schwerin, Ministerium für Volksbildung für das Jahr 1946," BArch DR 2/280, January 1, 1947, no. 33.

25. Donald A. Norman, *Learning and Memory* (San Francisco: W.H. Freeman and Co., 1982), 2, 11–12, 24, 81.

26. I have borrowed the concept of a "second history" from Anne Michaels, *Fugitive Pieces* (London: Bloomsbury, 1996), 20.

27. Dorothea Uebrig, Deutsch Kl. 10a, no title, SM Do 88/555, 16 February 1948; and Geschichte, Kl. 11a, "Absolutismus-Bismarcks Außenpolitik," SM Do 88/550, Fall 1949.

28. Otto Dieß, fifth grade, "Mein Freund (Hausaufsatz)," Oscheersleben, December 15, 1944, SM/Do 93/174.

29. Otto Dieß, fifth grade, "Wiederaufbau in Stadt und Land," Oscheersleben, October 1945, SM/Do 93/174.

30. Ibid.

31. For a discussion of both these plans, see Rolf Steininger, *Deutsche Geschichte seit 1945*, vol. 1, *1945–1947* (Frankfurt AM: Fischer, 1996), 42–45 and 85–101.

32. Vera Müller, thirteen years old, "Angestellt bei Beschuss," [ca. 1945 or 1946], LAB/STA 134/13, 179, no. 21.

33. Kiaz Kiewicz, "Kämpfe um Berlin," [ca. 1946], LAB/STA 134/13, 180/1, no. 539.

34. "Lehrplan für die Grund- und Oberschulen," 154.

35. Vera Müller, thirteen years old, "Angestellt bei Beschuss," [ca. 1945 or 1946], LAB/STA 134/13; 179, no. 20–21.

36. Ibid.

37. Ibid, no. 21.

38. The thirteen-year-old Brigitte Reimann, later a well-known author in the GDR, wrote a friend that she had selected the essay topic "A twenty Mark bill tells of its trip through the city and countryside" for her admission exam into upper secondary school. Brigitte Reimann, letter to Veralore, Burg bei Magdeburg, June 11, 1947, *Aber wir schaffen es, verlaß Dich drauf!" Briefe an eine Freundin im Westen.* 2nd ed. (Berlin: Aufbau Taschenbuch Verlag, 1999), 6–7.

39. Edgar Günther, "Nr. 7, Klassenaufsatz. Ein alter Geldschein erzählt seine Lebengeschichte," [1946], SM/Do 89/109, carton 09/24/02/14.

40. Ibid.

41. Ibid.

42. Ibid.

43. Selma Leydesdorff, Luisa Passerini, and Paul Thompson, "Introduction," in *Gender and Memory*, ed. Selma Leydesdorff, Luisa Passerini, and Paul Thompson (International Yearbook of Oral History and Life Stories, Oxford: Oxford University Press, 1996), 12.

44. Annett Gröschner, *"Ich schlug meiner Mutter die brennenden Funken ab": Berliner Schulaufsätze aus dem Jahr 1946* (Berlin: Kontext, 1996), 9–15.

45. See my discussion of gender and Soviet zone school essays in Benita Blessing, "The Antifascist Narrative: Memory Lessons in the Schools of the Soviet Occupation Zone, 1945–1949," in *Children and War*, ed. James Merten (New York: New York University Press, 2002).

46. Christel Novak, Brigitte Klingeberger, Sonja Krakau, Bärbel Lisurek, Ruth Jenner, Ingrid Sorgenfrei, and Siegrid Neumann, "Wiederaufbau Prenzlauer Berg," girls' middle school, LAB/STA 134/13, 181/1, no. 103–117.

47. Ibid., no. 114.

48. Ibid., no. 117.

49. Gerhard Krüger, born May 11, 1946, 13. Volksschule Prenzlauer Berg, fifth grade, "Ein Jahr Wiederaufbau," LAB/STA 134/13, 181/1, no. 114.

50. Vera Rietz, eighth grade, 26. Volksschule, Prenzlauer Berg, [1946], LAB/STA 134/13, 181/2, no. 445.

51. Helga Hoch, fifth grade, 10. Volksschule, [1946], LAB/STA 134/13, 181/1, no. 49.

52. This was the case in all zones. For the point of view in the Western zones, see for instance Erich Kästner's essays for the Munich youth newspaper *Pinguin* in Erich Kästner, *Der tägliche Kram: Chansons und Prosa 1945–1948*. 2nd ed. (Munich: Deutscher Taschenbuch Verlag, 1999); or the Hamburg youth newspaper *Benjamin: Zeitschrift für junge Menschen*, Hamburg, 1947 ff.

Chapter 6 "Vati's home!": From Defeated Nazi to Antifascist Hero

1. Adelheid zu Castell, "Die demographischen Konsequenzen des Ersten und Zweiten Welkriegs für das Deutsche Reich, die Deutsche Demokratische Republik und die Bundesrepublik Deutschland," in *Zweiter Weltkrieg und sozialer Wandel*, ed. Waclaw Dlugoborski (Göttingen: Vandenhoeck and Ruprecht, 1981), 121.

2. Barbara Willenbacher, "Zerrüttung und Bewährung der Nachkriegs-Familie," in *Von Stalingrad zur Währungsreform: Zur Sozialgeschichte des Umbruchs in Deutschland*, ed. Martin Broszat, Klaus-Dietmar Henke, and Hans Woller (Munich: Oldenburg Verlag, 1988), 595–618.

3. Staatliche Zentralverwaltung für Statistik, *Statisches Jahrbuch der Deutschen Demokratischen Republik* (Berlin [East]: VEB Deutscher Zentralverlag, 1956), 22.

4. The GDR statistical administration did not collect information on families in which the head of the household was female but the male continued to work, suggesting that this situation was extremely rare. Staatliche Zentralverwaltung für Statistik, *Statistisahes Jahrbuch der Deutschen Demokratischen Republik*, 23. See also zu Castell, "Die demographischen Konsequenzen des Ersten und Zweiten Welkriegs," 121.

5. Thurnwald, *Gegenwartsprobleme Berliner Familien.*

6. zu Castell, "Die demographischen Konsequenzen," 121.

7. Ina Merkel, "Leitbilder und Lebensweisen von Frauen in der DDR," in *Sozialgeschichte der DDR*, ed. Hartmut Kaelble, Jürgen Kocka, and Hartmut Zwahr (Stuttgart: Klett-Cotta, 1994), 361–365.

8. Robert R. Shandley, *Rubble Films: German Cinema in the Shadow of the Third Reich* (Philadelphia: Temple University Press, 2001), 17–18, 20, 25–26.

9. Ibid., 118–119.

10. Ingelore König, Dieter Wiedmann, and Lother Wolf, *Zwischen Marx und Muck: DEFA-Filme für Kinder* (Berlin: Henschel, 1996), 71–73. See also "Filmwesen," *DDR Handbuch.*

11. Barton Byg, "DEFA and the Traditions of International Cinema," in *DEFA: East German Cinema, 1946–1992* ed. Seán Allan and John Sandford (New York: Berghahn, 1999), 33.

12. König et al., *Zwischen Marx und Muck*, 72–73; Poster: "Irgendwo in Berlin" poster. Deutsches Historisches Museum, ed. *Plakate der SBZ/DDR: Politik, Wirtschaft, Kultur* (Munich: Sauer, 1999), Inv. Nr.—90/6647 SBZ/GDR [CD-ROM].

13. Peter Kast, "Irgendwo in Berlin—allüberall in Deutschland, *Vorwärts*, December 20, 1946.

14. Merkel, "Leitbilder und Lebensweisen von Frauen in der DDR," 364.

15. Ibid., 362–363.

16. *Plakate der SBZ/DDR* [CD-ROM] Inv. Nr. P—90/1741, Stengel, Gerhard. Hg. KPD Leipzig.

17. Cf. Shandley, *Rubble Films*, 124–125.

18. Alfred Geyer, "Enttrümmerung des Hauses," LAB/STA XVIII, 37, n.d., 404; Wolfgang Sachs, "Wiederaufbau in der Kastanienallee," May 1946, LAB/STA XIX, 5, 489.

19. Ingrid Höll, "Ein jahr Wiederaufbau im Bezirk Prenzlauer Berg. Frauen räumen auf und putzen Steine," LAB/STA 13413, 182/1, XX, 57, n.d.[1946], fo. 71.

20. Vera Rietz, "Reinlichkeit und Gemütlichkeit!" n.d. LAB/STA XVIII, 67, 445.

21. "In Berlin soll kein Kind mehr frieren: Ein Gespräch mit Frau Stadtrat Schirmer-Pröscher," *Neues Deutschland*. December 16, 1948, no. 293, 6.

22. Thurnwald, *Gegenwartsprobleme Berliner Familien*, 26–37.

23. König et al., *Zwischen Marx und Muck*, 74–76.

24. *Plakate der SBZ/DDR* [CD-ROM] Inv. Nr "Vermeide hemmungloses Leben," Artist Unger, [?] n.d. [1945/1948], Erfurt, Inv. Nr. P —73/357.

25. Byg, "DEFA and the Traditions of International Cinema," 32.

26. David Forgacs, Sarah Lutton, and Geoffrey Nowell-Smith, *Roberto Rossellini: Magician of the Real* (London: British Film Institute, 2000), 150–151.

27. Thurnwald, *Gegenwartsprobleme Berliner Familien*, 295–296.

28. Ibid., 343.

29. Ibid., 289.

30. Giesela Mischok, born March 12, 1935, grade 5, 32. Volksschule, "Wiederaufbau in Bezirk Prenzlauer Berg," 8 May 1946, LAB/STA 134/13, 182/2, no. 381.

31. Hannelore Adler, "Unsere Wohnung," n.d., LAB/STA 134/13, 181/2, XVIII, 16, 371.
32. Thurnwald, *Gegenwartsprobleme Berliner Familien*, 260–264.
33. Christel Novak, Brigitte Klingeberger, Sonja Krakau, Bärbel Lisurek, Ruth Jenner, Ingrid Sorgenfrei, Siegrid Neumann, "Wiederaufbau Prenzlauer Berg," [May 1946], LAB/STA 134/13, 182/1, no. 105.
34. "Generation ohne Väter," in *Benjamin: Zeitschrift für junge Menschen*, vol. 1 (Hamburg: 1947), 14.
35. Thurnwald, *Gegenwartsprobleme Berliner Familien*, 191.
36. Ibid., 192.
37. Reimann, October 10, 1947, *Aber wir schaffen es*, 16.
38. In this context, the use of Russian carries a slightly pejorative connotation.
39. Dieter Kirchhöfer has examined a similar phenomenon among eastern German families after 1989. Routines and value systems changed radically after the *Wende* of 1989/1990, but family members tended to adapt themselves to new situations without altering their relationships to one another significantly. Dieter Kirchhöfer, "Veränderungen in der sozialien Konstruktion von Kindheit," in *Kindheit, Jugend und Bildungsarbeit im Wandel: Ergebnisse der Transformationsforschung*, ed. Heinz-Elmar Tenorth, vol. 37th Supplement (Weinheim and Basel: Beltz, 1997), 25–26.

Similarly, in an enlightening comparative study of eastern and western German parent-child relationships in 1993, Peter Büchner, Burkhard Fuhs, and Heinz-Hermann Krüger concluded that relationships between family members throughout Germany were almost identical in spite of fundamental transformations in everyone's daily routine. Peter Büchner, Burkhard Fuhs, and Heinz-Hermann Krüger, "Transformation der Eltern-Kind-Beziehungen? Facetten der Kindbezogenheit des elterlichen Erziehungsverhaltens in Ost-und Westdeutschland," in *Kindheit, Jugend und Bildungsarbeit im Wandel*, ed. Heinz-Elmar Tenorth, 35–52.

Chapter 7   Reestablishing Traditions

1. "Pädagogisches Manifest des Kulturbundes zur demokratischen Erneuerung Deutschlands (Entwurf)," [ca. 1946], DIPF/BBF/Archiv NL Heinrich Deiters, fo. 42.
2. Erich Kästner, "Und dann fuhr ich nach Dresden," in *Der tägliche Kram*, 112. First published in *Neue Zeitung*, November 1946.
3. Manfred Jäger, *Kultur und Politik in der DDR: 1945–1990* (Cologne: Edition Deutschland Archiv, 1995), 20.
4. See Alexander Dymschiz, *Ein unvergeßlicher Frühling. Literarische Porträts und Erinnerungen* (Berlin [East]: Dietz, 1970).
5. Jürgen Rühle, *Deutschland Archiv* 12 (1980): 1306 ff., cited in Manfred Jäger, *Kultur und Politik in der DDR*, 8.

6. See Wolfgang Emmerich, *Kleine Literaturgeschichte der DDR*. 2nd ed. (Leipzig: Gustav Kiepenhauer, 1996), 72 and Manfred Jäger, *Kultur und Politik in der DDR*, 19.

7. See J. M. Bernstein, *The Philosophy of the Novel: Lukacs, Marxism and the Dialectics of Form* (Minneapolis: University of Minnesota Press, 1984).

8. Manfred Jäger, for instance, has reminded western scholars that Lukács lived in "faraway Budapest" and was treated somewhat skeptically by party functionaries, thus preventing him from having direct personal influence on cultural policy. Manfred Jäger, *Kultur und Politik in der DDR*, 21–22.

9. Wolfgang Emmerich, *Kleine Literaturgeschichte der DDR*. 2nd ed. (Leipzig: Gustav Kiepenhauer, 1996), 526–527.

10. David Bathrick, *The Powers of Speech: The Politics of Culture in the GDR* (Lincoln: University of Nebraska Press, 1995), 173–177.

11. Interpretations of the French Revolution also changed with the self-perception of the state. See the following essays in *Bilder einer Revolution: Die Französische Revolution in den Geschichtsschulbüchern der Welt*, ed., Rainer Riemenschneider.

    (Frankfurt AM: Diesterweg, 1994), 329–345; Horst Diere, "Die französische Revolution von 1789 bis 1795 und die Zeit Napoleons im Geschichtslehrbuch der DDR," 329–345; Matthias Middell, "La Révolution française enseignée dans les écoles de l'ex RDA et le changements d'hier à aujourd'hui," 347–355; Axel Koppetsch, "Die Französische Revolution: Dauer und Wandel ihrer Darstellung in deutschen Schulbüchern seit 1980," 357–376.

12. Heinrich Mann, "Die französische Revolution und Deutschland," *Aufbau* 3 (November 1945): 210.

13. Hans Hopp, "Der Wiederaufbau von Neubrandenburg," *Deutsche Architektur* 6 (1955): 294 and 297.

14. Bathrick, *The Powers of Speech*, 167–173.

15. Ibid., 168.

16. Ibid., 175. This process took on more concrete forms in the 1950s, when Marxist-Leninist thought began to redefine traditional disciplinary categories of scholarship. During this shift, the field of ethnography began to shift its emphasis from linguistics to material products, such as utensils and currency. No longer fixated on the idea of *Volk* as a spiritual identity, ethnographers looked to issues of labor and economy in their research. This change in focus permanently altered the way in which the GDR population defined itself and accorded the school a more central role in the education of enlightened workers. The new socialist nationhood excluded biological definitions of the *Volk* and centered instead upon "working class struggles." Grimms' fairy tales could be incorporated into the antifascist canon, because they now recounted stories of the victories of labor—albeit with continued emphasis on typical gender divisions. For examples of how this narrative was incorporated into GDR historiographical traditions, see the series Günter Wendel, ed. "Beiträge zur Wissenschaftsgeschichte," for the Arbeitskreises

Wissenschaftsgeschichte beim Ministerium für Hoch- und Fachschulwesen der DDR, especially *Wissenschaft im kapitalistischen Europa: 1871–1917* (Berlin [East]: Deutscher Verlag der Wissenschaften, 1983).

17. Jakob [Jacob] and Wilhelm Grimm, *Die Kinder- und Hausmärchen der Brüder Grimm*, ed. Walther Pollatschek, 5th ed. (Berlin [East]: Kinderbuchverlag, 1955), 21.

18. Jan Herman Brinks, *Die DDR-Geschichtswissenschaft auf dem Weg zur deutschen Einheit: Luther, Friedrich II und Bismarck als Paradigmen politischen Wandels* (Frankfurt and New York: Campus, 1992), 297–300.

19. Edgar Wolfrum, *Geschichtspolitik in der Bundesrepublik Deutschland. Der Weg zur bundesrepublikanischen Erinnerung 1948–1990* (Darmstadt: Wissenschaftliche Buchgesellschaft, 1990), 1–38.

20. Brinks, *Die DDR-Geschichtswissenschaft auf dem Weg zur deutschen Einheit*, 309.

21. Rudolf Bonna, *Die Erzählung in der Geschichtsmethodik von SBZ und DDR* (Bochum: Universitätsverlag Brockmeyer, 1996).

22. For the international historical context of Kerscheinsteiner's work, see Hermann Röhrs, *Die Reformpädagogik: Ursprung und Verlauf unter internationalem Aspekt*, 5th rev. ed. (Weinheim: Deutscher Studien-Verlag, 1998).

23. Rudolf Bonna concluded that reform pedagogues in the Soviet zone had to compromise in the first postwar years with Soviet-style Marxist-Leninists and gradually lost influence. The focus here on the Marxist-Leninist question is too narrow. Educational philosophies in the Soviet zone and GDR resulted from conflict and compromise, not from a single battle with only one victor. More actors participated in discussions of the school than Bonna acknowledged. As a consequence, more was at stake than dogmatic adherence to a Marx, Engels, or Lenin. Bonna, *Die Erzählung in der Geschichtsmethodik*, 140–141.

24. "Lehrpläne für die Grund- und Oberschulen in der Sowjetischen Besatzungszone Deutschlands, Geschichte, 8. Klasse," July 1, 1946, in *Zwei Jahrzehnte Bildungspolitik in der Sowjetzone Deutschlands*, vol. 1, ed. Siegfried Baske and Martha Engelbert (Heidelberg: Quelle and Meyer, 1966), 28.

25. Ibid., 29 and 31.

26. Ibid., 31.

27. Paul Wandel, "Introduction," in Deutsche Zentralverwaltung für Volksbildung in der Sowjetischen Besatzungszone Deutschlands, *Lehrpläne für die Grund- und Oberschulen in der Sowjetischen Besatzungszone Deutschlands: Geschichte* (Berlin[East]/Leipzig: 1 July 1946), 2.

28. Emphasis in original. Ibid., 3.

29. Ibid.

30. For a later example of May 10, 1933 commemorations in the GDR, see the booklet published by the Humboldt faculty: Gesellschaftswissenschaftliche Fakultät, *Nie wieder Faschismus und Krieg: die Mahnung der faschistischen Bücherverbrennung am 10. Mai 1933* (Berlin [East]: Humboldt University, 1983).

31. The first discussion of commemorating May 10 on a Soviet zone-wide level that I found was in the proposed minutes for an educational minister conference. There may have been other, earlier local or regional celebrations. "Zur Vorlage in der Ministerkonferenz am 22. und 23.4.1947," Volkmann, Abteilung Kunst und Literatur, April 3, 1947, BArch DR 2/54, no. 36–38.

32. See Gabriele Baumgartner and Dieter Hebig, ed. *Biographisches Handbuch der SBZ/DDR*, s.v. "Volkmann, Herbert," in *Enzyklopädie der DDR* [CD-ROM], Digitale Bibliothek, vol. 32 (Berlin: Directmedia, 2000 [Munich: Saur, 1996/1997]), 17429.

33. "10. Mai 1948, Tag des freien Wortes," Soviet zone, 1948, in *Das politische Plakat der DDR, 1945–1970* [CD-ROM], Inv. Nr. P 94/2483.

34. "10. Mai, Tag des freien Buches," Dresden, 1949, in *Das politische Plakat* [CD-ROM], Inv. Nr. P 94/2484.

35. Norman Naimark, *The Russians in Germany: A History of the Soviet Zone of Occupation, 1945–1949* (Cambridge, MA: Belknap Press of Harvard University Press, 1995), 434–435.

36. Paul Wandel, "Ministerbesprechung" March 18–19, 1947, Berlin, BArch DR 2/53, no. 41.

37. Ibid.

38. Max Kreuziger, "Ministerbesprechung," Berlin, March 18–19, 1947, BArch DR 2/53, no. 41. Franz Albrecht and Gerd Hohendorf, "Max Kreuziger, 1880 bis 1953. 'Wir wollen vorwärts!' " in *Wegbereiter der neuen Schule*, ed. Gerd Hohendorf, Helmut König and Eberhard Meumann (Berlin [East]: Volk und Wissen, 1989), 150–156.

39. Harald Müller, "Demokratische Lehrerausbildung," *Horizont*, June 9, 1946, LAB/STA 120/3226, no. 25.

40. Correspondence from Otto Winzer to Allied educational commission, June 27, 1946, LAB/STA 120/3226, no. 22.

41. Correspondence from Otto Winzer to *Horizont*, LAB/STA 120/3226, June 27, 1946, no. 20.

42. "Für eine fortschrittliche demokratische Nationalbildung," unnamed newspaper article, [ca. 1946], LAB/STA 120/3228, no. 53.

43. Hans Siebert [?], "Arbeitsgemeinschaft Anti-Nazi Deutsche Lehrer," [ca. 1944], DIPF/BBF/Archiv NL Hans Siebert, L.3 fo. 19, 1.

44. "Schulchronik Bergfelde," November 28, 1947, SM/Do 92/31 b.

45. "Protokollbuch der Lehrerbesprechungen der 2. Volksschule Berlin Prenzlauer Berg," April 1947, SM/Do 92/55. Beethoven died on March 26, 1827; the minutes listed the anniversary as being April 26.

46. Manfred Hettling, "Shattered Mirror. German Memory of 1848: From Spectacle to Event," in *1848: Memory and Oblivion*, ed. Charlotte Tacke (Brussels: Peter Lang, 2000), 79–98.

47. "Berlins historischer Novembertag 1948," *Neues Deutschland*, December 1, 1948, 4.

48. Edgar Wolfrum, *Geschichtspolitik in der Bundesrepublik Deutschland*, 39–49.

49. "Schulratskonferenz," Dresden-Wachwitz, February 3–5, 1948 BArch DR 2/489, no. 97.

50. "Ministerkonferenz in Stuttgart," February 19–20, 1948, BArch DR 2/4009, no. 177.

51. "Niederschrift über die Sitzung der Geschichtskommission," Abteilung Volksbildung, Hauptschulamt, Berlin, October 1, 1947, LAB/STA 120/177, no. 8.

52. Ibid.

53. "Schulratkonferenz," Dresden-Wachwitz, February 3–5, 1948, BArch DR 2/489, no. 98.

54. "Entwurf, Die neue deutsche Schule," [1945 or 1946], DIPF/BBF/Archiv NL Karl Sothmann, 0.4.0.1., fo. 13.

55. Karl-Heinz Günther, Franz Hofmann, Gerd Hohendorf, Helmut König, and Hernz Schttennauer, eds., *Geschichte der Erziehung*, 9th ed. (Berlin [East]: Volk und Wissen, 1969 [1966]), 295.

56. "Schulratskonferenz," Dresden-Wachwitz, February 3–5, 1948 BArch DR 2/489, no. 99.

57. "Ministerkonferenz," Berlin, December 2–3, 1947, BArch DR 2/4008, no. 42–50.

58. Vice president Weinert, Berlin, "Ministerkonferenz," December 2–3, 1947, BArch DR 2/4008, no. 50.

59. "Revolutionsfeier 1848–1948 in Schule Waltersdorf," (grades five–eight), March 16, 1948, Museum für Kindheit und Jugend—Schulmuseum, SM/Do 88/1165; "Gedenkblatt zur Schuljahresfeier," Leipzig, [Spring 1948], SM/Do 88/955.

60. Observation of the 1848 lesson taught by an "älterer Herr" (older gentlemen) in Freiberg, "Tagung der Bezirksschulräte," Meißen, April 4 and 25, 1946, BArch DR 2/488, no. 216.

61. "1948 Hundert Jahreskampf um die Demokratie in Deutschland," Halle-Saale, in *Das politische Plakat der DDR* [CD-ROM], Inv. Nr. P—94/153.

62. Wernfried Faschina, "Die Rolle der deutschen Studenten in der Revolution 1848," in *Geschichte in der Schule* 1, nr. 1/2 (July/October 1948): 18–26.

63. "Argumente gegen unsere Schulreform," "Referat Grundschule, Material für die Ministerkonferenz in Stuttgart," gez. Ellrich, Berlin, January 12, 1948, BArch DR 2/73, no. 34–35.

64. Ibid.

65. Michael J. Hogan, *The Marshall Plan: America, Britain, and the Reconstruction of Western Europe, 1947–1952* (New York: Cambridge University Press, 1987).

66. Alexander Abusch, "Der Schnitt durch die Nation," *Junge Generation: Zeitschrift für Fragen der Jugendbewegung*, Berlin, 2. Jg., Heft 6, 1948 [reprint from *Weltbühne* III/26]: 210.

67. Minutes of meeting for the departmental directors of the Hauptschulamt in Berlin, November 22, 1948, LAB/STA 120/201, no. 220.

68. Ernst Wildangel, Minutes of meeting for the departmental directors of the Hauptschulamt in Berlin, November 22, 1948, LAB/STA 120/201, no. 230.

69. English Goethe Society website: <http://www.sas.ac.uk/igs/HPARCHIVEEGS.htm#EGSARC>, accessed March 1, 2006.

70. For an analysis of Nazism as a secular religion, see George Mosse, *The Nationalization of the Masses: Political Symbolism and Mass Movements in Germany from the Napoleonic Wars through the Third Reich* (New York: H. Fertig, 1975).

71. Karl König, "Der Weg zu Goethe," Schwerin, n.d., DIPF/BBF/Archiv, NL Marie Torhorst, fo. 68.

72. "Die Lebensleistung Goethes besteht vor allem darin, daß er inmitten der Zerrissenheit Deutschlands die nationale Einheit im Sprachlichen und im Geistigen darstellte." "Der Wegweiser, Goethe Festtage der Deutschen Nation 1749–1949," Weimar, August 24–29, 1949 DIPF/BBF/Archiv NL Torhorst, fo. 68, no. 3.

73. "Deutschlands Kultusminister im Gespräch," *Deutschlands Stimme*, February 29, 1948, DIPF/BBF/Archiv NL Marie Torhorst, fo. 32.

74. Ibid.

75. Volk und Wissen, ed. "Preface" in *Goethe: Eine Auswahl für die Schule.* Berlin [East]/Leipzig: Volk und Wissen, 1949.

76. Ibid.

77. "Goethe: In seinem Geiste arbeiten. Wählt Liste 6, Die Liste des Kulturbundes," Böhlitz-Ehrenberg, *Das politische Plakat der DDR*, [CD-ROM], Inv. No. P 94/2081.

78. Brinks, *Die DDR-Geschichtswissenschaft*, 110.

79. Comment from Minister Grimme, "Konferenz der deutschen Erziehungsminister, Protokoll der Tagung am 19. und 20. Februar 1948 in Stuttgart-Hohenheim," DIPF/BBF/Archiv NL Marie Torhorst, fo. 32, n.p.

80. "Einladung zur Feierstunde zum Beginn unserer Arbeit am 25. November 1945 . . . im UFA Theater am Friedrichshain," Volkshochschule Prenzlauer Berg, DIPF/BBF Sammlungsgut nach 1945.

81. Brigitta Rehter [?] Nordgernersleben [?],February 15, 1947, SM/Do 87/1153.

82. *Goethe: Eine Auswahl für die Schule* (Berlin [East]/Leipzig: Volk und Wissen Verlag, 1949), DIPF/BBF/Archiv, NL Torhorst, fo. 68.

83. Brinks included a discussion of Engels' ambivalent stance toward Goethe in Brinks, *Die DDR-Geschichtswissenschaft*, 108–109.

84. Naimark, *The Russians in Germany*, 438 and Hans Siebert, "Westliche gegen östliche Kultur?" Presentation to FDGB, London, 1946, DIPF/BBF/Archiv NL Hans Siebert, fo. 137, no. 3.

85. A collection of brochures and pamphlets on Goethe and the Goethe Year are included in Marie Torhorst's file DIPF/BBF/Archiv, fo. 68.

86. Many politicians in the Soviet zone viewed the Bauhaus movement skeptically, in particular since many Bauhaus artists in exile did not return after

the war. Andreas Schätzke, "Rückkehr aus dem Exil. Zur Remigration bildender Künstler in der SBZ/DDR," in *Kunstdokumentation SBZ/DDR 1945–1990. Aufsätze—Berichte—Materialien*, ed. Günter Feist, Eckhart Gillen and Beatrice Vierneisel (Cologne: DuMont, 1996), 97.

87. Walter Wolf, *Wie gelang es dem illegalen deutschen Volksfrontkomittees im Konzentrationslager Buchenwald neue Vorstellungen zu erarbeiten, was in Deutschland nach der Zerschlagung der faschistischen Diktatur geshehen muß?* 1974, DIPF/BBF/Archiv, NL Walter Wolfs, 0.4.0.3., fo. 42.

88. Correspondence from Commissariat der Fuldaer Bischofkonferenz, Berlin Wichmannstr, to Marschall der Sowjetunion V.D. Sokolowkij, Cher der Sow. Militäradministration Deutschlands, October 14, 1948, BArch DR 2/4699, 15–17.

89. Ibid., andWandel to Marquardt, December 28, 1948.

90. STA/LAB 120/2425, "Rededisposition zur Hilfsaktion 'Rettet die Kinder!' " Hauptausschuß 'Opfer des Faschismus, Berlin, n.d. [fall 1945], 44–45.

91. LAB/STA 120/2425, "Bericht betr. Aktion 'Rettet die Kinder,' von Bildende Kunst," January 3, 1946, 74–76.

92. Victor Klemperer, *"Und so ist alles schwankend."* *Tagebücher Juni bis Dezember 1945*, December 9, 1945 (Berlin: Aufbau, 1996), 200–201.

93. Ibid., December 17, 1945, 209.

94. Ibid., December 26, 1945, 214.

95. "Was schenke ich meinem Kinde," Berlin, 1949, *Plakate der SBZ/DDR* DHM CD-ROM, Inv. P 90/2376.

96. "Weihnachten 1947," *Plakate der SBZ/DDR*, DHM CD-ROM, Dresden, 1947, Inv. P 90/6222.

97. "Sachsen produziert wieder," *Plakate der SBZ/DDR*, DHM CD-ROM, [Saxony], February 1947, Inv. P 90/3333

98. "Dresdner Weichnachtsmesse 1945," Horst Naumann, CD-Rom SBZ/GDR, P 90/6964.

99. K.S. [Karl Sothmann?], "Die Tagung der Kultusminister," unknown newspaper excerpt, [1948], DIPF/BBF/Archiv NL Hans Löffler 0.4.34, pt. 1.

100. Minister Bäuerle, "Konferenz der deutschen Erziehungsminister, Protokoll der Tagung am 19. und 20. Februar 1948 in Stuttgart-Hohenheim," DIPF/BBF/Archiv NL Marie Torhorst, fo. 32.

Conclusion: Redemption through Reconstruction and Beyond

1. Brigitte Reimann, May 31, 1950, *"Aber wir schaffen es, verlaß Dich drauf!"* in *Briefe an eine Freundin im Westen*, 2nd ed. (Berlin: Aufbau Taschenbuch Verlag, 1999), 32.

2. Ibid., December 28, 1950, 91–92.

3. Ibid., 92.

4. See, for example, Heidemarie Kühn [Kemnitz], "Mädchenbildung im Schulsystem der DDR," *Geschichte der Mädchen- und Frauenbildung*. Elke Kleinau and Claudia Opitz, eds., vol. 2. (Frankfurt AM: Campus, 1996), 434–445. My appreciation to Heidi Kemnitz for her insightful comments on this topic.

5. Sterling Fishman, "Colonizing Your Own People: German Unification and the Role of Education," in *The Educational Forum* 60.1 (Fall 1995): 30–31.

6. Helga A. Welsh, "Deutsche Zentralverwaltung für Volksbildung," in *SBZ Handbuch*, 231.

7. Heinrich Deiters, correspondence to directory of ministry Viktor Fadrus in Vienna, June 15, 1948, DIPF/BBF/Archiv, NL Heinrich Deiters, 0.4.05, fo. 1.

8. Heinrich Deiters, correspondence to E. Löffler, cultural minister in Stuttgart, June 21, 1950, DIPF/BBF/Archiv, NL Heinrich Deiters 0.4.05, fo. 2.

9. Ibid., fo. 54, 9.

10. Ibid., fo. 13.

11. Fishman, "Colonizing Your Own People," 24–25.

12. This second comment was posted on February 15, 2006 by Ruud van Dijk on the H-German discussion board (<http://h-net.msu.edu/cgi-bin/log-browse.pl?trx=lx&list=H-German&user=&pw=&month=0602>) sponsored by H-Net in response to comments I made about the current state of GDR historiography. <http://h-net.msu.edu/cgi-bin/logbrowse. pl?trx=vx&list=H-German&month= 0602&week= c&msg= 50xyt PlHXLGGnqB8Q7Opaw&user=&pw=> (accessed March 3, 2006).

13. My special thanks to Norman Naimark for his critical reading of my interpretation here.

14. Jutta-B. Lange-Quassowski, *Neuordnung oder Restauration? Das Demokratiekonzept der amerikanischen Besatzungsmacht und die politische Sozialisation der Westdeutschen: Wirtschaftsordnung—Schulstruktur— Politische Bildung* (Opladen, Germany: Leske, 1979); Brian Puaca, "Learning Democracy: Education Reform in Postwar West Germany, 1945–1965;" "The Pen is Mightier than the Sword?"

15. Gail Schmunk Murray, a contributor to the subject "Children and War" for the list-serve H-childhood, co-sponsored by H-Net and the Society for the History of Children and Youth, referred to the children on the American homefront affected by the war as "children-turned adults." Comment posted 31 March 1999 at: http:// h-net.msu.edu/cgi-bin/logbrowse. pl?trx= vx&list=H-Childhood&month= 9903&week= e&msg= GMAiP6OB6xdmbbI2nBb%2bYg&user=&pw=accessed March 3, 2006.

16. Wolfgang B., born 32, Volksschule, reprinted in Annett Gröschner, "*ich schlug meiner Mutter die brennenden funken ab*": *Berliner Schulaufsätze aus dem Jahr 1946* (Berlin: Kontext, 1996), 260 [LAB/STA 134/13, 181/1, no. 159].

17. See for instance Kurt Ortmann, thirteen years old, LAB/STA 134/13, 182/2, no. 413.

18. Siegfried Baske, "Schulen und Hochschulen," in *Handbuch der deutschen Bildungsgeschichte*, vol. 6/2 (Munich: Beck, 1998), 159–202.

19. Siegfried Baske, "Schulen und Hochschulen," in *Handbuch der deutschen Bildungsgeschichte*, vol. 6/2 ed. Christoph Führ and Carl-Ludwig Furck, 159–201.

20. Dorothee Wierling, *Geboren im Jahr Eins: Der Jahrgang 1949 in der DDR. Versuch einer Kollektivbiographie* (Berlin: Christoph Links, 2002), 99.

21. Katrin M., "Das erste Schritt zum neuen Mitenander," Premnitz, November 4, 1984, *Schüleraufsätze*, ed. Rudolf Chowanetz and Walter Lewerenz (Berlin [East]: Neues Leben, 1986), 86.

22. Claudia Ragalyi, "Auch meine Familie war davon betroffen," Rathenow, n.d., *Schüleraufsätze*, ed. Rudolf Chowanetz and Walter Lewerenz, 87–88.

23. Bärbel Michel, "Eine große Familie," Probstzella, *Schüleraufsätze*, ed. Rudolf Chowanetz and Walter Lewerenz, 63.

24. Kristin Zimmerman, "Meine Mutter," Saalfeld, *Schüleraufsätze*, ed. Rudolf Chowanetz and Walter Lewerenz, 62.

25. Jens Oerter, eighth grade, "Wie Eulen auf den Ästen," *Schüleraufsätze*, ed. Rudolf Chowanetz and Walter Lewerenz, 70.

26. Steffen Koch, eighth grade, "Alle Jahre wieder ..." *Schüleraufsätze*, ed. Rudolf Chowanetz and Walter Lewerenz, 68. For information on military training in schools, see Gert Geißler and Ulrich Wiegmann, "Wehrerziehung und Schule," in *Handbuch der deutschen Bildungsgeschichte*, Band VI/2, ed. Christoph Führ and Carl-Ludwig Furck, 359–375.

27. Quoted in Jeannette Z. Madarász, *Conflict and Comprise in East Germany, 1971–1989: A Precarious Stability* (New York: Palgrave, 2003), 40.

28. Reimann, August 3, 1950, 52; September 17, 1950, *Aber wir schaffen es, verlaß Dich drauf!* 61.

29. Ibid., October 7, 1950, 65–66.

30. Ibid., December 28, 1950, 91–92.

31. Brigitte Reimann, November 1, 1968, *Alles schmeckt nach Abschied*, 217.

32. Reimann, April 21, 1972, *Aber wir schaffen es, verlaß Dich drauf!* 183.

Bibliography

Albisetti, James. *Schooling German Girls and Women Secondary and Higher Education in the Nineteenth Century*. Princeton: Princeton University Press, 1988.

Albrecht, Franz. "Ernst Wildangel—1891 Bis 1951—'Du bist Lehrer, Du kannst unsere Schulen in Gang bringen!' " *Wegbereiter Der neuen Schule*. Gerd Hohendorf, Helmut König, and Eberhard Meumann, eds. Berlin [East]: Volk und Wissen, 1989. 267–273.

Albrecht, Franz, and Gerd Hohendorf. "Max Kreuziger, 1880 Bis 1953. 'Wir wollen Vorwärts!' " *Wegbereiter Der Neuen Schule*. Gerd Hohendorf, Helmut König, and Eberhard Meumann, eds. Berlin [East]: Volk und Wissen, 1989. 150–156.

Alt, Robert. *Das Bildungsmonopol*. Berlin [East]: Akademie-Verlag, 1978.

———. *Erziehung und Gesellschaft: Pädagogische Schriften*. Karl-Heinz Günther, Helmut König, and Rudi Schulz, eds. Berlin [East]: Volk und Wissen, 1975.

———. "Über unsere Stellung zur Reformpädagogik." *Pädagogik* 11:5/6 (1956): 345–367.

Alter, Peter. "Kulturnation und Staatsnation—Das Ende einer langen Debatte?" *Die Intellektuellen und die Nationale Frage*. Ed. Gerd Langguth. Frankfurt AM: Campus, 1997. 33–44.

Ambrosoli, Luigi. "Associazioni e Sindacati Degli Insenati Secondari dal 1945 Ad Oggi." *Storia Della Scuola e Storia D'Italia Dall'Unita Ad Oggi*. A. Santoni Rugui, G. Vigo, T. Tomasi Ricuperati, G. Talamo, D. Ragazzini, and G. Bonnetta. Bari: De Dorato, 1982. 207–234.

Anderson, Benedict. *Imagined Communities: Reflections on the Origin and Spread of Nationalism*. 2nd ed. London: Verso, 1991.

Anderson, W. *Holy Places of Britain*. London: Ebury Press, 1983.

Apple, Michael W. *Cultural Politics and Education*. New York: Teachers College Press, 1996.

Applegate, Celia. *A Nation of Provincials: The German Idea of Heimat*. Berkeley: University of California Press, 1990.

Ash, Timothy Garton. *The File: A Personal History*. London: Flamingo, 1997.

Aufbau. Berlin [East]: 1945 ff.

Bahne, Siegfried. "Sozialfaschismus in Deutschland. Zur Geschichte eines politischen Begriffs." *International Review of Social History* 10 (1965): 211–245.

Barnard, Henry. *School Architecture, or Contributions to the Improvement of School Houses in the United States.* 2nd ed. Ann Arbor: University of Michigan Press; reprint 2001 [1848].

Baske, Siegfried. "Allgemeinbildende Schulen." *Handbuch der deutschen Bildungsgeschichte.* Christoph Führ and Carl-Ludwig Furck, eds. Vol. 6, *1945 bis zur Gegenwart.* Bk. 2, *Die Deutsche Demokratische Republik und neue Bundesländer.* Munich: Beck, 1998. 159–202.

———. "Grund- und Rahmenbedingungen." *Handbuch der deutschen Bildungsgeschichte.* Christoph Führ and Carl-Ludwig Furck, eds. Vol. 6, *1945 bis zur Gegenwart.* Bk. 2, *Die Deutsche Demokratische Republik und neue Bundesländer.* Munich: Beck, 1998. 3–25.

———. "Pädagogische Wissenschaft." *Handbuch der deutschen Bildungsgeschichte.* Christoph Führ and Carl-Ludwig Furck, eds. Vol. 6, *1945 bis zur Gegenwart.* Bk. 2, *Die Deutsche Demokratische Republik und neue Bundesländer.* Munich: Beck, 1998. 137–157.

Baske, Siegfried, and Martha Engelberg, eds. *Zwei Jahrzehnte Bildungspolitik in der Sowjetzone Deutschlands.* Vol. 1, *1945–1958.* Heidelberg: Quelle and Meyer, 1966.

———, eds. *Zwei Jahrzehnte Bildungspolitik in der Sowjetzone Deutschlands.* Vol. 2. *1959–1965.* Heidelberg: Quelle and Meyer, 1966.

Bathrick, David. *The Powers of Speech: The Politics of Culture in the GDR.* Lincoln: University of Nebraska, 1995.

Batzli, Stefan, Friddin Kissling, and Rudolf Zihlmann, eds. *Menschenbilder-Menschenrechte. Islam und Okzident: Kulturen im Konflikt.* Zürich: Unionsverlag, 1994.

Baumgartner, Gabriele, and Dieter Hebig. "Biographisches Handbuch SBZ/GDR 1945–1990." *Enzyklopädie der DDR.* [CD-ROM]. Vol. 32. Berlin: Directmedia, 2000.

Bebel, August. *Die Frau und der Sozialismus.* 1895. Hannover: Fackelträger, 1974.

Benjamin: Zeitschrift für Junge Leute. Hamburg: 1947.

Benner, Dietrich, Jürgen Schriewer, and Heinz-Elmar Tenorth, eds. *Erziehungsstaaten: Historisch-Vergleichende Analysen ihrer Denktraditionen und nationaler Gestalten.* Weinheim: Deutscher Studien Verlag, 1998.

Berg, Stefan, Floriam Gless, Horand Knaup, Jürgen Leinemann, Paul Lersch. "Das Rote Gespenst." *Der Spiegel.* Vol. 10. 1999. 22–33.

Bernstein, J. M. *The Philosophy of the Novel: Lukacs, Marxism and the Dialectics of Form.* Minneapolis: University of Minnesota Press, 1984.

Blackburn, Gilmer W. *Education in the Third Reich: A Study of Race and History in Nazi Textbooks.* Albany: State University of New York Press, 1985.

Bonna, Rudolf. *Die Erzählung in der Geschichtsmethodik von SBZ und DDR.* Bochum: Universitätsverlag Brockmeyer, 1996.

Borneman, John. *Belonging in the Two Berlins: Kin, State, Nation.* Cambridge: Cambridge University Press, 1992.

Boswell, Laird. *Rural Communism in France, 1920–1939*. Ithaca: Cornell University Press, 1998.

Bowen, James. *A History of Western Education*. Vol. 3. London: Methuen & Co., 1981.

Boyd, William. *The History of Western Education*. 10th ed. London: Adam & Charles Black, 1995.

Brandt, Peter. *Antifaschismus und Arbeiterbewegung: Aufbau-Ausprägung-Politik in Bremen 1945/46*. Hamburg: Hans Christian, 1976.

Braun, Günter. "Daten Zur Demokgraphischen und Sozialen Struktur Der Bevölkerung." *SBZ Handbuch: Staatliche Verwaltung, Parteien, Gesellschaftliche Organisationen und ihre Führungskräfte in der Sowjetischen Besatzungszone Deutschlands 1945–1949*. Martin Brozat and Hermann Weber, eds. Munich: Oldenbourg, 1993. 1069–1074.

Breuilly, John. *Nationalism and the State*. New York: St. Martin's Press, 1982.

Brinks, Jan Herman. *Die DDR-Geschichtswissenschaft auf dem Weg zur deutschen Einheit: Luther, Friedrich II und Bismarck als Paradigmen Politischen Wandels*. Frankfurt AM: Campus, 1992.

Broszat, Martin, and Hermann Weber, eds. *SBZ Handbuch: Staatliche Verwaltung, Parteien, Gesellschaftliche Organisationen und ihre Führungskräfte in der Sowjetischen Besatzungszone Deutschlands 1945–1949*. Munich: Oldenbourg, 1993.

Brubaker, Rogers. *Citizenship and Nationhood in France and Germany*. Cambridge, MA: Harvard University Press, 1992.

Bruner, Jerome. *The Culture of Education*. Cambridge, MA: Harvard University Press, 1996.

Büchner, Peter, Burkhard Fuhs, and Heinz-Hermann Krüger. "Transformation der Eltern-Kind-Beziehungen? Facetten der Kindbezogenheit des elterlichen Erziehungsverhaltens in Ost- und Westdeutschland." *Kindheit, Jugend und Bildungsarbeit im Wandel: Ergebnisse der Transformationsforschung*. Ed. Heinz-Elmar Tenorth. Vol. 37th Supplement. Weinheim and Basel: Beltz, 1997. 32–52.

Bundesministerium des Innern. "DDR Handbuch." *Enzyklopädie der DDR*. [CD-ROM.] Vol. 32. Berlin: Directmedia, 2000.

Bundesministerium für gesamtdeutsche Fragen, ed. *Die Sowjetische Besatzungszone Deutschlands in den Jahren 1945–1954: Eine chronologische Übersicht*. Bonn: Deutscher Bundesverlag, 1956.

———, ed. *Die Sowjetisierung der deutschen Länder Brandenburg / Mecklenburg / Sachsen / Sachsen-Anhalt / Thüringen*. [Bonn]: Bundesministerium für gesamtdeutsche Fragen, 1950.

———, ed. *Das Schulwesen in der Sowjetzone*. Bonn: Bundesministerium für gesamtdeutsche Fragen, 1958.

Büngel, Dr. Werner. *Die Entwicklung des Brandenburgisch-Preußischen Staates bis 1786*. Lehrhefte für den Geschichtsunterricht in der Oberschule nr. 3. Berlin and Leipzig: Volk und Wissen, 1946.

Burnett, John, ed. *Destiny Obscure: Autobiographies of Childhood, Education and Family From the 1820s to the 1920s*. London and New York: Routledge, 1994.

Byg, Barton. "DEFA and the Traditions of International Cinema." Seán Allan and John Sandford, eds. *DEFA: East German Cinema, 19946–1992*. New York: Berghahn, 1999, 22–41.

Cambi, Franco. *Storia della pedagogia*. 1995. 6th ed. Milan: Laterza, 1999.

Caron, Jean-Claude. "Young People in School: Middle and High School Students in France and Europe." *A Story of Young People in the West: Stormy Evolution to Modern Times*. Levi Giovanni and Schmitt Jean-Claude, eds. Vol. 2. Cambridge, MA: Belknap Press of Harvard University Press, 1997. 117–173.

Castle, Terry. "Contagious Folly: An Adventure and Its Skeptics." *Questions of Evidence: Proof, Practice and Persuasion across the Disciplines*. James Chandler, Arnold I. Davidson, and Harry Harootunian, eds. Chicago: University of Chicago Press, 1994. 10–42.

Cheval, René. "Die Bildungspolitik in der Französischen Besatzungszone." *Umerziehung und Wiederaufbau: Die Bildungspolitik der Besatzungsmächte in Deutschland und Österreich*. Ed. Manfred Heinemann. Stuttgart: Klett-Cotta, 1981. 190–210.

Clark, Burton R., and Guy R. Neave, eds. *Encyclopedia of Higher Education*. Vol. 1. Oxford: Pergamon, 1992.

Clark, Martin. *Modern Italy: 1871–1982*. London and New York: Longman, 1984.

Clemens, Gabriele, ed. *Kulturpolitik im besetzten Deutschland 1945–1949*. Stuttgart: Franz Steiner, 1994.

Colley, Linda. *Britons: Forging the Nation 1707–1837*. New Haven: Yale University Press, 1992.

Confino, Alon. "Collective Memory and Cultural History." *American Historical Review* 102.5 (1997): 1386–1403.

————. *The Nation as a Local Metaphor: Württemberg, Imperial Germany, and National Memory, 1871–1918*. Chapel Hill: University of North Carolina Press, 1997.

Crane, Susan A. "Writing the Individual Back into Collective Memory." *American Historical Review* 102.5 (1997): 1372–1385.

Cutler, William. "Cathedral of Culture: The Schoolhouse in American Educational Thought and Practice Since 1820." *History of Education Quarterly* 29.1 (1989): 1–40.

D'Attore, Pier Paolo. "The European Recovery Program in Italy: Research Problems." *The Role of the United States in the Reconstruction of Italy and West Germany, 1943–1949*. Ed. Ekkehart Krippendorff. Berlin [West]: Free University, 1981. 77–105.

Dähn, Horst. "Kirchen und Religionsgemeinschaften." *SBZ Handbuch: Staatliche Verwaltung, Parteien, Gesellschaftliche Organisationen und ihre Führungskräfte in der Sowjetischen Besatzungszone Deutschlands 1945–1949*.

Martin Broszat and Hermann Weber, eds. Munich: Oldenbourg, 1993. 819–827.

Dann, Otto. *Nation und Nationalismus in Deutschland 1770–1990*. Munich: C.H. Beck, 1993.

Davey, Thomas. *A Generation Divided: German Children and the Berlin Wall*. Durham: Duke University Press, 1987.

Davis, Fanny. *The Ottoman Lady: A Social History from 1718 to 1918*. New York: Greenwood, 1986.

Davis, Kathleen Southwell. "Das Schulbuchwesen als Spiegel der Bildungspolitik von 1945 bis 1950." *Umerziehung und Wiederaufbau: Die Bildungspolitik der Besatzungsmächte in Deutschland und Österreich*. Ed. Manfred Heinemann. Stuttgart: Klett-Cotta, 1981. 153–171.

de Bruyn, Günter. *Zwischenbilanz: Eine Jugend in Berlin*. 1992. Frankfurt AM: Fischer, 1994.

de Fort, Ester. *La Scuola Elementare Dall Unita' alla Caduta del Fascismo*. Bologna: Mulino, 1996.

de Gouges, Olympe. "Introduction" by Olivier Blanc. *Ecrits Politiques*. Paris: Côté-femmes, 1993.

Deiters, Heinrich. *Bildung und Leben: Erinnerungen eines deutschen Pädagogen*. Cologne: Böhlau, 1989.

Deutsche Wissenschaft Erziehung und Volksbildung 7 (1941): 314–334.

Deutsche Zentralverwaltung für Volksbildung in der Sowjetischen Besatzungszone Deutschlands. *Lehrpläne für die Grund- und Oberschulen in der Sowjetischen Besatzungszone Deutschlands: Geschichte*. Berlin, 1946.

Dewey, John. *Democracy and Education*. New York: Macmillan, 1916.

Dickinson, Edward Ross. *The Politics of German Child Welfare from the Empire to the Federal Republic*. Cambridge, MA: Harvard University Press, 1996.

Die deutsche demokratische Schule im Aufbau. Berlin[East]/Leipzig: Volk und Wissen, 1949.

die neue schule [sic]. Berlin [East]: 1946 ff.

Die Zeit 1946ff.

Diederich, Jürgen, and Heinz-Elmar Tenorth. *Theorie der Schule: Ein Studienbuch zu Geschichte, Funktionen und Gestaltung*. Berlin: Cornelsen, 1997.

Diederich, Martin, and Friedrich Blage. *Das Schulbuch in der Sojwetzone: Lehrbücher im Dienst totalitärer Propaganda*. Bonn: Bundesministerium für gesamtdeutsche Fragen, 1958.

Diefendorf, Jeffry M., ed. *Rebuilding Euorpe's Bombed Cities*. London: Macmillan, 1990.

Dienst, Karl. "Bildungspolitik und Kirchen." *Handbuch der deutschen Bildungsgeschichte*. Christoph Führ and Carl-Ludwig Furck, eds. Vol. 6, *1945 bis zur Gegenwart*. Bk. 2, *Die Deutsche Demokratische Republik und neue Bundesländer*. Munich: Beck, 1998. 54–68.

Diere, Horst. "Die Französische Revolution von 1789 bis 1795 und die Zeit Napoleons im Geschichtslehrbuch der DDR." *Die Französische Revolution in den Geschichtsschulbüchern der Welt.* Ed. Rainer Riemenschneider. Frankfurt, AM: Diesterweg, 1994. 329–345.

Dietrich, Gerd. *Politik und Kultur in der SBZ 1945–1949.* Bern: Peter Lang, 1993.

Diner, Dan. "On the Ideology of Antifascism." *New German Critique* 67 (1997): 123–132.

Drewek, Peter. " Begriff, System und Ideologie Der *'Einheitsschule.'* Ein Kommentar zu Gerhart Neuners Beitrag über *'Das Einheitsprinzip'* im DDR-Bildungswesen." *Zeitschrift für Pädagogik* 4 (1997): 639–657.

Dymschiz, Alexansssder. *Ein unvergeßlicher Frühling. Literarische Porträts und Erinnerungen.* Berlin [East]: Dietz, 1970.

Ebbert, Birgit. *Erziehung zu Menschlichkeit und Demokratie: Erich Kästner und Seine Zeitschrift Pinguin im Erziehungsgefüge der Nachkriegszeit.* Frankfurt AM: Lang, 1994.

Eder, Ferdinand, Gisela Felhofer, and Satu with Muhr-Arnold. "Schule als Lebenswelt." *Kindliche Lebenswelten: Eine sozialwissenschaftliche Annäherung.* Liselotte Wilk and Johann Bacher, eds. Opladen: Leske and Budrich, 1994. 197–252.

Ehnert, Gunter. "Alte Parteien in der 'Neuen Zeit:' Vom Bund Demokratischer Sozialisten zum SPD-Bezirksverband in Thüringen 1945." *Von der SBZ zur DDR: Studien zum Herrschaftssystem in der Sowjetischen Besatzungszone und in der Deutschen Demokratischen Republik.* Ed. Hartmut Mehringer. Munich: Oldenbourg, 1995. 13–42.

Eley, Geoff, ed. *Society, Culture, and the State in Germany, 1870–1930.* Ann Arbor: University of Michigan Press, 1996.

Eley, Geoff, and Ronald Grigor Suny. "Introduction: From the Moment of Social History to the Work of Cultural Representation." *Becoming National.* Geoff Eley and Ronald Grigor Suny, eds. New York: Oxford University Press, 1996. 3–37.

Ellrich, Karl. "Die Entwicklung des Grundschulwesens in der Sowjetischen Besatzungszone seit 1945." *Die deutsche demokratische Schule im Aufbau,* ed. Schulabteilung der Deutschen Verwaltung für Volksbildung in der Sowjetischen Besatzungszone. Leipzig: Volk und Wissen, 1949. 9–17.

Emmerich, Wolfgang. *Kleine Literatureschichte der DDR.* 2nd ed. Leipzig: Gustav Kiepenhauer, 1996.

Fait, Barbara, "Dokumentation, Mecklenburg," and "Führungskräfte in Staat, Politik und Gesellschaft." *SBZ Handbuch: Staatliche Verwaltung, Parteien, Gesellschaftliche Organisationen und Ihre Führungskräfte in der Sowjetischen Besatzungszone Deutschlands 1945–1949.* Munich: Oldenbourg, 1993. 103–125.

Feist, Günter, Eckhart Gillen, and Beatrice Vierneisel, eds. *Kunstdokumentation SBZ/GDR 1945–1990.* Cologne: DuMont Buchverlag, 1996.

Fichte, Johann Gottlieb. *Reden an die deutsche Nation.* 1808. 5th ed. Hamburg: Meiner, 1978.

Fischer, Friedhelm. "German Reconstruction as an International Activity." *Rebuilding Europe's Bombed Cities.* Ed. Jeffry M. Diefendorf. London: Macmillan, 1990. 131–144.

Fishman, Sterling. "Colonizing Your Own People: German Unification and the Role of Education." *The Educational Forum* 60.1 (Fall 1995): 24–31.

———. *The Struggle for German Youth: The Search for Educational Reform in Imperial Germany.* New York: Revisionist Press, 1976.

Fishman, Sterling, and Lothar Martin. *Estranged Twins: Education and Society in the Two Germanys.* New York: Praeger, 1987.

Fivush, Robyn. "Constructing Narrative, Emotion, and Self in Parent-Child Conversations about the Past." *The Remembering Self: Construction and Accuracy in the Self-Narrative.* Ulrich Neisser and Robyn Fivush, eds. Cambridge: Cambridge University Press, 1994. 136–157.

Foitzik, Jan. "Sowjetische Militäradministration in Deutschland." *SBZ Handbuch: Staatliche Verwaltung, Parteien, Gesellschaftliche Organisationen und ihre Führungskräfte in der Sowjetischen Besatzungszone Deutschlands 1945–1949.* Martin Broszat and Hermann Weber, eds. Munich: R. Oldenbourg, 1993. 7–69.

———. "Vereinigung der Verfolgten des Naziregimes." *SBZ Handbuch: Staatliche Verwaltung, Parteien, Gesellschaftliche Organisationen und ihre Führungskräfte in der Sowjetischen Besatzungszone Deutschlands 1945–1949.* Martin Broszat and Hermann Weber, eds. Munich: Oldenbourg, 1993. 748–759.

Forgacs, David, Sarah Lutton, and Geoffrey Nowell-Smith, *Roberto Rossellini: Magician of the Real.* London: British Film Institute, 2000.

Frauen und Geschichte Baden-Württemberg, Eds. *Frauen und Nation.* Tübingen: Silberburg, 1996.

Friedländer, Ernst. "Nationalismus." *Die Zeit.* February 6, 1947.

Friemel, Rudolf. "Trennung der Geschlechter oder gemeinschaftliche Erziehung?" *Pädagogisches Magazin* 327 (1908): 8–12.

Führ, Christoph, and Carl-Ludwig Furck, eds. *Handbuch der deutschen Bildungsgeschichte* Vol. 6, *1945 bis zur Gegenwart.* Bk. 2, *Die Deutsche Demokratische Republik und neue Bundesländer.* Munich: Beck, 1998.

———, eds. *Handbuch der deutschen Bildungsgeschichte.* Vol. 6, *1945 bis zur Gegenwart.* Bk. 1, *Die Bundesrepublik Deutschland.* Munich: Beck, 1998.

Fulbrook, Mary. *Anatomy of a Dictatorship: Inside the GDR 1949–1989.* Oxford and New York: Oxford University Press, 1995.

———. *The People's State: East Germany Society from Hitler to Honecker.* New Haven: Yale University Press, 2005.

Füssl, Karl-Heinz. *Die Umerziehung der Deutschen: Jugend und Schule unter den Siegermächten des zweiten Weltkrieges, 1945–1955.* Paderborn: Ferdinand Schöningh, 1994.

———. "Zwischen NS-Traumatisierung und Demokratie: Die Erziehungspolitik der USA in der deutschen Nackriegsgeschichte (1945–1952)." *Paedagogica Historica* 33.1 (1997): 221–246.

Füssl, Karl-Heinz, and Gregory Paul Wegner. "Education Under Radical Change: Education Policy and the Youth Program of the United States in Postwar Germany." *History of Education Quarterly* 36 (1996): 1–18.

Gedi, Noa, and Yigal Elam. "Collective Memory—What Is It?" *History and Memory* 8.1 (1996): 30–50.

Geißler, Gert. "Schulämter und Schulreformer in Berlin nach Kriegsende 1945." *Reformpädagogk in Berlin— Tradition und Wiederentdeckung.* Ed. Wolfgang Keim. Frankfurt AM: Peter Lang, 1998. 137–168.

———. "Schulreform von Oben: Bemerkungen zum Schulpolitischen Herrschaftssystem in der SBZ/DDR." *Erinnerung für die Zukunft II: Das DDR-Bildungssystem als Geschichte und Gegenwart.* Ed. Petra Gruner. Ludwigsfeld-Struveshof: Pädagogisches Landesinstitut Brandenburg, 1997. 49–60.

———. "Schulreform zwischen Diktaturen? Pädagogik und Politik in der frühen Sowjetischen Besatzungszone Deutschlands." *Bildung und Erziehung in Europa.* Ed. Dietrich Benner. Vol. 32nd Supplement. Weinheim: Beltz, 1994. 49–63.

Gellner, Ernest. "Nationalism." *Nationalism.* John Hutchinson and Antony D. Smith, eds. Oxford: Oxford University Press, 1994.

———. *Nations and Nationalism.* Blackwell: Oxford, 1983.

Gernert, Dörte. "Mädchenerziehung im allgemeinen Volksschulwesen." *Geschichte der Mädchen—und Frauenbildung.* Elke Kleinau and Claudia Opitz, eds. Vol. 2. Frankfurt AM: Campus, 1996. 85–98.

Geschichte in der Schule. Berlin [East]: 1948.

Gesellschaftswissenschaftliche Fakultät. *Nie Wieder Faschismus und Krieg: Die Mahnung der Faschistischen Bücherverbrennung am 10. Mai 1933.* Berlin [East]: Humboldt University, 1983.

Gibas, Monika. "Vater Staat snd seine Töchter: Offizielle propagierte Frauenleitbilder der DDR und ihre Sozialisationswirkungen." *Parteiauftrag: Ein neues Deutschland: Bilder, Rituale und Symbole der frühen DDR.* Ed. Dieter Vorsteher. Berlin: Deutsches Historisches Museum, 1996. 311–312.

Gibbs Jr., Raymond W. "Process and Products in Making Sense of Tropes." *Metaphor and Thought.* Ed. Andrew Ortony. 2nd ed. Cambridge: Cambridge University Press, 1993. 252–276.

Gill, Ulrich. *Der Freie Deutsche Gewerkschaftsbund: Geschichte-Organisation-Funktionen-Kritik.* Opladen: Leske and Budrich, 1989.

Gillis, John R. *Commemorations.* Princeton, NJ: Princeton University Press, 1994.

Goguel, Rudi. *Cap Arcona: Report über den Untergang der Häftlingsflotte in der Lübecker Bucht am 3. Mai 1945.* Frankfurt AM: Röderberg, 1972.

Gotschlich, Helga, ed. *Aber nicht im Gleichschritt: Zur Entstehung der Freien Deutschen Jugend.* Berlin: Metropol, 1997.Griese, Christiane. "Auf 'Spurensuche.' Überlegungen zu Schwerpunkten und Perspektiven der bildungsgeschichtlichen DDR-Forschung." *Paedagogica Historica* 31.2 (1995): 483–491.

Griese, Christiane, and Helga Marburger. "Sozialistischer Patriotismus und Proletarischer Internationalismus." *Pädagogische Rundschau* 51 (1997): 165–178.

Grimm, Jacob [Jakob]. "Einleitung." *Deutsches Wörterbuch.* Jacob Grimm. Vol. 1. Leipzig: S. Hirzel, 1854. LI–LXII.

Grimm, Jacob [Jakob], and Wilhelm. *Die Kinder- und Hausmärchen der Brüder Grimm.* 3 vols. Hans Siebert and Walther Pollatschek, eds. 5th ed. Berlin [East]: Kinderbuchverlag, 1955 [1952].

Gröschner, Annett. *"Ich schlug meiner Mutter die brennenden Funken ab": Berliner Schulaufsätze aus dem Jahr 1946.* Berlin: Kontext, 1996.

Grossmann, Atina. "Eine Frage des Schweigens? Die Vergewaltigung deutscher Frauen durch Besatzungssoldaten." *Sozialwissenschaftliche Informationen* 24.2 (1995): 109–119.

Grunenberg, Antonia. *Antifaschismus—Ein deutscher Mythos.* Reinbek bei Hamburg: Rowohlt, 1993.

Gruner, Petra. "Die Neulehrer: Ein Schlüsselsymbol der DDR-Gesellschaft?" Ph.D. diss. Humboldt University—Berlin, 1998.

———, ed. *Erinnerung für die Zukunft II: Das DDR-Bildungssystem als Geschichte und Gegenwart.* Conference Protocol Pädagogisches Landesinstitut Brandenburg December 1995. Ludwigsfeld-Struveshof: Pädagogisches Landesinstitut Brandenburg, 1997.

Günter, Karl-Heinz, and Gottfried Uhlig, eds. *Dokumente zur Geschichte des Schulwesens in der Deutschen Demokratischen Republik 1945–1955.* Vol. 6. *Monumenta Paedagogica.* Berlin [East]: Volk und Wissen, 1970.

Günther, Karl-Heinz, Franz Hofmann, Gerd Hohendorf, Helmut König, and Heinz Schuffenhauer, eds. *Geschichte der Erziehung.* 1966. 9th ed. Berlin [East]: Volk und Wissen, 1969.

Günther, Karl-Heinz, and Gottfried Uhlig. *Zur Entwicklung des Volksbildungswesens auf dem Gebiet der Deutschen Demokratischen Republik 1946–1949. Monumenta Paedagogica III.* Berlin [East]: Volk und Wissen, 1968.

Haberl, Othmar Nikola. "Die Sowjetunion und der Marshall-Plan." *The Role of the United States in the Reconstruction of Italy and West Germany, 1943–1949.* Ed. Ekkehart Krippendorff. Berlin [West]: Free University, 1981. 330–346.

Hacker, Jens. "Über die Tabuisierung der nationalen Frage im intellektuellen Diskurs." *Die Intellektuellen und die nationale Frage.* Ed. Gerd Langguth. Frankfurt AM: Campus, 1997. 314–329.

Häder, Sonja, and Heinz-Elmar Tenorth, eds. *Bildungsgeschichte einer Diktatur.* Weinheim: Deutscher Studien Verlag, 1997.

Hahn, Hans J. "Education System: FRG." *Encyclopedia of Contemporary German Culture.* Ed. John Sandford. London: Routledge, 1999. 172–174.

———. "Education System: GDR." *Encyclopedia of Contemporary German Culture.* Ed. John Sandford. London: Routledge, 1999. 175–176.

Harp, Stephen. *Learning to Be Loyal: Primary Schooling as Nation Building in Alsace and Lorraine, 1850–1940.* DeKalb: Northern Illinois University Press, 1998.

Hartmann, Silvia. *Fraktur oder Antiqua: Der Schriftstreit von 1881 bis 1941.* New York: Peter Lang, 1998.

Haupt, Heinz-Gerhard, ed. *Luoghi quotidiani nella storia d'Europa.* Rome: Laterza, 1993.

Haus der Geschichte der Bundesrepublik Deutschland, ed. *Ungleiche Schwestern? Frauen in Ost- und Westdeutschland.* Berlin: Nicolai, 1998.

Haydock, Michael D. *City under Siege: The Berlin Blockade and Airlift, 1948–1949.* Washington, DC: Brassey's, 1999.

Hearnden, Arthur. *The British in Germany: Educational Reconstruction after 1945.* London: Hamilton, 1978.

———. *Education in the Two Germanies.* Oxford: Blackwell, 1974.

Heinemann, Manfred. *Umerziehung und Wiederaufbau: Die Bildungspolitik der Besatzungsmächte in Deutschland und Österreich.* Stuttgart: Klett-Cotta, 1981.

Heitzer, Heinz. *DDR: Geschichtlicher Überblick.* 4th ed. Berlin [East]: Dietz, 1987.

Herb, Guntram H., and David H. Kaplan, eds. *Nested Identities: Nationalism, Territory, and Scale.* Lanham, MD: Rowman and Littlefield, 1999.

Herbst, Andreas, Winfried Ranke, and Jürgen Winkler. *So Funktionierte die DDR.* Vol. 3, *Lexikon der Funktionäre.* Reinbek bei Hamburg: Rowohllt, 1994.

———. *"So funktionierte die DDR. Lexikon der Organisationen und Institutionen."* Enzyklopädie der DDR. [CD-ROM.] Digitale Bibliothek. Vol. 32. Berlin: Directmedia, 2000.

Herbst, Jürgen. *Requiem for a German Past: A Boyhood among Nazis.* Madison, WI: University of Wisconsin Press, 1999.

Herf, Jeffrey. *Divided Memory: The Nazi Past in the Two Germanys.* Cambridge, MA: Harvard University Press, 1997.

Hettling, Manfred. "Shattered Mirror. German Memory of 1848: From Spectacle to Event." *1848: Memory and Oblivion.* Ed. Charlotte Tacke. Brussels: Peter Lang, 2000. 79–98.

Heym, Stefan. *Five Days in June.* Buffalo: Prometheus Books, 1978.

Hielscher, Almut. " 'Bei den Wessis ist jeder für sich': Vier Thüringer Schüler über ihre Kindheit in der DDR." *Der Spiegel* 10 (1999): 28.

Hobsbawm, Eric. *The Age of Extremes: A History of the World, 1914–1991.* New York: Pantheon, 1994.

Hoffmann, Dietrich. *Erziehung und Erziehungswissenschaft in der BRD und der DDR.* Vol. 1, *Die Teilung der Pädagogik.* Weinheim: Deutscher Studien Verlag, 1994.

Hoffmann, Stanley, and Charles Maier, eds. *The Marshall Plan: A Retrospective.* Boulder: Westview Press, 1984.

Hofmann, J., and D. Säuberlich. "Nationale Frage und Nationsentwicklung in der Politik der SED von der Mitte der 50er Jahre bis 1963." *Jahrbuch für Geschichte* 31 (1984): 41–70.

Hogan, Michael J. *The Marshall Plan: America, Britain, and the Reconstruction of Western Europe, 1947–1952.* New York: Cambridge University Press, 1987.

Hohendorf, Gerd, Helmut König, and Eberhard Meumann. "Vorwort." *Wegbereiter der neuen Schule.* Gerd Hohendorf, Helmut König, and Eberhard Meumann, eds. Berlin [East]: Volk und Wissen, 1989. 11–12.

———, eds. *Wegbereiter der neuen Schule.* Berlin [East]: Volk und Wissen, 1989.

Hohlfield, Brigitte. *Die Neulehrer in der SBZ/DDR 1945–1953.* Weinheim: Deutscher Studien Verlag, 1992.

Hohmann, Joachim S. *Deutschunterricht in SBZ und DDR 1945–1962.* Frankfurt AM: Peter Lang, 1997.

Hopp, Hans. "Der Wiederaufbau von Neubrandenburg." *Deutsche Architektur* 6 (1955): 293–299.

Horstkemper, Marianne. "Die Koedukationsdebatte um die Jahrhundertwende." *Geschichte der Mädchen- und Frauenbildung* Eds. Elke Kleinau and Claudia Opitz. Vol. 2. Frankfurt AM: Campus, 1996. 203–218.

Howard, Marc Alan. "Die Ostdeutschen als ethnische Gruppe? Zum Verständnis der neuen Teilung des geeinten Deutschland." *Berliner Debatte* 4.5 (1995): 119–131.

Hübner-Funk, Sibylle. *Loyalität und Verblendung. Hitlers Garanten der Zukunft als Träger der zweiten deutschen Demokratie.* Potsdam: Land Berlin-Brandenburg, 1998.

Hudemann, Rainer. "Kulturpolitik in der französischen Besatzungszone— Sicherheitspolitik oder Völkerverständigung? Notizen zur wissenschaftlichen Diskussion." *Kulturpolitik im besetzten Deutschland 1945–1949.* Ed. Gabriele Clemens. Stuttgart: Franz Steiner, 1994. 185–199.

Huschner, "Vereinheitlichung und Differenzierung in der Schulentwicklung der SBZ und DDR. Zweig und Klassen mit verstärktem alt- bzw. neusprachlichem Unterricht im Schulsystem der SBZ/DDR (1945 bis Anfang der siebzer Jahre)." *Zeitschrift für Pädagogik* 2 (March/April 1997): 279–297.

Jäger, Manfred. *Kultur und Politik in der DDR: 1945–1990.* Cologne: Edition Deutschland Archiv, 1995.

Jarausch, Konrad H., ed. *Dictatorship as Experience: Towards a Socio-Cultural History of the GDR.* New York: Berghahn Books, 1999.

———. "Introduction: Reshaping German Identities: Reflections on the Post-Unification Debate." *After Unity: Reconfiguring German Identities.* Ed. Konrad H. Jarausch. Providence: Berghahn, 1997. 1–24.

Jarausch, Konrad, and Hannes Siegrist, eds. *Amerikanisierung und Sowjetisierung in Deutschland 1945–1970.* Frankfurt AM: Campus, 1997.

———. "Amerikanisierung und Sowjetisierung: Eine vergleichende Fragestellung zur deutsch-deutschen Nachkriegsgeschichte." *Amerikanisierung und Sowjetisierung in Deutschland 1945–1970.* Eds. Konrad Jarausch and Hannes Siegrist. Frankfurt AM: Campus, 1997. 11–46.

Jarausch, Konrad H., Hinrich C. Seeba, and David P. Conradt. "The Presence of the Past: Culture, Opinion, and Identity in Germany." *After Unity: Reconfiguring German Identities.* Ed. Konrad H. Jarausch. Providence: Berghahn, 1997. 25–60.

Jeannette Z. Madarász. *Conflict and Comprise in East Germany, 1971–1989: A Precarious Stability.* New York: Palgrave, 2003.

Jeismann, Karl-Ernst, and Peter Lundgreen, eds. *Handbuch der deutschen Bildungsgeschichte* Vol. 3, *1800–1870. Von der Neuordnung Deutschlands bis zur Gründung des Deutschen Reiches.* Munich: Beck, 1987.

Johnson, Norris Brock. "School Spaces and Architecture: The Social and Cultural Landscape of Educational Environments." *Journal of American Culture* 5 (1982): 79–88.

Judt, Tony. "Europas Nachkriegsgeschichte Neu Denken." *Transit: Europäische Revue* (Vienna) 15 (1998): 3–11.

Jünger, Friedrich Georg. *Sämtliche Gedichte.* 1900. Vol. 1. Stuttgart: Klett-Cotta, 1985.

Kaelble, Hartmut, Jürga Kocka, and Hartmut Zwahr, eds. *Sozialgeschichte der DDR.* Stuttgart: Klett-Cotta, 1994.

Kaiser, Monika. "Sowjetischer Einfluß auf die Ostdeutsche Politik und Verwaltung 1945–1970." *Amerikaniserung und Sowjetisierung in Deutschland 1945–1970.* Konrad Jarausch and Hannes Siegrist, eds. Frankfurt AM: Campus, 1997. 111–133.

Kammeier-Nebel, Andrea. "Frauenbildung im Kaufmannsmilieu spätmittelalterlicher Städte." *Geschichte der Mädchen- und Frauenbildung.* Eds. Elke Kleinau and Claudia Opitz. Vol. 1. Frankfurt: Campus, 1996. 78–90.

Kant, Hermann. *Die Aula.* 1965. Munich: Rütlen & Loening, 1966.

Kästner, Erich. *Der tägliche Kram: Chansons und Prosa 1945–1948.* 1948. 2nd ed. Munich: Deutscher Taschenbuch Verlag, 1999.

Kedourie, Elie. *Nationalism.* 1960. 4th ed. Oxford: Blackwell, 1994.

Kehm, B., and U. Teichler. "Germany, Federal Republic of." *Encyclopedia of Higher Education.* Burton R. Clark and Guy R. Neave, eds. Vol. 1. Oxford: Pergamon, 1992. 240–260.

Kempf, Frau Dr. R. "Die soziale und wirtshcaftliche Lage in ihrer Bedeutung für das Problem der gemeinsamen Erziehung." *Dritter Deutscher Kongreß für Jugendbildung und Jugendkunde.* Ed. Arbeiten des Bundes für Schulreform. Leipzig/Berlin: BG Teubner, 1914. 17–38.

Kenny, Michael. "A Place for Memory: The Interface between Individual and Collective History." *Comparative Studies in Society and History* 41.3 (1999): 420–437.

Key, Ellen. *Das Jahrhundert des Kindes.* Berlin: S. Fischer, 1904.

Kienbaum, Jutta. "Kindliche Sozialisation in unterschiedlichen Kulturen: Eine Vergleichsstudie an deutschen und sowjetischen Kindern." *Kinder und Kindheit: Soziokulturelle Muster—Sozialisationstheoretische Perspektiven.* Michael-Sebastian Honig, Hans Rudolf Leu, and Ursula Nissen, eds. Munich: Juventa, 1996. 117–128.

Kirchhöfer, Dieter. "Veränderungen in der sozialien Konstruktion von Kindheit." *Kindheit, Jugend und Bildungsarbeit im Wandel: Ergebnisse der Transformationsforschung.* Ed. Heinz-Elmar Tenorth. Vol. 37th Supplement. Weinheim and Basel: Beltz, 1997. 15–34.

Kitzmüller, Erich, Heinz Kuby, and Lutz Niethammer. "Der Wandel der nationalen Frage in der Bundesrepublik Deutschland. Nationalstaat und Nationalökonomie." *Das Parlament* 33 (1973). 3–31.

Kleinau, Elke, and Claudia Opitz, eds. *Geschichte der Mädchen- und Frauenbildung.* Vol. 1 and 2. Frankfurt AM: Campus, 1996.

Klemperer, Victor. *I Will Bear Witness: A Diary of the Nazi Years 1933–1941*, trans. Martin Chalmers. New York: Random House, 1988.

———. *I Will Bear Witness: A Diary of the Nazi Years 1942–1945*, trans. Martin Chalmers. New York: Random House, 1989.

———. *LTI: Notizbuch eines Philologen.* Berlin [East]: Aufbau, 1947.

———. *Und so ist alles schwankend: Tagebücher Juni bis Dezember 1945.* 3rd ed. Berlin: Aufbau, 1995. Kleßmann, Christoph. *Die doppelte Staatsgründung: Deutsche Geschichte 1945–1955.* Göttingen: Vandenhoeck and Ruprecht, 1982.

Kleßmann, Christoph, and Georg Wagner, eds. *Das gespaltene Land: Leben in Deutschland 1945 bis 1990, Texte und Dokumente.* Munich: Beck, 1993.

Klier, Freya. *Lüg Vaterland: Erziehung in der DDR.* Munich: Kinder, 1990.

Kluchert, Gerhard, and Achim Leschinsky. "Glaubensunterricht in der Säkularität: Religionspädagogische Entwicklungen in Deutschland seit 1945." *Christenlehre und Religionsunterricht: Interpretation zu ihrer Entwicklung 1945–1990.* Ed. Comenius-Institut. Weinheim: Deutscher Studien Verlag, 1998. 1–113.

Kocka, Jürgen, ed. *Historische DDR-Forschung.* Berlin: Akademie, 1993.

Köhler, Helmut. *Was die Schulstatistik der SBZ-DDR Erfragte. Analyse und Dokumentation des Erhebungsprogramms 1945–1989.* Berlin: Max-Planck-Institut für Bildungsforschung, 1999.

Köhler, Tilo, and Rainer Nitsche, eds. *Stunde Eins oder Die Erfindung von Ost und West.* Berlin: Transit, 1995.

Kohli, Martin. "Die DDR als Arbeitsgesellschaft? Arbeit, Lebenslauf und Soziale Differenzierung." *Sozialgeschichte der DDR .* Harmut Kaelble, Jürgen Kocka, and Harmut Zwahr, eds. Stuttgart: Klett-Cotta, 1994.

König, Helmut. "Robert Alt: 1905 bis 1978—Erziehung und Gesellschaft—Grundthema seiner wissenschaftlichen Arbeit." *Wegbereiter der neuen Schule.* Gerd Hohendorf, Helmut König, and Eberhard Meumann, eds. Berlin [East]: Volk und Wissen, 1989. 31–38.

König, Ingelore, Dieter Wiedemann, and Lother Wolf, ed. *Alltagsgeschichten: Arbeiten mit DEFA-Kinderfilmen.* Munich: KoPäd, 1998.

———. *Zwischen Marx und Muck: DEFA-Filme für Kinder.* Berlin: Henschleg, 1996.

Koppetsch, Axel. "Die Französische Revolution: Dauer und Wandel ihrer Darstellung in deutschen Schulbüchern seit 1980." *Bilder einer Revolution: Die Französische Revolution in den Geschichtsschulbüchern der Welt.* Ed. Rainer Riemenschneider. Frankfurt AM: Diesterweg, 1994. 357–376.

Koshar, Rudy. "Building Pasts: Historic Preservation and Identity in Twentieth-Century Germany." *Commemorations.* Ed. John R. Gillis. Princeton: Princeton Universty Press, 1994. 215–238.

Koshar, Rudy. *Germany's Transient Pasts: Preservation and National Memory in the Twentieth Century.* Chapel Hill: University of North Carolina Press, 1998.

Kotre, John. *White Gloves: How We Create Ourselves through Memory.* New York: Free Press, 1995.

Koven, Seth. "From Rough Lads to Hooligans: Boy Life, National Culture and Social Reform." *Nationalisms and Sexualities.* Andrew Parker, Mary Russo, Doris Sommer, and Patricia Yaeger, eds. New York: Routledge, 1992. 365–388.

Krenn, Ruth. *Die Industrielle Revolution in England. Arbeitshefte für den Geschichtsunterricht in der Grundschule. Material für die Hand des Lehrers.* Berlin[East]/Leipzig: Volk und Wissen, 1949.

Krippendorff, Ekkehart, ed. *The Role of the United States in the Reconstruction of Italy and West Germany, 1943–1949.* German-Italian Colloquium at the John F. Kennedy-Institut Für Nordamerikastudien. Berlin: Free University, 1981.

Küchler, Stefan. "DDR-Geschichtsbilder: Zur Interpretation des Nationalsozialismus, der jüdischen Geschichte und des Holocaust im Geschichtsunterricht der DDR." *Internationale Schulbuchforschung* 22 (2000): 31–48.

Kudlek, Martin. "Architecture." *Encyclopedia of Contemporary German Culture.* Ed. John Sandford. London: Routledge, 1999. 17–20.

Kuhn, Annette. "Die stille Kulturrevolution der Frau. Versuch einer Deutung der Frauenöffentlichkeit." *Kulturpolitik im besetzen Deutschland 1945–1949.* Ed. Gabriele Clemens. Stuttgart: Franz Steiner, 1994. 83–101.

Kühn [Kemnitz], Heidemarie. "Mädchenbildung im Schulsystem der DDR." *Geschichte der Mädchen- und Frauenbildung.* Elke Kleinau and Claudia Opitz, eds. Vol. 2. Frankfurt AM: Campus, 1996. 434–445.

Labrousse, Ernest. *Le mouvement ouvrier et les théories sociales en France de 1815 à 1848.* Paris: Centre de documentation universitaire, 1961.

Lacroix-Riz, Annie. "Politique scolaire et universitaire en Allemagne occupée." *Kulturpolitik im besetzten Deutschland 1945–1949.* Ed. Gabriele Clemens. Stuttgart: Franz Steiner, 1994. 131–184.

Ladd, Brian. *The Ghosts of Berlin: Confronting German History in the Urban Landscape.* Chicago: University of Chicago Press, 1997.

Lagemann, Ellen Condliffe. *An Elusive Science: The Troubling History of Education Research.* Chicago: University of Chicago Press, 2000.

Lamberti, Marjorie. "Radical Schoolteachers and the Origins of the Progressive Education Movement in Germany, 1900–1914." *History of Education Quarterly* 40.1 (2000): 22–48.

Landler, Eva. "Käte Agerth: 1888 bis 1974—Im Herzen immer jung Geblieben." *Wegbereiter der neuen Schule.* Gerd Hohendorf, Helmut König, and Eberhard Meumann, eds. Berlin [East]: Volk und Wissen, 1989. 13–19.

Lane, Barbara Miller. *Architecture and Politics in Germany, 1918–1945.* Cambridge, MA: Harvard University Press, 1968.

Langewiesche, Dieter. "Nation, Nationalismus, Nationalstaat: Forschungsstand und Foschungsperspektiven." *Neue Politische Literatur* 40 (1995): 190–236.

Langewiesche, Dieter, and Heinz-Elmar Tenorth, eds. *Handbuch der deutschen Bildungsgeschichte*. Vol. 5, *1918–1945: Die Weimarer Republik und die nationalsozialistische Diktatur*. Munich: Beck, 1989.

Langguth, Gerd, ed. *Die Intellektuellen und die nationale Frage*. Frankfurt AM: Campus, 1997.

Lemke, Michael. "Die Sowjetisierung der SBZ/GDR im Ost-Westlichen Spannungsfeld." *Aus Politik und Zeitgeschichte* B 6.97 (1997): 41–53.

Lentricchia, Frank, and Thomas McLaughlin, eds. *Critical Terms for Literary Study*. Chicago: University of Chicago Press, 1990.

Leonhard, Wolfgang. "Die 'Gruppe Ulbricht'—Strategie und Taktik der Machteroberung 1945/46." *Einheit oder Freiheit? Zum 40. Jahrestag der Gründung der SED*. Ed. Friedrich-Ebert-Stiftung. Bonn: Friedrich-Ebert-Stiftung, 1985. 9–30.

Leschinsky, Achim, and Gerhard Kluchert. *Zwischen zwei Diktaturen: Gespräche über die Schulzeit im Nationalsozialismus und in der SBZ/DDR*. Weinheim: Deutscher Studium Verlag, 1997.

Lexikon der Pädagogik. 3rd ed. Vol. 2 and 4. Freiburg: Herder, 1970.

Leydesdorff, Selma, Luisa Passerini, and Paul Thompson, eds. *Gender and Memory*. International Yearbook of Oral History and Life Stories. Oxford: Oxford University Press, 1996.

———. "Introduction." *Gender and Memory*. Selma Leydesdorff, Luisa Passerini, and Paul Thompson, eds. Oxford: Oxford University Press, 1996. 1–16.

Longhi, Giuseppe. "Introduzione." *Storia urbana* 73 (October–December 1995): 5–10.

Loriga, Sabina. "The Military Experience." *A Story of Young People in the West: Stormy Evolution to Modern Times*. Levi Giovanni and Jean-Claude Caron, eds. Cambridge, Mass.: Belknap Press of Harvard University Press, 1997. 11–36.

Lüdtke, Alf, ed. *Alltagsgeschichte: Zur Rekonstruktion historischer Erfahrungen und Lebensweisen*. Frankfurt AM: Campus, 1989.

———. "La République Démocratique Allemande comme Histoire. Réflexions Historiographiques." *Annales, Histoire, Sciences Sociales* 53.1 (1998): 3–39.

Luhmann, Niklas. "Zeit und Gedächtnis." *Soziale Systeme* 2.2 (1996): 307–330.

Lukas, Richard. *Zehn Jahre Sowjetische Besatzungszone. Politik, Wirtschaft, Kultur, Rechtwesen*. Mainz-Gonsenheim: Deutscher Fachschriften, 1955.

Lundgreen, Peter. "La scuola." *Luoghi quotidiani nella storia d'Europa*. Ed. Heinz-Gerhard Haupt. Rome: Laterza, 1993. 285–295.

Lynch, Kathleen. *The Hidden Curriculum: Reproduction in Education. A Reappraisal*. New York: Falmer, 1989.

MacClancy, Jeremy, ed. *Sport, Identity and Ethnicity*. Oxford: Berg, 1996.

Mählert, Ulrich. *Die Freie Deutsche Jugend 1945–1949: Von den "Antifaschistischen Jugendausschüssen" zur SED-Massenorganisation: Die Erfassung der Jugend in der Sowjetischen Besatzungszone*. Paderborn: F. Schöningh, 1995.

Maier, Charles. *Recasting Bourgeois Europe: Stabilization in France, Germany, and Italy in the Decade after World War I.* Princeton, NJ: Princeton University Press, 1975.

Mallmann, Klaus-Michael. *Kommunisten in der Weimarer Republik: Sozialgeschichte einer revolutionären Bewegung.* Darmstadt: Wissenschaftliche Buchgesellschaft, 1996.

Malycha, Andreas. *Auf dem Weg zur SED. Die Sozialdemokratie und Bildung einer Einheitspartei in den Ländern der SBZ.* Bonn: Dietz, 1996.

Mang, Karl. *Storia del mobile moderno* [Geschichte des modernen Möbels]. 1978. Rome: Laterza, 1991.

March, James G., and Johan P. Olsen. *Ambiguity and Choice in Organizations.* Oslo: Universitetsforlaget, 1976.

Marx, Karl, and Friedrich Engels. *Über Pädagogik und Bildungspolitik.* Ed. Karl-Heinz Günther and Helmut König. Berlin [East]: Volk und Wissen, 1976.

Massey, Doreen. *Space, Place and Gender.* Cambridge: Polity Press, 1994.

Mayer, Christine. "Die Anfänge einer institutionalisierten Mädchenerziehung an der Wende vom 18. zum 19. Jahrhundert." *Geschichte der Mädchen- und Frauenbildung.* Elke Kleinau and Claudia Opitz, eds. Vol. 1. Frankfurt AM: Campus, 1996. 373–392.

Maynes, M. J. "Childhood Memories, Political Visions, and Working-Class Formation in Imperial Germany: Some Comparative Observations." *Society, Culture, and the State in Germany, 1870–1930.* Ed. Geoff Eley. Ann Arbor: University of Michigan Press, 1996. 143–162.

McDougall, Alan. *Youth Politics in East Germany: The Free German Youth Movement, 1946–1968.* Oxford: Oxford University Press, 2004.

McLaughlin, Thomas. "Figurative Language." *Critical Terms for Literary Study.* Frank Lentricchia and Thomas McLaughlin, eds. Chicago: University of Chicago Press, 1990. 80–90.

Mebus, Sylvia. "Zur Entwicklung der Lehrerausbildung in der SBZ/DDR 1945 bis 1959 an Beispiel Dresdens: Pädagogik Zwischen Selbst- und Fremdbestimmung." Habilitation, Greifswald, Germany: University of Greifswald, 1997.

Mehringer, Hartmut, ed. *Von der SBZ zur DDR: Studien zum Herrschaftssystem in der Sowjetischen Besatzungszone und in der Deutschen Demokratischen Republik.* Munich: Oldenbourg, 1995.

Meinecke, Friedrich. *Weltbürgertum und Nationalstaat.* 1907. Ed. Hans Herzfeld. Munich: Oldenbourg, 1962.

Meltzer, Françoise. "Unconscious." *Critical Terms for Literary Study.* Frank Lentricchia and Thomas McLaughlin, eds. Chicago: University of Chicago Press, 1990. 147–162.

Merkel, Ina. "Leitbilder und Lebensweisen von Frauen in der DDR." Hartmut Kaelble, Jürgen Kocka, and Hartmut Zwahr, eds. *Sozialgeschichte der DDR.* Stuttgart: Klett-Cotta, 1994, 361–365.

———. *Utopie und Bedürfnis. Die Geschichte der Konsumkultur in der DDR.* Cologne: Böhlau, 1999.

Meyers Enzyklopädisches Lexikon. Mannheim: Lexikonverlag, 1997.

Michaels, Ann. *Fugitive Pieces.* London: Bloomsbury, 1996.

Michaud, Eric. "Soldiers of an Idea: Young People Under the Third Reich." *A Story of Young People in the West: Stormy Evolution to Modern Times.* Levi Giovanni and Schmitt Jean-Claude, eds. Vol. 2. Cambridge, MA: Belknap Press of Harvard University Press, 1997. 257–280.

Middell, Matthias. "La Révolution Française enseignée dans les Écoles de L'Ex RDA et les Changements d'Hier à Aujourd'Hui." *Bilder einer Revolution: Die Französische Revolution in den Geschichtsschulbüchern der Welt.* Ed. Rainer Riemenschneider. Frankfurt AM: Diesterweg, 1994. 347–355.

Miller, Daniel, Peter Jackson, Nigel Thrift, Beverly Holbrook, and Michael Rowlands. *Shopping, Place and Identity.* London and New York: Routledge, 1998.

Mitchell, W. J. T. *Iconology: Image, Text, Ideology.* Chicago: University of Chicago Press, 1986.

———. "Representation." *Critical Terms for Literary Study.* Frank Lentricchia and Thomas McLaughlin, eds. Chicago: University of Chicago Press, 1990. 11–22.

Möbus, Gerhard. *Unterwerfung durch Erziehung: Zur politischen Pädagogik im Sowjetischen Besetzten Deutschland.* Mainz: V. Hase und Koehler, 1965.

Moeller, Robert G. *Protecting Motherhood. Women and the Family in the Politics of Postwar West Germany.* Berkeley: University of California Press, 1993.

Möhle, H. "German Democratic Republic." *Encyclopedia of Higher Education.* Burton R. Clark and Guy R. Neave, eds. Vol. 1. Oxford: Pergamon, 1992. 231–240.

Monmonier, Mark. *How to Lie with Maps.* 2nd ed. Chicago: University of Chicago Press, 1996.

Montessori, Maria. *Formazione dell'uomo.* Milan: Garzanti, 1949.

———. *La scoperta del bambino* (First published as *Il metodo della pedagogia scientifica,* 1909.) Milan: Garzanti, 1950.

Mort, Frank. *Cultures of Consumption: Masculinitės and Social Space in Late Twentieth-Century Britain.* London and New York: Routledge, 1996.

Moses, A. D. "The Forty-Fivers: A Generation Between Fascism and Democracy." *German Politics and Society* 17.1 (1999): 94–126.

Mosse, George. *The Nationalization of the Masses: Political Symbolism and Mass Movements in Germany from the Napoleonic Wars through the Third Reich.* New York: H. Fertig, 1975.

Motyl, Alexander J., ed. *Thinking Theoretically about Soviet Nationalities.* New York: Columbia University Press, 1992.

Müller, Werner. "Sozialistische Einheitspartei Deutschlands." *SBZ Handbuch: Staatliche Verwaltung, Parteien, Gesellschaftliche Organisationen und ihre Füehrungskräefte in der Sowjetischen Besatzungszone Deutschlands 1945–1949.* Martin Broszat and Hermann Weber, eds. Munich: Oldenbourg, 1993. 481–514.

Naimark, Norman. "Der Nationalismus und die osteuropäische Revolution 1944–1947." *Transit: Europäische Revue* (Vienna) 15 (1998): 40–59.

Naimark, Norman M. *The Russians in Germany: A History of the Soviet Zone of Occupation, 1945–1949.* Cambridge, MA: Belknap Press of Harvard University Press, 1995.

Neisser, Ulrich. "Self-Narratives: True and False." *The Remembering Self: Construction and Accuracy in the Self-Narrative.* Ulrich Neisser and Robyn Fivush, eds. Cambridge: Cambridge University Press, 1994. 1–18.

Neisser, Ulrich, and Robin Fivush, eds. *The Remembering Self: Construction and Accuracy in the Self-Narrative.* Cambridge: Cambridge University Press, 1994.

Niekus-Moore, Cornelia. *The Maiden's Mirror: Reading Material for German Girls in the Sixteenth and Seventeenth Centuries.* Wiesbaden: Otto Harrassowitz, 1987.

Nietzsche, Friedrich. "On the Uses and Disadvantages of History for Life." *Untimely Meditations.* Cambridge: Cambridge University Press, 1997. 57–124.

Noll, Dieter. *Aufbau: Berlin 1945–58. Bibliographie einer Zeitschrift.* Berlin [East]: Aufbau, 1978.

Nora, Pierre, ed. *Les Lieux de Mémoire.* 7 vols. Paris: Gallimard, 1992.

Norden, Albert. *Die Nation und Wir.* Vol. 1. Berlin [East]: Dietz, 1964.

Norman, Donald A. *Learning and Memory.* San Francisco: W.H. Freeman and Co., 1982.

Nothnagle, Alan L. *Building the East German Myth: Historical Mythology and Youth Propaganda in the German Democratic Republic, 1945–1989.* Ann Arbor: University of Michigan Press, 2002.

Novoa, Antonio. "The Restructuring of the European Educational Space: Changing Relationships among States, Citizens, and Educational Communities." *Educational Knowledge: Changing Relationships between the State, Civil Society, and the Educational Community.* Ed. Thomas Popkewitz. Albany: State University of New York Press, 2000. 31–57.

Oelkers, Jürgen, ed. *Zeitschrift für Pädagogik.* Vol. 28 (Beiheft). Weinheim and Basel: Beltz, 1992.

Offer, Daniel. "The Altering of Reported Experiences." *The Journal of the American Academy of Child and Adolescent Psychiatry* 39.6 (2000): 735–742.

Organisation for Economic Co-Operation and Development. *Education and the Economy in a Changing Society.* Paris: OECD, 1989.

Ortony, Andrew, ed. *Metaphor and Thought.* 2nd ed. Cambridge: Cambridge University Press, 1993.

pädagogik [sic]. Berlin [East]: 1946 ff.

Parker, Andrew, Mary Russo, Doris Summer, and Patricia Yaeger. "Introduction." *Nationalisms and Sexualities.* Andrew Parker et al., eds. New York: Routledge, 1992. 1–18.

Parsons, Talcott. *Structure and Personality.* London: Free Press of Glencoe, 1964.

Passerini, Luisa. "Youth as a Metaphor for Social Change: Fascist Italy and America in the 1950s." *A Story of Young People in the West: Stormy Evolution to Modern Times.* Levi Giovanni and Jean-Claude Caron, eds. Cambridge, MA: Belknap Press of Harvard University Press, 1997.

Paul, Jürgen. "Reconstruction of the City Centre of Dresden: Planning and Building during the 1950s." *Rebuilding Euorpe's Bombed Cities.* Ed. Jeffry M. Diefendorf. London: Macmillan, 1990. 170–189.

Penny III, H. Glenn. "The *Museum für Deutsche Geschichte* and German National Identity." *Central European History* 28.3 (1995): 343–372.

Pike, David. *The Politics of Culture in Soviet-Occupied Germany.* Stanford: Stanford University Press, 1992.

Pilarczyk, Ulrike. "Veränderung des schulischen Raum-, Zeit- und Rollengefüges im Prozeß der Politisierung der DDR-Schule." *Kindheit, Jugend und Bildungsarbeit im Wandel: Ergebnisse der Transformationsforschung.* Ed. Heinz-Elmar Tenorth. Vol. 37th Supplement. Weinheim and Basel: Beltz, 1997. 115–143.

Pingel, Falk. "National Socialism and the Holocaust in West German School Books." *Internationale Schulbuchforschung* 22 (2000): 11–29.

Pope, Kenneth S. "Memory, Abuse and Science: Questioning Claims about the False Memory Syndrome Epidemic." *American Psychologist* 51.9 (1996): 957–974.

———. "Science as Careful Questioning: Are Claims of a False Memory Syndrome Epidemic Based on Empirical Evidence?" *American Psychologist* 52.9 (1996): 997–1006.

Popkewitz, Thomas, ed. *Educational Knowledge: Changing Relationships between the State, Civil Society, and the Educational Community.* Albany: State University Press of New York, 2000.

Popkewitz, Thomas S. "Globalization/Regionalization, Knowledge, and the Eduational Practices. Some Notes on Comparatives Strategies for Educational Research." *Educational Knowledge: Changing Relationships between the State, Civil Society, and the Educational Community.* Ed. Thomas S. Popkewitz. Albany: State University Press of New York, 2000. 3–27.

Puaca, Brian. "Learning Democracy: Education Reform in Postwar West Germany, 1945–1965." Ph.D. diss. University of North Carolina—Chapel Hill, 2005.

———. " 'We learned what democracy really meant.' The Berlin Student Parliament and School Reform in the 1950s." *History of Education Quarterly* 45.4 (Winter 2005): 615–624.

———. "Missionaries of Goodwill: Deutsche Austauschlehrer und—schüler und die Lehren der amerikanischen Demokratie in den fünzigen Jahren." Arnd Bauerkämper, Konrad H. Jarausch, and Marcus M. Payk, eds. Göttingen: Vanderhoeck & Ruprecht, 2005. 305–331.

———. "The Pen is Mightier than the Sword? Student Newspapers and Democracy in Postwar West Germany." Unpublished ms. Christopher Newport University, 2005.

Quassowski, Jutta-B. Lange. *Neuordnung oder Restauration. Das Demokratiekonzept der amerikanischen Besatzungsmacht und die Politische Sozialisation der Westdeutschen: Wirtschaftsordnung—Schulstruktur—Politische Bildung.* Opladen: Leske and Budrich, 1979.

Rabinbach, Anson. "Introduction: Legacies of Antifascism." *New German Critique* 67 (1997): 3–17.

Radde, Gerd. *Fritz Karsen: Ein Berliner Schulreformer der Weimarer Zeit.* Berlin: Colloquium, 1973.

Rasche, Klaus. "1945–1990: Architettura e Urbanistica nella Germania Orientale. L'esempio di Dresda." *Storia Urbana* 73 (1995): 131–152.

Raschke, Eva-Christine. "Schulbauten 1928 bis 1988." *Köln—Seine Bauten 1928–1988.* Ed. Heribert Hall. Cologne: J.P. Bachem, 1990.

Reese, William J. *Power and the Promise of School Reform: Grassroots Movements during the Progressive Era.* Boston: Routledge and Paul Kegan, 1986.

Reimann, Brigitte. *"Aber wir schaffen es, verlaß Dich drauf!" Briefe an eine Freundin im Westen.* 2nd ed. Berlin: Aufbau Taschenbuch Verlag, 1999.

———. *Alles schmeckt nach Abschied. Tagebücher 1964–1970.* 3rd ed. Berlin: Aufbau, 1998.

———. *Franziska Linkerhand.* 1974. Unedited version. Berlin: Aufbau, 1998.

———. *Ich bedaure nichts. Tagebücher 1955–1963.* 5th ed. Berlin: Aufbau, 1997.

Renza, Louis. "Influence." *Critical Terms for Literary Study.* Frank Lentricchia and Thomas McLaughlin, eds. Chicago: University of Chicago Press, 1990. 186–202.

Reuter, Lutz R. "Administrative Grundlagen und Rahmenbedingungen." *Handbuch der deutschen Bildungsgeschichte.* Vol. 6, *1945 bis zur Gegenwart.* Bk. 2, *Die Deutsche Demokratische Republik und neue Bundesländer.* Munich: Beck, 1998.

———. "Rechtliche Grundlagen und Rahmenbedingungen." *Handbuch der deutschen Bildungsgeschichte.* Christoph Führ and Carl-Ludwig Furck, eds. Vol. 6, *1945 bis zur Gegenwart.* Bk. 2, *Die Deutsche Demokratische Republik und neue Bundesländer.* Munich: Beck, 1998. 26–36.

Richter, J. G. "Von der darstellung der Rede durch die Schrift." *Berlinische Monatschrift* 1 (1792): 203–315.

Riemenschneider, Rainer, ed. *Bilder einer Revolution: Die Französische Revolution in den Geschichtsschulbüchern der Welt.* Frankfurt AM: Dieterweg, 1994.

Riepl-Schmidt, Maja. "Die 'Wertkonservativen' Erziehungskonzepte zur Vorbereitung einer 'Mit Vernunft getragenen Nation' der Therese Huber (1764–1829)." *Frauen und Nation.* Ed. Frauen und Geschichte Baden-Württemberg. Tübingen: Silberburg, 1996. 90–103.

Rodden, John. *Repainting the Little Red Schoolhouse: A History of Eastern German Education, 1945–1995.* New York: Oxford University Press, 1999.

Röhrs, Hermann. *Die Reformpädagogik: Ursprung und Verlauf unter internationalem Aspekt.* 5th rev. ed. Weinheim: Deutscher Studien Verlag, 1998.

Roseman, Mark. *Generations in Conflict: Youth Revolt and Generation Formation in Germany 1770–1968.* Cambridge: Cambridge University Press, 1995.

———. "Introduction: Generation Conflict and German History." Ed. Mark Roseman. Cambridge: Cambridge University Press, 1995. 1–46.

Rosenholtz, Susan J. *Teacher's Workplace: The Social Orgnization of Schools.* White Plains, NY: Longman, 1989.

Rosenzweig, Beate. *Erziehung zur Demokratie? Amerikanische Besatzungs- und Schulreformpolitik in Deutschland und Japan.* Stuttgart: Franz Steiner, 1998.

Ross, Corey. *The East German Dictatorship: Problems and Perspectives in the Interpretation of the GDR.* London: Arnold, 2002.

Rousseau, Jean-Jacques. *Emile, or on Education.* 1762. New York: Basic Books, 1979.

Rueck, Peter. "Die Sprache der Schrift. Zur Geschichte des Frakturverbots." *Homo Scribens. Perspektiven der Schriftlichkeitsforschung.* Jürgen Baurmann, Hartmut Guenther, and Ulrich Knoop, eds. Tübingen: Niemeyer, 1993. 231–272.

Rühle, Jürgen. *Deutschland Archiv* 12 (1980).

Ruigui, Santoni. *Storia Della Scuola e Storia d'Italia dall'Unita' Ad Oggi:* Bari: De Dorato, 1982.

Rust, Val D. *Education in East and West Germany: A Bibliography.* New York: Garland, 1984.

Rust, Val D., and Diane Rust. *The Unification of German Education.* New York: Garland, 1995.

Sachse, Petra. *Tanz Den Lenin/Right on Red.* Film documentary. Munich: Discovery Channel, 1999.

Sächsische Volkszeitung (Saxony): 1946.

Sandford, Gregory W. *From Hitler to Ulbricht: The Communist Reconstruction of East Germany, 1945–46.* Princeton: Princeton University Press, 1983.

Sandford, John, ed. *Encyclopedia of Contemporary German Culture.* London: Routledge, 1999.

Schacter, Daniel L. *Searching for Memory: The Brain, the Mind and the Past.* New York: Basic Books, 1996.

Schaffer, Gordon. *Russian Zone.* London: Co-Operative Press, 1947.

Schätzke, Andreas. "Rückkehr aus dem Exil. Zur Remigration bildender Künstler in der SBZ/DDR." *Kunstdokumentation SBZ/DDR 1945–1990. Aufsätze—Berichte—Materialien.* Günter Feist, Eckhart Gillen, and Beatrice Vierneisel, eds. Cologne: DuMont, 1996. 96–109.

Schleunes, Karl. *Schooling and Society: The Politics of Education in Prussia and Bavaria 1750–1900.* Oxford: Berg, 1989.

Schmidt-Thomsen, Jörn-Peter, Helga Schmidt-Thomsen, and Manfred Scholz. *Schulen.* Vol. 5. Bk. c. *Berlin und seine Bauten.* Berlin: Ernst & Sohn, 1991.

Schmidt, Walter. "The Nation in German History." *The National Question in Europe in Historical Context.* Mikulás Teich and Roy Porter, eds. Cambridge: Cambridge University Press, 1993. 148–180.

Scholz, Manfred. "Schulen Nach 1945." *Berlin und seine Bauten.* Jörn-Peter, Schmidt-Thomsen and Helga Schmidt-Thomsen, Vol. 5. Bk. c, Schulen. Berlin: Ernst & Sohn, 1991. 197–326.

Schön, Donald A. "Generative Metaphor and Social Policy." *Metaphor and Thought.* Ed. Andrew Ortony. 2nd ed. Cambridge: Cambridge University Press, 1993. 137–163.

Schöttler, Peter. "Mentalitäten, Ideologien, Diskurse: Zur Sozialgeschichtlichen Thematisierung der 'Dritten Ebene.'" *Alltagsgeschichte: Zur Rekonstruktion historischer Erfahrungen und Lebensweisen.* Ed. Alf Lüdtke. Frankfurt AM: Campus, 1989. 85–136.

Schulabteilung der Deutschen Verwaltung für Volksbildung in der Sowjetischen Besatzungszone, ed. *Die Deutsche Demokratische Schule im Aufbau.* Leipzig: Volk und Wissen, 1949.

Schulz, Rudi. "Bibliographie der Arbeiten von Robert Alt." *Jahrbuch für Erziehungs- und Schulgeschichte.* Vol. 5/6. Berlin [East]: Akademie, 1966. 427–432.

Schulze, Edeltraud. *DDR-Jugend: Ein statistisches Handbuch.* Berlin: Akademie, 1995.

Scott, Tom. *Regional Identity and Economic Change: The Upper Rhine, 1450–1600.* New York: Oxford University Press, 1997.

Shandley, Robert R. *Rubble Films: German Cinema in the Shadow of the Third Reich.* Philadelphia: Temple University Press, 2001.

Siegel, Mona L. "Lasting Lessons: War, Peace, and Patriotism in French Primary Schools, 1914–1939." Ph.D. diss. University of Wisconsin—Madison, 1996.

Siegfried Dubel. *Die Situation der Jugend im kommunistischen Herrschaftssystem der Sowjetischen Besatzungszone Deutschlands.* Bonn: Bundesministerium für gesamtdeutsche Fragen, 1957.

Smith, Anthony D. *The Ethnic Origins of Nations.* Oxford: Basil Blackwell, 1986.

———. *National Identity.* Reno: University of Nevada Press, 1991.

———. *Nationalism in the Twentieth Century.* New York: New York University Press, 1979.

Smith, Bonnie. *Ladies of the Leisure Class: The Bourgeoises of Northern France in the 19th Century.* Princeton: Princeton University Press, 1981.

Snyder, Benson R. *The Hidden Curriculum.* New York: Knopf, 1971.

Sontheimer, Kurt, and Wilhelm Bleek. *Die DDR. Politik, Gesellschaft, Wirtschaft.* 4th ed. Hamburg: Hoffmann und Campe, 1975.

Staatliche Zentralverwaltung für Statistik. *Statisches Jahrbuch der Deutschen Demokratischen Republik.* Berlin [East]: VEB Deutscher Zentralverlag, 1956.

Stachura, Peter D. *Nazi Youth in the Weimar Republic.* Santa Barbara, CA: Clio, 1975.

Stahl, Walter, ed. *Education for Democracy in West Germany: Achievements, Shortcomings, Prospects.* New York: Praeger, 1961.

Staritz, Dietrich. *Die Gründung der DDR: Von der Sowjetischen Besatzungsherrschaft zum sozialistischen Staat.* 1984. 3rd ed. Munich: Deutscher Taschenbuch Verlag, 1995.

Staritz, Dietrich, and Siegfried Suckut. "Einleitung." *SBZ Handbuch: Staatliche Verwaltung, Parteien, gesellschaftliche Organisationen und ihre Führungskräfte in der Sowjetischen Besatzungszone Deutschlands 1945–1949.* Martin Broszat and Hermann Weber, eds. Munich: Oldenbourg, 1993. 435–439.

———. *Geschichte der DDR: 1949–1985.* Frankfurt AM: Suhrkamp, 1985.

————. "National-Demokratische Partei Deutschlands (NDPD)." *SBZ Handbuch: Staatliche Verwaltung, Parteien, gesellschaftliche Organisationen und ihre Führungskräfte in der Sowjetischen Besatzungszone Deutschlands 1945–1949.* Martin Broszat and Hermann Weber, eds. Munich: Oldenbourg, 1993. 574–594.

Stegmann, Natali. " 'Je mehr Bildung, desto polnischer.' Die Nationalisierung polnischer Frauen in der Provinz Posen (1870–1914)." Ed. Frauen und Geschichte Baden-Württemberg. *Frauen und Nation.* Tübingen: Silberburg, 1996. 165–177.

Steiner, Rudolf. *Die Erziehung des Kindes vom Geschichtspunkte der Geisteswissenschaften.* Berlin: Besant-Zweig of the Theosophical Society, 1907.

Steininger, Rolf. *Deutsche Geschichte seit 1945.* Vol. 1, *1945–1947.* Frankfurt AM: Fischer, 1996.

————. *Deutsche Geschichte seit 1945.* Vol. 2: 1948–1955. Frankfurt AM: Fischer, 1996.

Sternberg, Robert J., Roger Trangeau, and Georgia Nigro. "Metaphor, Introduction, and Social Policy: The Conveyence of Macroscopic and Microscopic Views," in Metaphor and Thought, 2nd ed., ed. Andrew Orthony. Cambridge: Cambridge University Press, 1993. 277–303.

Stenhell, Zeev, Mario Sznajder, and Asheri Maia, eds. *The Birth of Fascist Ideology: From Cultural Rebellion to Political Revolution.* Princeton: Princeton University Press, 1994.

Storbeck, Dietrich. *Soziale Strukturen in Mitteldeutschland: Eine sozialstatische Bevölkerungsanalyse im gesamtdeutschen Vergleich.* Berlin[West]: Duncker and Humblot, 1964.

Sunker, Heinz, and Hans-Uwe Otto. *Political Formation and Social Education in German National Socialism.* 1991. London: Falmer, 1996.

Teich, Mikulás, and Roy Porter, eds. *Nationalism and Nation-State in Germany.* Cambridge: Cambridge University Press, 1993.

Tenorth, Heinz-Elmar, ed. *Kindheit, Jugend und Bildungsarbeit im Wandel: Ergebnisse der Transformationsforschung.* Vol. 37. Beiheft. *Zeitschrift für Pädagogik.* Weinheim and Basel: Beltz, 1997.

————. "Politisierung des Schulalltags im historischen Vergleich—Grenzen von Indoktrination." *Erinnerung für die Zukunft II: Das DDR-Bildungssystem als Geschichte und Gegenwart.* Ed. Petra Gruner. Ludwigsfeld-Struveshof: Pädagogisches Landesinstitut Brandenburg, 1997. 37–48.

Tenorth, Heinz-Elmar, Sonja Kudella, and Andreas Paetz. *Politisierung im Schulalltag der DDR: Durchsetzung und Scheitern einer Erziehungsambition.* Weinheim: Deutscher Studien Verlag, 1996.

Tent, James. *Mission on the Rhine: "Reeducation" and Denazification in American-Occupied Germany.* Chicago: University of Chicago Press, 1982.

Thurnwald, Hilde. *Gegenwartsprobleme Berliner Familien: Eine Soziologische Untersuchung an 498 Familien.* Berlin [West]: Weidmannsche Verlagsbuchhandlung, 1948.

Timm, Albrecht. *Das Fach Geschichte in Forschung und Lehre in der Sowjetischen Besatzungszone Deutschlands seit 1945.* Bonn: Bundesministerium für gesamtdeutsche Fragen, 1965.

Trommler, Frank, and Joseph McVeigh, eds. *The Relationship in the Twentieth Century.* Vol. 2. *America and the Germans. An Assessment of a Three-Hundred Year History.* Philadelphia: University of Pennsylvania Press, 1985.

Tyack, David, and Elisabeth Hansot. *Learning Together: A History of Coeducation in American Schools.* New Haven: Yale University Press, 1990.

Uhlig, Christa. "Zur Erarbeitung der bildungspolitischen Programmatik für Nachkriegsdeutschland in der UdSSR: Konzepte und Personen." *Kindheit, Jugend und Bildungsarbeit im Wandel: Ergebnisse der Transformationsforschung.* Ed. Heinz-Elmar Tenorth. Vol. 37th Supplement. Weinheim/Basel: Beltz, 1997. 411–432.

Uhlig, Gottfried. *Der Beginn der Antifaschistisch-Demokratischen Schulreform, 1945–1946. Monumenta Paedagogica* II. Berlin [East]: Akademie, 1965.

———, ed. *Dokumente zur Geschichte des Schulwesens in der Deutschen Demokratischen Republik.* Vol. 1, *1945–1955. Monumenta Paedagogica.* Berlin [East]: Volk und Wissen, 1970.

Ulrich, Bernd. *Die Augenzeugen: Deutsche Feldpostbriefe in Kriegs- und Nachkriegszeit 1914–1933.* Essen: Klartext, 1997.

Vavrus, Frances. "Governmentality in an Era of 'Empowerment': The Case of Tanzania." *Educational Knowledge: Changing Relationships between the State, Civil Society, and the Educational Community.* Ed. Thomas S. Popkewitz. Albany: State University of New York Press, 2000. 221–242.

Vivian, Malcolm, John Seaborne, and R. Lowe, eds. *The English School: Its Architecture and Organization.* Vol. 2. London: Routledge and Kegan, 1977.

Volk und Wissen, ed. *Goethe: Eine Auswahl für die Schule.* Berlin [East]/Leipzig: Volk und Wissen, 1949.

von Beyme, Klaus. "Reconstruction in the German Democratic Republic." *Rebuilding Euorpe's Bombed Cities.* Ed. Jeffry M. Diefendorf. London: Macmillan, 1990. 190–208.

von Ditfurth, Christian. *Ostalgie, Oder, Linke Alternative: Meine Reise durch die PDS.* Cologne: Kiepenheuer & Witsch, 1998.

von Goethe, Johann, and Friedrich Schiller. *Xenien* Nr. 95–96, 1796.

von Hähling, Heinrich, and Weihbischof von Paderborn. *Die Koedukation oder, Die gemeinschaftliche Erziehung der Knaben und Mädchen, Besonders in der Volksschule. Die richtigen Grundsätze in einer brennenden Frage.* 2nd ed. Paderborn: Bonifacius-Druckerei, 1924.

Warnecke, Heinz. "Wilhelm Heise: 1897 Bis 1949—Der neue Lehrer soll ein politischer Mensch sein." *Wegbereiter der neuen Schule.* Gerd Hohendorf, Helmut König, and Eberhard Meumann, eds. Berlin [East]: Volk und Wissen, 1989. 130–137.

Waterkamp, Dietrich. *Einheitsprinzip Im Bildungswesen Der DDR.* Cologne: Böhlau, 1985.

Weber, Hermann. *Die DDR: 1945–1990.* 2nd ed. Munich: Oldenbourg, 1993.

———. "Freie Deutsche Jugend (FDJ)." *SBZ Handbuch: Staatliche Verwaltung, Parteien, gesellschaftliche Organisationen und ihre Führungskräfte in der*

Sowjetischen Besatzungszone Deutschlands 1945–1949. Martin Broszat and Hermann Weber, eds. Munich: Oldenbourg, 1993. 665–690.

Wegner, Gregory P. "Germany's Past Contested: The Soviet-American Conflict in Berlin over History Curriculum Reform, 1945–48." *History of Education Quarterly* 30.1 (1990): 1–16.

———. "The Power of Tradition in Education: The Formation of the History Curriculum in the Gymnasium of the American Sector in Berlin, 1945–1955." Ph.D. diss. University of Wisconsin—Madison, 1988.

Weidenfeld, Werner. "Was ist nationale Identität?" *Die Intellektuellen und die nationale Frage.* Ed. Gerd Langguth. Frankfurt AM: Campus, 1997. 45–62.

Welch, Steven R. *Subjects or Citizens? Elementary School Policy and Practice in Bavaria 1800–1918.* Melbourne: Universty of Melbourne, 1998.

Wells, C. J. *German: A Linguistic History to 1945.* New York: Oxford University Press, 1985.

Welsh, Helga A., Andreas Pickel, and Dorothy Rosenberg. "East and West German Identities: United and Divided?" *After Unity: Reconfiguring German Identities.* Ed. Konrad H. Jarausch. Providence: Berghahn, 1997. 103–136.

———. "Deutsche Zentralverwaltung für Volksbildung." *SBZ Handbuch: Staatliche Verwaltung, Parteien, gesellschaftliche Organisationen und ihre Führungskräfte in der Sowjetischen Besatzungszone Deutschlands 1945–1949.* Martin Broszat and Hermann Weber, eds. Munich: Oldenbourg, 1993. 229–238.

Wendel, Günter, ed. *Wissenschaft im Kapitalistischen Europa: 1871–1917.* Beiträge zur Wissenschaftsgeschichte. Berlin [East]: Deutscher Verlag der Wissenschaften, 1983.

Wernet-Tietz, Bernhard. "Demokratische Bauernpartei Deutschlands." Eds. Martin Broszat and Hermann Weber. *SBZ Handbuch: Staatliche Verwaltung, Parteien, gesellschaftliche Organisationen und ihre Führungskräfte in der Sowjetischen Besatzungszone Deutschlands 1945–1949.* Munich: Oldenbourg, 1993, 584–594.

Westphal, Siegrid. "Reformatorische Bildungskonzepte für Mädchen und Frauen—Theorie und Praxis." *Geschichte der Mädchen- und Frauenbildung.* Vol. 1. Elke Kleinau and Claudia Opitz, eds. Frankfurt AM: Campus, 1996. 135–151.

White, Steven F. "Carleton Washburne: L'influenza Deweyana nella scuola italiana." *Scuola e Città* 40.2 (1989): 49–57.

Wierling, Dorothee. *Geboren im Jahr Eins. Der Jahrgang 1949 in der DDR: Versuch einer Kollektivbiographie.* Berlin: Christoph Links, 2002.

Wilk, Liselotte, and Johann Bacher, eds. *Kindliche Lebenswelten: Eine sozialwissenschaftliche Annäherung.* Opladen: Leske and Budrich, 1994.

Willenbacher, Barbara, "Zerrüttung und Bewährung der Nachkriegs-Familie." Martin Broszat, Klaus-Dietmar Henke, and Hans Woller, eds. *Von Stalingrad zur Währungsreform: Zur Sozialgeschichte des Umbruchs in Deutschland.* Munich: Oldenburg, 1988, 595–618.

Winkeler, Rolf. "Das Scheitern einer Schulreform in der Besatzungszeit: Analyse der Ursachen am Beispiel der Französisch Besetzten Zone Württembergs und Hohenzollerns von 1945 Bis 1949." *Umerziehung und Wiederaufbau: Die Bildungspolitik der Besatzungsmächte in Deutschland und Österreich.* Ed. Manfred Heinemann. Stuttgart: Klett-Cotta: 1981. 211–228.

Winkler, Heinrich August. "Nationalism and Nation-State in Germany." *The National Question in Europe in Historical Context.* Mikulás Teich and Roy Porter, eds. Cambridge: Cambridge University Press, 1993. 181–195.

Wolff, Sylvia. "Der Umgang mit der Reformpädagogik in der SBZ/DDR 1945–1949." Diplom-Arbeit. Berlin: Free University, 1996.

Wolfrum, Edgar. *Geschichtspolitik in der Bundesrepublik Deutschland. Der Weg zur bundesrepublikanischen Erinnerung 1948–1990.* Darmstadt: Wissenschaftliche Buchgesellschaft, 1990.

Wyneken, Gustav. *Schule und Jugendkultur.* 2nd ed. Jena: Eugen Diederichs, 1914.

Young, Alan. *The Harmony of Illusions: Inventing Post-Traumatic Stress Disorder.* Princeton: Princeton University Press, 1995.

Zarusky, Jürgen. *Die deutschen Sozialdemokraten und das sowjetsiche Modell. Ideologische Auseinandersetzung und außenpolitische Konzeptionen 1917–1933.* Munich: Oldenbourg, 1992.

zu Castell, Adelheid. "Die demographischen Konsequenzen des Ersten und Zweiten Weltkrieges für das Deutsche Reich, die Deutsche Demokratische Republik und die Bundesrepublik Deutschland." *Zweiter Weltkrieg und Sozialer Wandel.* Ed. Waclaw Dlugoborski. Göttingen: Vandenhoeck and Ruprecht, 1981. 117–137.

Zymek, Bernd. "Schulen." *Handbuch der deutschen Bildungsgeschichte.* Dieter Langewiesche and Heinz-Elmar Tenorth, eds. Vol. 5, *1918–1945: Die Weimarer Republik und die nationsozialistische Diktatur.* Munich: Beck, 1989. 155–208.

Index

Note: Page numbers in **bold** refer to illustrations

1848 (Revolution, etc.), 11, 28, 42, 58,
    78, 128, 159, 169, 171, 172, 173,
    174, 186, 190
abortions, 101
Ackermann, Anton (co-founder of
    Kulturbund), 22, 27
adults, 6, 38, 70, 80, 87, 91, 95, 96, 99,
    105, 106, 111, 112, 117, 130,
    136, 142, 147, 150, 153, 179,
    183, 188, 191, 195
    see also family, fathers, mothers
age categories/differences, *see* adults,
    children, generation
Agerth, Käte, 33, 78, 109
Allied Command, 9, 24, 39, 56, 75,
    77, 82, 104, 124, 126, 156, 167,
    171, 196
    see also Western allies
Alt, Robert, 34, 35
Altlehrer (teachers employed before
    1933 or 1945), 48, 94, 189, 192
American zone, 40, 54, 55, 100
antifascism, 4, 8, 11, 16, 20–8, 31, 33,
    35, 109, 122, 126, 128, 137, 142,
    143, 144, 153, 154, 158, 159,
    160, 161, 162, 163, 169, 173,
    179, 181, 185, 186, 187, 189,
    190, 192, 196, 197, 198, 199
    see also antifascism democratization
antifascist democratization, 1, 2–5, 9,
    13, 23, 25, 27, 31, 33, 37, 39, 42,
    50, 51, 52, 54, 55, 60, 61, 62, 70,

72, 76, 83, 84, 88, 95, 98, 101,
    102, 104, 105, 106, 109, 117,
    120, 121, 123, 124, 140, 141,
    144, 145, 159, 161, 162, 163,
    164, 166, 167, 169, 170, 175,
    177, 185, 188, 189, 190, 191,
    195, 196, 197, 198, 202
    see also antifascism,
antifascist education, *see* antifascist
    democratization
anti-Semitism, 167
autograph books, girls', 117

Bach, Johann Sebastian (1685–1750),
    17, 175, 183
Basic Law (constitution of Federal
    Republic), 193
Basic Treaty, 193
Bebel, August, (1840–1913), 108
Beethoven, Ludwig van (1770–1827),
    17, 74, 169, 175, 183
Belgium, 25, 29
Berlin, 16, 19, 25, 30, 34, 44, 56, 59,
    68, 69, 77, 81, 92, 113, 114, 142,
    144, 150, 171
    East, 1, 16, 25, 30, 31, 72, 92, 117,
        174, 199
    West, 16, 25, 30, 72, 92, 174, 199
Berlin Blockade, 69
"Berlin Film" genre, 144
black market, 87, 105, 144, 149, 152
Bonn, 25, 29

borders, 14–16, 25, 30, 35, 53, 72, 73, 74, 188, 190, 194
boys, 10, 65, 68, 82, 87, 93, 96, 102, 103, 106, 107, 108, 109, 110, 111, 112, 113, 114, 115, 117, 120, 123, 126–7, 129, 133, 134, 144, 145, **146**, 147, 148, 149, 150, 152, 153, 155, 156, 158, 181, **182**, 187, 191, 195, 200
see also gender, girls
Brandenburg, 51, 59, 68, 70, 76, 190
Brecht, Bertolt, 160
British zone, 4, 54, 55, 100
see also American zone, French zone, Great Britain, Western zones
de Bruyn, Günter, 76, 190
buildings, 65, 67, 69, **71**, 74, 77, 125–6, 144, 145, 150, 153, 161, 183
church, 74, 80, 168,2 183, 190
school, 9–10, 43, 47, 51, 61, 62, 63, 64, 65, 67, 68, 69, 72, 73, 74, 76, 78, 80, 83, 84, 86, 87, 88, 110, 112, 113, 135, 137, 169, 190, 191

Campe, Joachim Heinrich (1746–1818), 106, 109
Cap Arcona, 34
see Alt, Robert
CDU (Christian Democratic Union), 20, 21, 22, 41, 42, 45, 47, 115
celebrations, 47
children, 10–11, 37, 38, 39, 53, 54, 61, 65, 70, 77, 80, 82, 83, 84, 87, 88, 91, 92, 93, 95, 96, 98, 103, 104, 109, 110, 112, 116, 125, 126, 128, 129, 133, 141, 142, 144, **146**, 148, 149, 150, 151, 152, 153, 154, 155, 156, 157, 158, 179, 180, 187, 188, 191, 195, 197, 199, 200
see also families, pupils, students, young people, youth
Christianity, *see* religion
Christmas, 11, 85, 96, **97**, 178–84, **182**, **184**, 186, 200

churches, 88, 106, 108, 114, 141, 168
see also buildings, religion
Catholic, 4, 22, 168, 179
Protestant, 168
class issues and differences (social), 3, 5, 16, 23, 26, 32, 33, 35, 46, 58, 74, 128, 132–3, 171, 188, 189, 196, 200
classicism, 160, 161, 178, 192
coeducation, 10, 33, 42, 92, 103, 106, 107, 108, 109, 110, 111, 112, 113, 114, 115, 133, 191, 195, 196, 198, 200
cold war, 111, 194, 197
collective consciousness, 122, 123, 130, 137, 190, 197
see also forgetting, memory, remembering
Comintern (Communist International), 21
commemoration, 11, 58, 78, 159, 164, 165, 167, 168, 169, 173, 175, 178, **184**, 186, 200
communism, 1, 13, 18, 20, 21, 22, 24, 26, 27, 28, 31, 33, 34, 35, 39, 43, 44, 45, 46, 47, 78, 97, 98, 102, 105, 109, 112, 160, 163, 165, 171, 174, 178, 180, 192, 194, 197
see also KPD
comprehensive schools, *see Einheitsschule*
confession, *see* religion
constitution
of Federal Republic of Germany, *see* Basic Law
construction, 63, 64, 67, 69, 70, 124, 125–6, 143, 188, 190, 191, 196
see also rebuilding, reconstruction
consumerism, 96, **97**, 179, 180, 181, **182**, 183, **184**
corporal punishment, 102, 103, 106, 177
culture, 6, 10, 11, 15, 17, 23, 24, 28, 29, 31, 35, 38, 41, 59, 69, 74–5, 76, 78, 123, 128, 143, 156, 158, 159, 160, 161, 162, 165, 168, 174, 175, 176, 178, 185, 186, 188, 189, 193, 195, 200

currency reform, 30, 69, 74, 174, 192, 197
curriculum, 6, 22, 26, 31, 37, 42, 44, 46, 51, 55, 62, 74, 109, 121, 123, 124, 126, 130, 141, 162, 163, 164, 165, 168, 170, 171, 185, 189, 191

Day of German Literature, 164, 165, 166, 167, 186
DEFA (*Deutsche Film Aktiengesellschaft*), 143, 144, 146, 149, 150, 151, 157
see also film
Deiters, Heinrich, 114, 192, 193
democracy, 3, 4, 13, 21, 26, 27, 33, 60, 62, 105, 116, 121, 123, 127, 128, 130, 163, 164, 167, 170, 173, 189, 195
see also antifascist democratization
denazification, 1, 2, 4, 38, 49, 51, 54, 75, 85, 104–5, 109, 121, 130, 195
see also antifascist democratization, democracy, reeducation
Dewey, John, 34, 46
dictatorship, 1, 13, 31, 160, 166, 193
division, 2, 8–9, 11, 14, 19, 25, 26, 28–31, 37, 38, 45, 60, 69, 72, 73, 75, 106, 114, 120, 168, 169, 173, 174, 176, 185, 186, 188, 189, 190, 191, 198
DVV (*Deutsche Verwaltung für Volksbildung*), 9, 24, 41, 43, 44, 77, 79, 101, 108, 109, 110, 143, 164, 176, 179

East Germany, *see* German Democratic Republic
education, 1, 5, 11, 31, 33, 34, 35, 37, 38, 43, 46, 61, 69, 74, 83, 106, 107, 110, 113, 114, 121, 149, 163, 167, 185, 193, 195, 199
see also educators, *Einheitsschule*, reeducation, schools, "new school"
educational materials, 3, 16, 31, 51, 52, 54, 55, 57, 59, 60, 84, 126, 159, 189

educational journals, 6, 19, 58, 59, 173, 177
educators, 5, 6, 11, 13, 14, 16, 21, 22, 23, 25, 28, 29, 30, 31–5, 37, 38, 39, 41, 42, 44, 46, 47, 48, 49, 50, 51, 52, 53, 54, 56, 57, 62, 63, 67, 68, 69, 72, 73, 77, 79, 81, 83, 85, 88, 92, 93, 94, 95, 99, 101, 102, 103, 104, 105, 108, 109, 110, 111, 112, 117, 121, 122, 123, 124, 126, 127, 129, 133, 140, 142, 143, 144, 145, 148, 151, 152, 162, 163, 167, 168, 169, 171, 173, 174, 177, 183, 186, 188, 189, 190, 192, 193, 196, 197
see also education
leaving the profession in Soviet zone and GDR (*Lehrerfluktuation*), 49–50
Einheitsschule, 1, 8, 10, 26, 27, 31, 35, 37, 38, 41, 42, 45, 60, 92, 111, 114, 125, 168, 170, 173, 188, 189, 200
see also education, "new school," reeducation, schools
England, *see* Great Britain
Enlightenment, 4, 106, 160, 161, 192
essays
pupils', 3, 6, 7, 8, 9–10, 11, 51, 52, 53, 59, 80, 81, 82, 86, 87, 88, 104, 105, 113, 116, 117, **118–19**, 121, 124, 128, 129, 130, 132, 133, 134, 135, **136**, 137, **138–139**, 140, 148, 153, 154, 155, 195, 196, 198, 199, 200
university students', 156
teacher candidates', 6–8, 101, 177, 192
teachers', 84, 192, 196
experience, 3, 4, 5, 6, 8, 9, 10, 12, 15, 18, 28, 30, 32, 33, 34, 35, 41, 45, 49, 53, 59, 60, 63, 84, 86, 91, 93, 94, 99, 104, 111, 115, 116, 120, 121, 123, 128, 129, 130, 131, 132, 133, 135, 137, 140, 141, 155, 156, 157, 160, 189, 190, 195, 201, 202, 202
Europe, 142

families, 10–11, 39, 52, 53, 60, 61, 63,
    64, 76, 86, 88, 42, 96, 99, 117,
    122, 123, 126, 128, 130, 133, 134,
    137, 141, 142, 143, 144, 145, 148,
    150, 151, 152, 154, 155, 156,
    157, 158, 179, 180, 181, 183,
    192, 199, 200
  see also adults, children, fathers,
    mothers
fascism, 14, 15, 20, 24, 26, 27, 35, 46,
    110, 125, 128, 190, 199
  see also National Socialist Party
fathers, 10–11, 86, 98, 117, 120, 131,
    133, 141, 142, 143, 144, 145, 146,
    147, 149, 150, 152, 153, 154, 155,
    156, 157, 158, 180, 181, 182,
    199, 200
  see also adults, families, mothers
Federal Republic of Germany, 1, 25, 45,
    169, 187, 193, 194, 195, 196,
    199, 201
Fénelon, François (1651–1751), 106
females, see gender, girls, women
film, 6, 58, 141, 143, 144, 145, 148,
    149, 150, 152
  see also DEFA
First World War, see World War I
food, 53, 54, 60, 78, 79, 102, 120, 147,
    149, 150, 151, 152, 157, 180,
    181, 200
forgetting, 62, 83, 84
  see also memory, remembering
founding, 28, 29, 174, 177, 188, 197
France, 16, 17, 18, 19, 25, 29, 172, 194
  see also French zone, Western zones
fraternization (between soldiers and
    civilians), 100, 101, 103, 106, 191
Free German Trade Union, see Freier
    Deutscher Gewerkschaftsbund, 111
Free German Youth (Freie Deutsche
    Jugend, FDJ), 45, 115, 201
Freie Deutsche Jugend (FDJ), see Free
    German Youth
Freier Deutscher Gewerkschaftsbund
    (FDGB), 47, 169
French Revolution, 5, 18, 24, 159, 160,
    188, 190

French zone, 54, 55, 107
Freunde der neuen Schule, see Friends of
    the New School
Friends of the New School, 47, 70, 125

gender, 7,10, 11, 33, 68, 81, 82, 91,
    92, 93, 94, 95, 102, 103, 106, 107,
    108, 109, 110, 112, 113, 115, 116,
    120, 123, 133, 143, 147, 148,
    149, 150, 181, 182, 183, 188,
    191, 196, 199, 200, 201
  see also boys, girls, men, women
generations, 10, 91, 94, 95, 106, 109,
    110, 120, 126, 145, 149, 168,
    197, 200, 201, 202
German language
  school instruction, 56, 62, 121, 122,
    123, 124, 126, 128, 130,
    160, 164
German Education Administration, see
    DVV
German Democratic Republic (GDR),
    1, 11, 20, 45, 47, 49, 108, 143,
    145, 160, 162, 163, 169, 187, 191,
    193, 194, 196, 197, 198, 199, 200,
    201, 202
  founding, 1, 11, 14, 22, 28, 177
German Educational Administration,
    see DVV
Germania Anno Zero (Germany Year
    Zero), 150, 152
Germany, 160
  postwar years: 1, 11, 15, 39, 51, 106,
    110, 116, 123, 128, 140, 141,
    142, 143, 144, 145, 149,
    151, 156, 160, 162, 174,
    187, 196
Germany Year Zero, see Germania Anno
    Zero
ghosts, 11, 126, 168, 185
girls, 7, 10, 65, 68, 70, 82, 93, 94, 96,
    97, 98, 99, 103, 104, 106, 107,
    108, 110, 111, 113, 114, 115, 116,
    117, 120, 123, 130, 133, 134, 148,
    149, 150, 153, 155, 181, 182,
    183, 187, 191, 199, 200
  see also boys, gender

Goethe, Johann Wolfgang von
(1749–1832), 16, 24, 26, 160,
161, 165, 169, 175, 176, 177,
178, 185, 192
Goethe-Year, 11, 174–8, 186
Great Britain, 16, 19, 25, 29, 31, 39,
40, 58, 104, 163, 172, 194
see also British zone, Western zones
Grimm, Brothers, (Jakob and Wilhelm),
161
Grundgesetz, see Basic Law

Haydn, Franz Josef, (1732–1809), 175
health issues, 54, 79, 86, 91, 99, 101,
102, 112, 143, 144, 145, 147, 148,
149, 154, 156, 157, 158, 191
Herder, Gottfried von (1744–1803),192
Heine, Heinrich, 169, 192
heritage, 11, 24, 35, 46, 59, 62, 128,
161, 165, 169, 170, 178, 183,
188, 192
antifascist, 11, 17, 23, 24, 35, 128,
160, 161, 162, 165, 169, 178,
185, 192
heroes, 11, 65, 128, 141, 144, **146**, 147,
151, 157, 158, 159, 160, 165, 169,
174, 175, 177, 181, 186, 196
see also fathers
history, 159, 188, 199
instruction, 58, 121, 122, 123, 124,
126, 128, 159, 163, 164, 171,
173, 197
and nation-building, 162, 167, 168,
169, 185, 189
historiography, about Soviet zone/GDR
and antifascism, 3–4, 11, 12, 14,
20, 31–2, 43, 45, 55, 58, 59, 123,
142, 160, 163, 168, 189, 192,
193, 195, 200
Hitler, Adolf, 17, 18, 21, 27, 43, 84,
132, 145, 156, 175, 179
Hitler Youth, 102, 127, 129, 167
holidays, see commemorations,
Christmas
Holocaust, 14, 17, 18, 20, 34, 132,
164, 178, 179
see also World War II

homes, 7, 11, 52, 53, 64, 86, 108, 109,
111, 117, 126, 131, 135, 137, 142,
144, 145, 147, 148, 153, 154, 155,
156, 157, 158, 183
homework, 7, 52, 53, 105, 155, 195
homosexuality, 99
humanism, see socialist humanism
Humboldt, Wilhelm von, 38
hunger, 53, 54, 148, 157

identity, 161, 189
see also antifascism, culture,
experience, gender, heritage,
nation-building, religion
ideology, 14, 16, 22, 23, 25, 30, 31, 33,
35, 45, 46, 49, 110, 112, 122,
123, 125, 127, 137, 167, 188,
189, 190, 192, 194, 196
indoctrination, 5, 123, 129, 162, 188,
189, 195
Irgendwo in Berlin, see Somewhere in Berlin

Jews, 18, 34, 164, 167, 199
Judaism, see Jews, religion

Karsen, Fritz 1885–1951, reform
pedagogue, later educational
reformer), 34
Kerscheinsteiner, Georg, 40, 163
Key, Ellen, 34
Klemperer, Victor, 15, 78, 80, 180
Koebbel, Eberhard "Tusk," 19
KPD, 2, 11, 20, 21, 22, 24, 26, 27, 28,
31, 32, 33, 34, 39, 42, 43, 44, 45,
46, 47, 98, 105, 109, 112, 147,
160, 178, 180, 181, 185, 192
see also communism, SED, socialism,
SPD
Kreuziger, Max (DVV representative for
the school division), 166
Kruspkaja, Nadešda Konstantinovna, 40
Kuckucks, die (1949), 149, 152

language instruction, 126
classical languages, 42, 114, 198
modern languages, 42, 114, 166, 198
Russian, 58–9, 166, 195, 197, 198

Law for the Democratization of the
German School, 26
LDP (Liberal Democratic Party), 20,
21, 22, 41, 42, 46, 47
legitimacy, 1, 38, 121, 135, 185, 193,
194
Literature, 58, 76, 164, 177, 115, 142,
157, 160, 161, 162, 164, 165,
166, 176
of the Soviet zone/GDR
London Conference (1948), 25, 29, 73
Lunačarskij, Anatolij Vassilevič, 40
Luxembourg, 25, 29

Makarenko, Anton Semëovi, 40
male, see boys, gender, men
Mann, Heinrich, 65
Mann, Thomas, 160, 165
maps, 16
Marshall Plan, 29, 174
Marquardt, Erwin (vice-president of the
DVV), 29
Marxist-Leninism, 22, 23, 32, 40, 46,
163, 181, 188, 197
mathematics instruction, 56, 80, 82,
114, 198
May 10, 1933, see Day of German
Literature
Mecklenburg-Vorpommern, 68, 73, 74,
127, 190
memory, 11, 61, 62, 70, 83, 84, 85, 86,
87, 88, 121, 122, 123, 127, 128,
129, 130, 131, 132, 133, 134, 135,
137, 140, 153, 157, 169, 190
see also collective consciousness,
experience, forgetting,
remembering
men, 11, 13, 33, 68, 92, 93, 94, 95,
102, 103, 106, 108, 110, 111,
116, 117, 120, 142, 143, 144,
145, 146, 147, 148, 149, 151,
152, 153, 200
see also women, gender
military, 111
conscription, 107, 108
Montesorri, Maria, 34
Morgenthau Plan, 130, 140
Moscow, see Soviet Union

mothers 7, 53, 81, 98, 102, 109, 110,
120, 127, 131, 142, 143, 144,
146, 147, 148, 149, 151, 152,
153, 154, 156, 157, 162, 180,
187, 200
see also adults, families, fathers
movies, see DEFA, film
myth, 4, 18, 28, 32, 35, 81, 95, 143, 145

narratives, 162
antifascist, 10, 11, 17, 35, 121, 122,
128, 130, 131, 133, 135, 140,
143, 171, 179, 190, 196, 199
gender differences, 133, 134, 135,
140
nation, 106, 107, 109, 120, 123, 186,
190, 191, 196, 198, 202
German, 10, 11, 21, 23, 24, 29, 35,
37, 61, 74, 75, 79, 86, 88, 89,
91, 95, 99, 104, 108, 114, 121,
125, 127, 128, 130, 140, 142,
143, 151, 159, 162, 163, 174,
176, 186, 189, 190
nationalism, 100, 116, 120, 121,
164, 165, 168, 200
nation-building (German and
antifascist), 9, 11, 19, 20, 42,
62, 70, 83, 86, 88, 91, 91, 94,
95, 98, 116, 120, 121, 122,
123, 128, 129, 143, 144, 149,
153, 158, 161, 162, 170, 177,
179, 181, 183, 185, 186, 187,
188, 189, 190, 191, 196
national consciousness, 1, 2, 4, 5, 11,
13, 18, 23, 24, 28, 29, 35, 60,
83, 92, 122, 163, 164, 165, 178,
190, 196, 197, 198, 202
National Socialism (NSDAP), 1, 4, 11,
13, 17, 18, 21, 23, 26, 31, 39, 40,
45, 49, 50, 52, 53, 57, 63, 69, 73,
85, 86, 88, 94, 96, 98, 102, 104,
106, 108, 110, 116, 122, 123, 125,
126, 127, 129, 131, 132, 137, 141,
142, 143, 145, 150, 151, 155, 156,
164, 165, 167, 170, 174, 179,
185, 190, 197, 199
National Socialist education, 2, 11,
26, 27, 31, 40, 52, 55, 56, 63,

95, 99, 102, 102, 104, 111,
124, 125, 132, 141, 145
National Socialists and NSDAP
members, 10, 11, 37, 48, 49,
84, 110, 125, 127, 128, 132,
133, 164, 165, 167
Nazi, *see* National Socialism
Nazism, *see* National Socialism
Netherlands, the, 25, 29
neorealism, 144, 150
"*neue schule*," *see* "new school"
Neulehrer ("new teachers"), 6, 37,
43, 48, 57, 58, 76, 78, 85,
86, 94, 124, 125, 127, 167,
189, 192
"new school," 1, 2–5, 10, 12, 14, 23,
24, 26, 28, 31, 37, 38, 41, 49, 51,
59, 60, 70, 84, 93, 101–2, 105,
121, 124, 126, 127, 128, 144,
165, 168, 173–4, 188, 189,
192, 193, 197, 198
see also education, *Einheitsschule*,
reeducation, schools
"new teachers," *see Neulehrer*
nineteenth century, 5, 8, 26, 37, 38,
63, 107, 108, 128, 170, 172,
173, 190
normal, normalcy, normalization, etc.,
61, 76, 77, 79, 99, 116, 137, 140,
155, 164, 180, 181, 191
NSDAP, *see* National Socialism

parents, 6, 11, 38, 50, 52, 53, 61, 83,
105, 112, 122, 126, 129, 130,
131, 140, 141, **146**, 180, 190,
191, 192
see also fathers, mothers
past, 11, 53, 61, 62, 69, 77, 83, 84, 87,
88, 95, 96, 106, 120, 121, 122,
124, 126, 127, 128, 129, 130,
133, 135, 137, 140, 143, 151,
158, 159, 162, 163, 167, 168,
169, 170, 175, 177, 183, 185,
186, 189, 190, 191, 196, 206
Paulskirche Revolution, 11, 78, 169,
170, 171, 172, 185
see also, 1848
pedagogues, *see* educators

pedagogy, *see* education
Pieck, Wilhelm (SED chair and GDR
president), 43
Poland, 16, 73
polytehnic schools, 197, 198
posters, 6, 39, 65, **66**, 95, 96, **97**, 98,
136, 145, **146**, 147, 149, 165,
173, 177, 181, **182**, 183
Potsdam Accords, 129, 140
POWs, *see* Prisoners of War
Prague Spring (1968), 197, 201
Prisoners of War (POWs), 142, 143,
144, **146**, 152, 154, 155, 156,
157, 158, 199
propaganda, 11, 40, 82, 104,
110, 167
prostitution, 100, 102, 191
see also semi-prostitution
Protestantism, *see* religion
pupils, 5, 10, 11, 15, 23, 37, 39, 44, 48,
50, 51, 52, 53, 54, 56, 57, 58, 59,
61, 63, 65, 67, 68, 70, 72, 73, 77,
78, 79, 80, 81, 82, 83, 87, 88, 92,
104, 105, 113, 117, 123, 124,
125, 129, 130, 137, 148, 153,
154, 155, 163, 164, 167, 168,
172, 173, 175, 188, 189, 190,
195, 196, 200, 202
see also children, students, young
people, youth

rape, 100, 101, 102, 191
see also prostitution, sexually
transmitted diseases,
semi-prostitution
rebuilding, 18, 20, 61, 64, 65, 95, 106,
108, 109, 129, 144, 145, 153,
154, 158
see also construction, reconstruction
reconstruction, 3, 9, 11, 21, 29, 62,
63, 67, 68, 69, 70, 72, 75, 79,
81, 83, 88, 105, 128, 137, 140,
142, 143, 144, 147, 148, 149,
150, 153, 159, 161, 174, 178,
181, 187, 188, 196
see also construction, rebuilding
redemption, 1, 4, 143, 144, 150, 151,
180, 187, 188

reeducation, 1, 11, 54, 75, 87, 95, 105,
    109, 110, 122, 123, 124, 126,
    127, 142, 143, 144, 167, 175,
    194, 195, 196
  see also education, *Einheitsschule*,
    "new school," schools
reform pedagogy, 33, 40, 46–7, 163
refugees, 76, 77, 92, 190
Reimann, Brigitte, 157, 158, 187, 200,
    201, 202
religion, 11, 115, 141, 165, 170, 175,
    178, 179, 180, 181, 188, 196, 199
  Catholicism, 178, 179, 181
  Judaism, 34
    Jewish education, 34
  Protestantism, 34, 178, 181
  school instruction, 22, 42, 80, 141,
    168, 170
remembering, 61, 77, 121, 122, 123,
    124, 126, 127, 129, 135, 137, 140
  see also collective consciousness,
    forgetting, memory
Renaissance, the, 159, 161
Republikflucht ("fleeing the [German
    Democratic] Republic"), 194
resistance, 18, 21, 24, 78, 19, 112, 143
reunification, 156, 157
  of Germany, see unification
Romanticism, 161
Rousseau, Jean-Jacques, 106
rubble, 3, 62, 64, 68, 69, 71, 81, 91,
    93, 95, 99, 101, 103, 104, 105,
    107, 109, 111, 113, 115, 117,
    119, 126, 137, 143, 145, 146,
    147, 148, 150, 181, 195
  See also "rubble films," "rubble
    women," "rubble youth"
"rubble films," 143, **146**,
  see also rubble, "rubble women,"
    "rubble youth"
"rubble women," 64, 81, **119–20**,
    147, 148
  see also rubble, "rubble films," "rubble
    youth"
"rubble youth," 95
  see also rubble, "rubble films," "rubble
    women"

rural areas, 55, 59, 60, 76–7, 107, 112,
    125, 129, 189, 190
Russia, *see* Soviet Union
Russian zone, *see* Soviet zone

Saxony, 15, 16, 25, 56, 46, 48, 55, 64,
    65, 67, 68, 73, 75, 78, 80, 84, 85,
    96, 97, 101, 102, 109, 110, 115,
    125, 171, 172, 180, 183, 184, 190
Saxony-Anhalt, 68, 130, 157
SBZ, *see* Soviet zone
Schiller, Friedrich (1759–1805), 16, 24,
    26, 160
schools, 1, 5, 11, 14, 25, 27, 31, 32, 37,
    38, 39, 40, 44, 45, 51, 53, 54, 59,
    80, 98, 108, 120, 122, 127, 128,
    135, 136, 140, 141, 144, 155,
    159, 167, 168, 169, 172, 177,
    178, 179, 183, 185, 187, 188,
    189, 190, 191, 194, 196, 197,
    198, 201, 202
  see also education, *Einheitsschule*,
    "new school," reeducation
  official beginning in Soviet zone,
    39, 123
  school boards, 5, 37, 104, 179
  school laws, 5, 41, 56, 112, 123
  secondary schools, middle and upper:
    41, 82, 92, 93, 94, 108, 113,
    116, 155
science instruction, 42, 109, 198
Second World War, *see* World War II
secularism, 11, 22, 33, 42, 107, 141,
    168, 170, 175, 198, 199
SED (Socialist Unity Party), 20, 21, 27,
    31, 32, 34, 41, 44, 45, 47, 48, 50,
    160, 167, 178, 185, 189, 192, 195,
    197, 200, 201
  see also communism, communist,
    KPD, socialism, SPD,
    socialist
semi-prostitution, 100, 102, 151, 191
sex, *see* boys, girls, gender, men,
    prostitution, rape, semi-
    prostitution, sexual education,
    sexuality, sexually transmitted
    diseases, women

sexual education, 99, 100, 101
  see also rape, sexually transmitted
      diseases
sexuality, 10, 91, 98, 99, 106, 109, 111,
      112, 120, 149, 150, 151
sexually transmitted diseases (STDs),
      99, 100, 101, 102, 149, 191
  see also rape, sexual education
Siebert, Johannes (Hans), 17, 19, 31
SMA(D), 9, 20, 21, 24, 25, 27, 28, 37,
      38, 39, 43, 45, 48, 50, 56, 63, 71,
      72, 79, 85, 101, 126, 143, 160,
      164, 166, 170, 171, 187
social studies instruction, *see* history
      instruction
socialism, 1, 2, 4, 12, 14, 19, 20, 21, 22,
      27, 30, 33, 57, 63, 67, 98, 108,
      109, 114, 115, 116, 142, 143, 145,
      149, 159, 161, 162, 168, 169, 175,
      178, 179, 180, 181, 187, 188, 192,
      193, 195, 196, 197, 198, 199,
      200, 201, 202
  see also socialist, SED, SPD
socialist
  see also SED, socialism, SPD
socialist humanism, 11, 24, 26, 28, 30,
      169, 170, 185, 187, 189, 196, 197
soldiers
  Allied, 152
  German (National Socialist), 142, **146**,
      151, 154
  Soviet, 100
Somewhere in Berlin (Irgendwo in Berlin,
      1946), 144, **146**, 147, 149, 150,
      152, 153, 157
Soviet Military Administration of
      Germany, *see* SMA(D)
Soviet Union (USSR), 1, 3, 11, 15, 16,
      19, 20, 23, 24, 28, 31, 34, 39, 40,
      42, 43, 46, 51, 67, 154, 156, 163,
      165, 166, 187, 189, 196, 199
  anti-Sovietism in Germany, 40, 56,
      72, 73, 87, 99, 101, 116, 129,
      158, 196
  pedagogical movements in, 39,
      40–1, 197
Socialist Unity Party, *see* SED

Soviet Occupation(al) Zone, *see* Soviet
      zone
Soviet zone (of Occupation), 1, 5, 10,
      13, 24, 28, 31, 32, 35, 37, 39, 40,
      41, 43, 48, 49, 50, 54, 55, 60, 67,
      69, 89, 100, 106, 108, 109, 112,
      114, 116, 120, 121, 128, 129, 142,
      143, 147, 156, 159, 163, 169, 170,
      173, 188, 189, 194, 195, 200, 202
Spanish civil war, 20, 23, 166
special schools (special classes), 198
SPD (Socialist Party of Germany), 2.
      20, 21, 22, 24, 26, 27, 28, 33, 34,
      39, 42, 45, 46, 47, 49, 98, 108,
      109, 112, 114, 142, 160, 161,
      178, 179, 180, 181, 192, 196,
      197, 198, 201
sports instruction, 198
Stalin, Josef (leader of the Soviet
      Union), 21, 31, 40, 63, 201
state, 1, 2, 4, 5, 10, 12, 14, 20, 23, 25,
      28, 46, 53, 57, 59, 70, 74, 80,
      106, 108, 109, 112, 122, 123,
      125, 127, 129, 133, 141, 143,
      144, 145, 147, 152, 159, 161,
      171, 174, 177, 185, 187, 188,
      189, 193, 194, 196, 197, 198,
      199, 201, 202
Steiner, Rudolf, 34
Stunde Null, see Zero Hour
students (university), 92, 115, 173
  see also children, pupils, young
      people, youth
suicide, 144, 151

teachers, *see* educators
teachers' union (*Gewerkschaft Unterricht
      und Erzieher*), 47, 111
textbooks, 6, 54, 55, 56–7, 70, 127,
      177, 189
  see also education
Thälmann, Ernst, 165
Thuringia, 16, 34, 41, 45, 48, 49,
      52, 64, 65, 66, 68, 114, 115,
      172, 178
totalitarianism, *see* dictatorship
Torhorst, Marie, 34, 35, 172, 176

traditions, 11, 40, 41, 42, 83, 87, 95,
107, 108, 109, 120, 142, 143, 158,
159, 160, 161, 163, 171, 178, 181,
183, 185, 188, 189, 195, 200
Truman, Harry S. (President of the
United States 1945–1953), 201
Trümmerfrauen, see "rubble women"
Trümmerjugend, see "rubble youth"
Tucholsky, Kurt, 165

Ulbricht, Walter, 35
"Ulbricht Group," *see* Ulbricht, Walter
unification
of Germany, pre-twentieth century,
170, 173, 188, 190
of post-World War II Germany, 8,
14, 25, 60, 73, 162, 168, 174,
176, 185, 189, 190, 193
of Germany, post-1989, 193
United Nations, 193
United States of America (USA), 16, 25,
29, 47, 104, 108, 111, 142, 174,
193, 194
see also American zone, Western zones
unity, 8, 11, 14, 25, 28–31, 35, 41, 42,
60, 73, 74, 78, 106, 108, 114, 120,
168, 174, 175, 176, 185, 188, 189,
191, 195, 197, 198, 199
see also unification
unity school, *see Einheitsschule*
USSR, *see* Soviet Union

Versailles Treaty, 16
venereal diseases (VD), *see* sexually
transmitted diseases
Vergangenheitsbewältigung (coping with
the past), 11, 96, 104, 133–4

Wandel, Paul (director of DVV), 27,
41, 43, 56, 57, 108, 166, 176
Weimar Republic, 5, 13, 21, 27, 32, 33,
34, 38, 40, 45, 46, 55, 63, 144,
158, 159, 174, 188, 189, 197
West Germany, *see* Federal Republic of
Germany
Western allies (France, Great Britain,
USA), 19, 24, 30, 40

Western zones, 3, 4, 5, 9, 10, 11, 23, 24,
25, 29, 30, 31, 38, 39, 40, 45, 49,
56, 60, 67, 74, 89, 100, 108, 112,
114, 130, 156, 161, 169, 170, 173,
176, 179, 185, 189, 195
see also American zone, British zone,
French zone
Wildangel, Ernst (director of Berlin's
central school administration), 17,
18–19, 69, 115, 116, 174
winters, 9, 51–2, 54, 73, 75, 77, 78, 79,
80, 81, 88, 97, 120, 140, 148, 154,
183, 195
Winzer, Otto (city councilor for education
of Greater Berlin), 15, 72, 167
women, 11, 68, 76, 81, 92, 93, 94, 95,
99, 100, 106, 107, 108, 109, 110,
113, 114, 116, 117, 120, 142,
143, 144, 145, 147, 148, 149,
151, 153, 179, 191, 200
see also men, mothers, "rubble
women"
World War I, 33, 98, 142, 143, 163
World War II, 2, 13, 14, 18, 22, 24, 33,
34, 37, 38, 63, 74, 77, 83, 87, 92,
98, 106, 117, 126, 128, 130, 131,
140, 141, 143, 153, 154, 155,
157, 158, 163, 164, 190, 196
see also Holocaust, Potsdam Accords

young people, 27, 28, 32, 35, 38, 53,
58, 63, 65, 82, 87, 91, 98, 104,
106, 113, 114, 115, 117, 120, 121,
126, 127, 128, 131, 137, 142, 144,
147, 151, 153, 166, 181, 188, 197
see also children, pupils, students, youth
Young Pioneers, 181, **182**
youth (especially symbolically), 27, 28,
32, 35, 65, 82, 87, 91, 92, 98, 142,
164, 191, 196
see also children, pupils, "rubble youth"
[*Trümmerjugend*], students,
young people

Zero Hour, 14
Zolotuchin, P. (director of Soviet Education
Department in Germany), 71–2

Printed in the United States
By Bookmasters